COASTS AND ESTUARIES

Management and Engineering

ADVANCED SERIES ON OCEAN ENGINEERING

ISSN 1793-074X

Editor-in-Chief
Philip L- F Liu (*School of Civil and Environmental Engineering,*
Cornell Univ., Hollister Hall Ithaca, NY)

*For the complete list of titles in this series, go to https://www.worldscientific.com/series/asoe

Advanced Series on Ocean Engineering — Volume 57

COASTS
AND ESTUARIES
Management and Engineering

V Sundar
S A Sannasiraj
K Murali
Indian Institute of Technology Madras, India

World Scientific

NEW JERSEY · LONDON · SINGAPORE · BEIJING · SHANGHAI · HONG KONG · TAIPEI · CHENNAI · TOKYO

Published by

World Scientific Publishing Co. Pte. Ltd.
5 Toh Tuck Link, Singapore 596224
USA office: 27 Warren Street, Suite 401-402, Hackensack, NJ 07601
UK office: 57 Shelton Street, Covent Garden, London WC2H 9HE

Library of Congress Cataloging-in-Publication Data
Names: Sundar, V. (Vallam), author. | Sannasiraj, S. A., 1968–   author. | Murali, K, author.
Title: Coasts and estuaries : management and engineering / V Sundar, S A Sannasiraj,
    K Murali, Indian Institute of Technology Madras, India.
Description: New Jersey : World Scientific, [2023] | Series: Advanced series on ocean engineering,
    1793-074X ; vol. 57 | Includes bibliographical references and index.
Identifiers: LCCN 2022025785 | ISBN 9789811261800 (hardcover) |
    ISBN 9789811261817 (ebook for institutions) | ISBN 9789811261824 (ebook for individuals)
Subjects: LCSH: Coastal engineering.
Classification: LCC TC205 .S846 2023 | DDC 627/.58--dc23/eng/20220818
LC record available at https://lccn.loc.gov/2022025785

British Library Cataloguing-in-Publication Data
A catalogue record for this book is available from the British Library.

For any available supplementary material, please visit
https://www.worldscientific.com/worldscibooks/10.1142/13011#t=suppl

Desk Editors: Balasubramanian Shanmugam/Amanda Yun

Typeset by Stallion Press
Email: enquiries@stallionpress.com

This book is dedicated to the parents of the authors, their family members, colleagues, students and friends.

Preface

The book *Coasts and Estuaries — Management and Engineering* was written with the sole purpose of introducing the readers to the spectrum of coastal parameters that influence the stability and existence of open beaches and estuaries. The importance of estuaries and tidal inlets is often overlooked compared to coastal erosion problems. The awareness of sustainable development in coastal and estuarine habitats is essential for the wholesome development of the environment and the communities dependent on the same for their livelihood. The coastlines and estuarine environments across the world are dynamic in nature owing to a host of parameters contributing to the nearshore processes. The effective and efficient management of these water-bound regions can be perfected through a thorough knowledge of the nearshore hydrodynamics and morphodynamics.

Several numerical and analytical tools are available to understand the physical processes in the vicinity of these water bodies, which requires to be supplemented with field measurements for model calibration and validation since they are extremely site-specific. This book encompasses the engineering principles involved in field data observation-measurement-collection-processing; prediction of wave climate and sediment transport using measured field data; numerical modelling involving calibration-validation of the hydrodynamic and morphodynamic processes and underlying physical processes and the application of sustainable engineering measures to combat the coast and estuarine related problems.

The book has three sections, the first section extends an elaboration on the need and framework of the existing management and engineering notions. The second section details the measurement of the various parameters such as wave climate (offshore and nearshore), shoreline changes, beach profile variation and sediment transport rates. The third section describes the aspects of wave prediction to arrive at design characteristics and modelling of the hydrodynamic and morphodynamic processes along open coasts and tidal inlets. The vast experience of the authors in field monitoring and numerical model development in coastal and estuarine dynamics has been presented through detailed case studies in all three sections of the book to enhance the reader's understanding of the subject. It attempts to enrich the reader's knowledge of physical and numerical modeling in Coastal Engineering to investigate a wide range of problems related to coastal processes, sediment dynamics, nearshore hydrodynamics and morphodynamics. Recent key research findings from various journals have been included. Ocean, coastal, harbour and civil engineers must understand the underlying physics of sediment transport, the probable causes for coastal erosion, and design aspects of appropriate coastal protection measures for making a well-informed decision. The subject covered in this book is therefore of societal importance and major interest to the coastal engineering community.

Prof. Sundar V.
Prof. Sannasiraj S.A.
Prof. Murali K.
Department of Ocean Engineering
Indian Institute of Technology Madras, India

About the Authors

Dr. V. Sundar is a Professor Emeritus in the Department of Ocean Engineering IIT Madras. He has contributed to 28 PhD and 14 MS theses and has received 10 distinguished awards amongst the scientific community to his credit, including an Honorary Doctorate from University of Wuppertal, Germany. He was the Chairman of the Asia Pacific division of International Association of Hydro-Environment Engineering and Research (IAHR) during 2007–2011.

He has 550 publications in conferences and journals to his credit. He has participated in over 100 conferences worldwide and has organized five such major conferences. He has served as a member of the international scientific committee for a number of international conferences. He is a member of the editorial board for about 10 international journals. He has successfully completed about 30 research projects of which 10 are international in nature.

On the research front, he has worked on topics related to breakwaters (like semi-circular, skirt type as well as rubble mound type), wave loads on structures, sediment transport under waves and currents and scour around structures. He has developed instrumentation to measure scour, which has been cited by several researchers.

He has completed close to 300 projects in the general area of Coastal Engineering and has delivered lectures to the media, educational institutions and several other agencies all over the world. He was instrumental in planning for tsunami mitigation measures and prepared the master plan for the tsunami-affected maritime states of Tamil Nadu and Kerala, India, which remain as the basic document for implement by the respective governments.

Dr. S. A. Sannasiraj is a Professor in the Department of Ocean Engineering, Indian Institute of Technology Madras and also served as its Head from 2017 to 2020. His area of specialization includes breaking waves, wind-wave modeling, numerical simulation of nonlinear wave-structure interaction and coastal erosion and protection. He has contributed to 11 PhD and 6 MS theses. Since 2003, he has completed 14 research projects sponsored by Department of Science & Technology, National Research Board, Indian Space Research Organization, European Union and National University of Singapore. To his credit, he has 72 peer-reviewed journal publications and has participated over 100 technical conferences at both national and international levels. He was awarded the DAAD fellowship from Germany during 2006 for tsunamic wave impact on coastal structures; Indian Maritime Award in 2005–06 for his work on the prediction of tsunami using data buoys; Endeavour India Executive Award from Australian Government (2007); Fulbright-Nehru senior research fellowship (2011) for modeling tsunami impact on coastal structures and the Institution Prize by Indian Engineering Congress for a technical paper in 2015. He has successfully executed more than 200 industrial and consultancy projects related to nature port and harbours, intake/outfall systems, design of coastal protection structures and wind-wave prediction. Further, he has organized about 20 short courses and workshops during the last 10 years for researchers, field engineers, faculties and senior managers.

Dr. K. Murali is a Professor in the Department of Ocean Engineering, Indian Institute of Technology, Madras and also served as its Head from 2020 to 2022. He specializes in Computational hydrodynamics, shallow water hydrodynamics and morphodynamics. He has supervised 13 PhD and 8 M.S. theses. Since 2003, he has completed numerous research projects sponsored by various agencies with a funding of about 15 Million Dollars. To his credit, he has 55 peer-reviewed journal publications and participated in over 100 technical conferences at both national and international levels. He was awarded: the IBM Blue challenge award and Endeavour Australia fellowship and several other awards. He has successfully executed more than 100 industrial and consultancy projects for nature ports and harbours, intake/outfall systems, design of coastal protection structures and wind-wave prediction. Further, he has organized six major conferences and several short courses and workshops during the last 20 years for researchers, field engineers, faculties and senior managers in the field of Ocean Engineering.

Acknowledgments

The authors would like to place on record their sincere gratitude to Central Water Commission (CWC), Ministry of Jal Shakti (MoJS) and Department of Science and Technology (DST), Government of India, for their financial support in field data collection.

Thanks to Dr. D. Kumaran Raju for his contribution to the chapter on Maritime Spatial Planning, with his expertise in the field of Remote Sensing. The services of Dr. J. Sriganesh and his survey team members Mr. A. Pradeep, Mr. S. Vasanthakumar and Mr. S. Tharaniraja for strenuously conducting field data measurements; V. Maheshvaran for collecting and providing data on the morphodynamics and Sai Chenthur for providing morphodynamic analysis, which form the crux of this book, are gratefully acknowledged. The contributions made by Ms. R. Sukanya, Ms. D. Monica and Mr. M. Dhananjayan in data analysis, processing as well as their support in drafting different chapters of this book demand a special mention and their efforts are appreciated. The authors would like to place on record their sincere thanks to all project staff who have immensely contributed for the successful outcome of this book.

Contents

Section B Observation of Coastal Environment 105

Section A

CMIS — Need and Framework

Chapter 1

Management and Engineering of Coasts and Estuaries: An Overview

Abstract

Vast lengths of coastlines around the world are experiencing threats that have led to the loss of infrastructure and livelihood of the community and, in several instances, loss of lives. Coastal erosion is caused by either nature or anthropogenic activities along the coastal belt. In addition to planning for combating the hazards, a rapid progress in coastal development has emphasized the need for coastal management information to understand the physical parameters that are responsible for shoreline stability, which is the focus of *Coasts and Estuaries: Management and Engineering*. This book encompasses monitoring parameters such as waves, currents, wind, tide, bathymetry, beach profile, shoreline, sediment characteristics in the near shore, as well as in case of estuarine coasts, riverine data such as river current, discharge, conductivity, temperature and depth. Three coastal sites along the southern Indian peninsula with different coastal features — Devaneri in Tamil Nadu, Ponnani in Kerala and Karaikal in Puducherry — were selected for this purpose, the salient details of which are reported in this chapter.

1.1. Purpose of the Book

There is insufficient information on the baseline or changes in waves and nearshore currents to provide an assessment of the effects of nearshore hydrodynamics on erosion. The local physical processes depends on local morphodynamics and coastal geomorphology. In several locations, shoreline stability risk is poorly defined under assumed wave and current climate conditions because of sparse field networks and relatively limited historical records. This provides a poor baseline for decision-makers to detect changes in observed records and assess future changes. Most monitoring systems are not capable of observing the variables particular for local site-specific changes. Monitoring and warning systems are in place to mainly observe coastal extremes, such as waves, storm surges and associated coastal floods.

Coastal erosion and deposition have a strong dependence on coastal structures, vegetation and the sediments properties and its transportation and settling. Even though physical and numerical models are available to describe these phenomena, they are hindered by accuracy issues as several parameters that dictate the driving forces are assumed. The ground truth of results from these simulation models is not easily verifiable against observations due to lack of high-quality observation data. Hence, accurate prediction of coastal erosion/deposition is still a major issue. Beach erosion rates can be predicted based on the data obtained from monitoring campaigns and via coastal line observations obtained remotely. Both methods, if combined, produce the most accurate results.

The Monitoring and Data Collection programme proposes monitoring (via onshore beach profiling through land surveying) to provide data relevant to the designers, decision-makers and modelling groups that are not available otherwise. For this monitoring, the necessary sensors and devices are planned to be deployed at strategic locations along southeast and southwest coasts of India, particularly for the stretches of the coast that are deemed vulnerable to erosion.

1.2. Parameters to be Measured to Plan for Coastal Protection

The most common parameters that dictate the stability of the coastal zone that need to be measured are provided in **Table 1.1**.

1.3. Site Selection

1.3.1. *Criteria for site selection*

The following criteria were considered for finalizing each of the sites:

- degree of vulnerability;
- need for reliable field data to effective planning of shore protection in the vicinity;
- type of activities and density of the dwelling units along the coast;
- accessibility to the site;
- safety of equipment to be erected;
- support from local public and concerned government department in maintaining the equipment;
- application of the data collected and scope for future projects;
- security arrangements as well as insurance.

Table 1.1.   Details of parameters to be measured and respective instruments.

S. No	Parameter	Characteristics	Instruments
1	Wave	Wave Height Wave Direction Time Period	Directional Wave Recorder (DWR)

2	Current	Ocean Current —
		    Velocity Profile
			Acoustic Doppler Current Profiler
			    (ADCP)

		River Current — Point Current	Current Meter (CM)
3	Tide	Tidal Amplitude	Tide Gauge (TG)
4	Wind	Wind Speed Wind Direction	Automatic Weather Station (AWS)
5	Bathymetry	Deep Water (-5 to -20 m)	Echo-sounder with DGPS
		Shallow Water (-1 to -5 m)	Water Scooter

		Shoreline to -1 m
		    contour
			Real-Time Kinematic Global
			    Positioning System (RTK-GPS)

6	Shoreline and Beach
	    Profile Changes
		HTL & LTL
		    Mapping
			RTK-GPS

7	Coastal Sediment Shore Sediments	Particle/Grain Size Specific Gravity	Grab Sampler NISKIN Water Sampler
	Seabed Sediments		Sieve Shaker/Hot Air Owen
	Suspended Samples	Salinity, Conductivity, TSS, TDS and Temperature	Membrane Filter Apparatus/ Conductivity Meter
8	Riverine Data	Salinity, Conductivity and Temperature	Conductivity, Temperature and Depth (CTD)

1.4.   Typical Instruments for the Designated Parameters

1.4.1.   *Waves*

1.4.1.1.   *General*

Waves are periodic undulations of the water surface due to the action of wind blowing over it, the characteristics of which dictate the loads on structures in the ocean. The ocean waves undergo deformation as they propagate towards the coast and govern the stability of shoreline.

To measure the wave characteristics, a bottom-mounted Directional Wave Recorder (DWR) was proposed for the wave observation. To achieve good-quality data, the field data needs to be collected covering the entire coastal stretch under consideration and a few locations in relatively deeper

Fig. 1.1(a).   Directional wave recorder.

waters (up to 15–30 m). This facilitates the validation of the numerical model, thereby, wave characteristics can be predicted by the model at any location along the coast (for any extended use of measurements). The wave data in 8–12 m water depth needs to be collected at specific locations based on vulnerability of the location to extreme events. A view of the DWR is shown in **Fig. 1.1(a)**. The DWR measures wave elevation from which its height, period and direction are derived.

1.4.1.2.   *Description of instrument*

The MIDAS DWR adopts the small amplitude wave theory to obtain the wave elevation time history from the measured pressure and flow signatures by strain gauge or high-accuracy piezo-resistive type pressure sensors. The electromagnetic current sensors housed in the DWR registers the magnitude and direction of currents. The DWR is handy and simple to operate at water depths less than 20 m and achieves a high degree of accuracy.

1.4.1.3.   *Data acquisition*

According to the linear wave theory, the accuracy in the measurement of waves in the deployed location depends upon the number of data points required to be sampled over the said data recording duration. The measured signals are then subjected to post-processing on board.

Table 1.2.   DWR specifications.

Description	Specifications
1. Sensor	
Type	Radar-based three sensors to measure Wave Direction, Height and Period
Material	Stainless Steel
Power	24–64 VDC
Frequency	10 GHz (X Band)
Temperature	$-30°$C to $+40°$C
Humidity	0–100%
Protection	IP67
2. Wave Height	
Range	0 to 20 m or better
Accuracy	± 1 cm
Interval	1 min or better
Processing	Standard Wave Analysis Program
3. Wave Period	
Range	1–50 s or better
Accuracy	± 50 ms
Interval	1 min or better
Processing	Standard Wave Analysis Program
4. Wave Direction	
Range	$0°$–$360°$
Accuracy	± 1
Period	1.6–50 s
Interval	1 min or better
Processing	Standard Wave Analysis Program
5. Processing Unit	
Protection	IP67
Communication	RS485, RS232, LAN, USB
Temperature	$-20°$C to 65$°$C
Power	DC supply
6. Software	Software for Processing Wave and Tide Data

1.4.1.4.   *Recommended specification of the instrument*

The specifications of the DWR deployed herein are provided in **Table 1.2**.

1.4.1.5.   *Deployment of the instrument*

The DWR need to be fitted with mooring frame, which in turn needs to be supported by four counterweight bars of minimum 20 kg each. This mooring frame should be connected to an anchor weight of about 50–80 kg. It is always advisable to deploy DWR with an anchor and counterweights to avoid drifting of the DWR in the event of extreme waves. The mooring frame with DWR should be connected to a marker buoy for the identification of the instrument and the anchor weight also needs to be connected to the marker buoy, so the instrument can be identified easily. The complete set-up of DWR (**Fig. 1.1(b)**) needs was then deployed in 15 m water depth. **The DWR was setup with 2 Hz sampling rate of 2048 sample collection with a Wave Burst Interval of 60 min.**

1.4.1.6.   *Typical results of wave observation by DWR*

The DWR was deployed in the ocean at a water depth of 15 m off the Karaikal coast during the period from 13 October 2017 to 19 April 2018. The significant wave height, H_s, maximum wave height, H_{max} and peak wave period, T_p, along with its direction derived from the acquired data, are presented in **Figs. 1.1c–1.1(e)**.

1.4.2.   *Ocean current*

1.4.2.1.   *General*

An ocean current is a continuous movement of a mass of water in the ocean from one location to another. Ocean currents are created by a gradient due to wind, temperature, density, wave heights, salinity and tides. An Acoustic Doppler Current Profiler (ADCP) was considered to acquire data regarding the direction and magnitude of currents. ADCP provides the velocity profile along the water column. Coastal currents are of significance in shallow waters, so the data collection needs to be done in water depths less than 6 m. A view of the ADCP is shown in **Fig. 1.2.**

The ADCP aims for the measurements of ocean current characteristics:

✓ **Velocity of Ocean Current**
✓ **Direction of Ocean Current**

1.4.2.2.   *Description of instrument*

To acquire the distribution of current over the water depth, the ADCP utilizes the Doppler effect, wherein sound waves get scattered or reflected from the particles within the medium. The ADCP is fixed either rigidly to a floating vessel or to a fixed structure like piers or piles.

DWR Mooring Frame

DWR Fitted with Mooring Frame

DWR Counter Weight and Ropes

Anchor Weight

Final Setup

DWR Deployment from boat

DWR Marker Buoy after Deployment

Deployment Plan

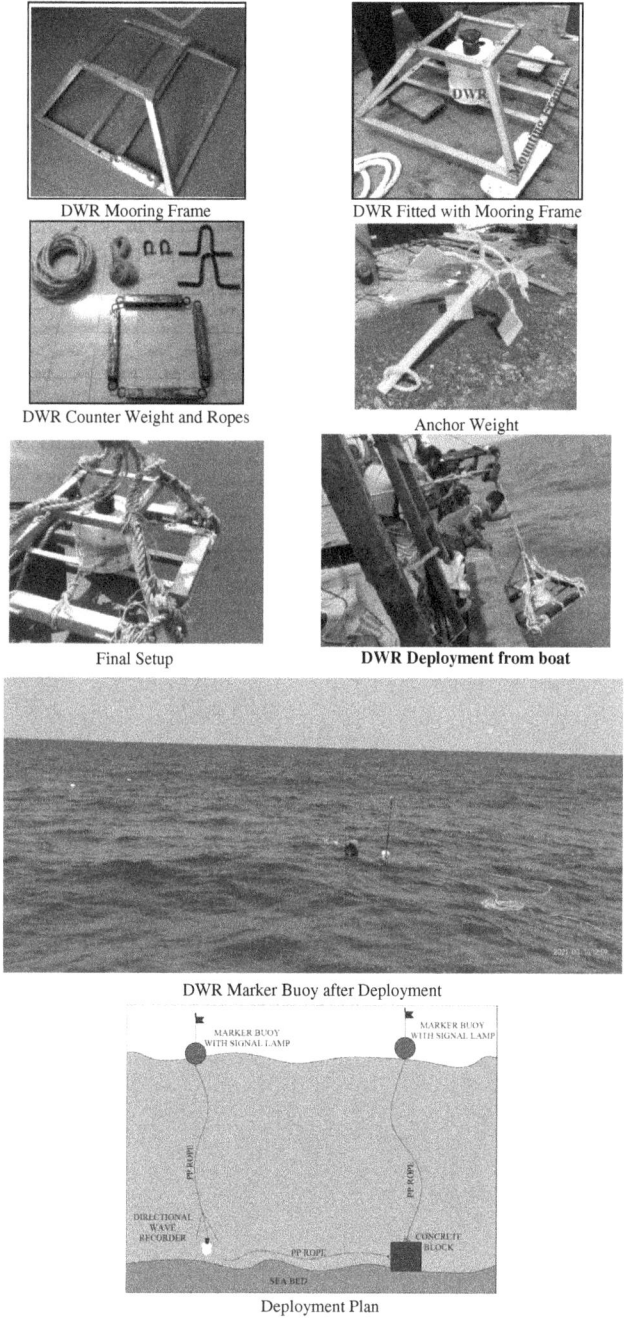

Fig. 1.1(b).   Deployment of DWR.

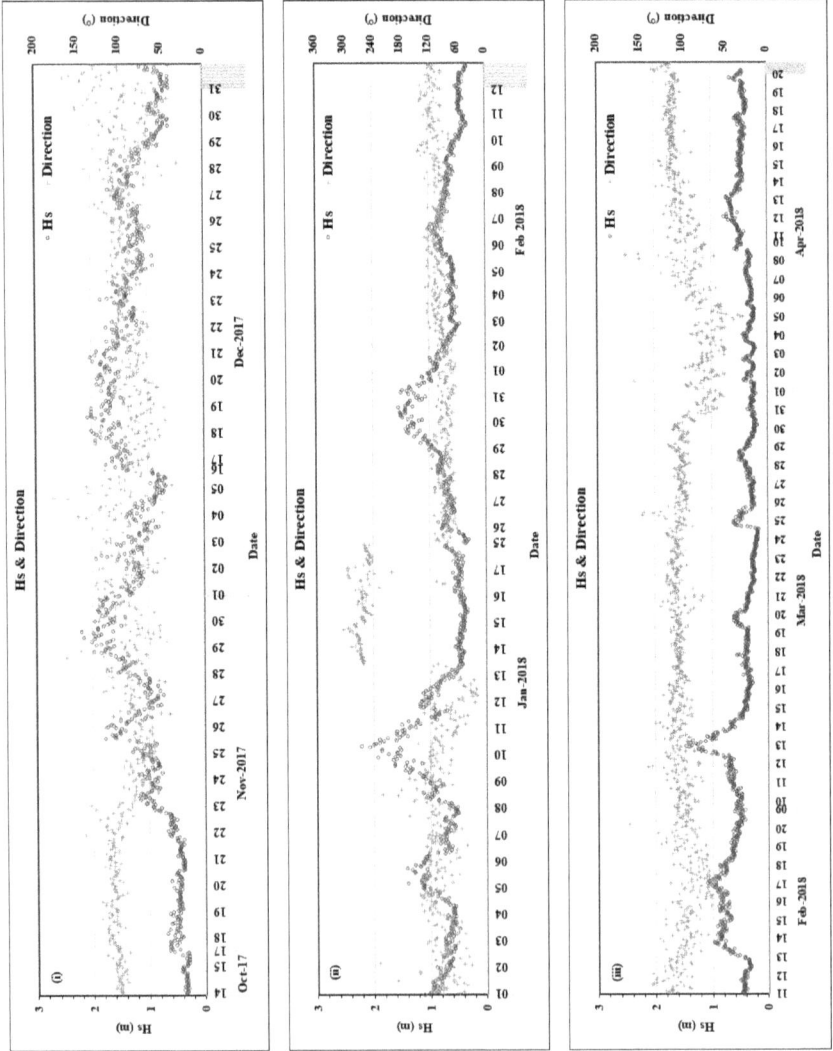

Fig. 1.1(c). Karaikal DWR data — H_S and direction *vs* date and month.

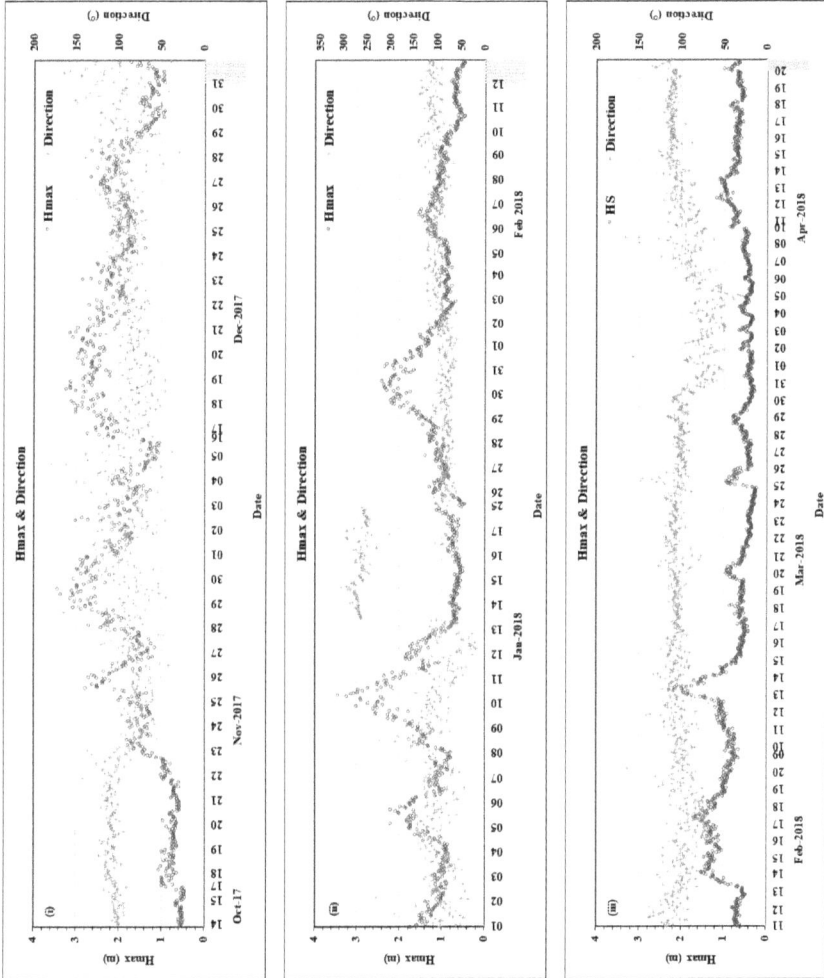

Fig. 1.1(d). Karaikal DWR data — H_{max} and direction vs date and month.

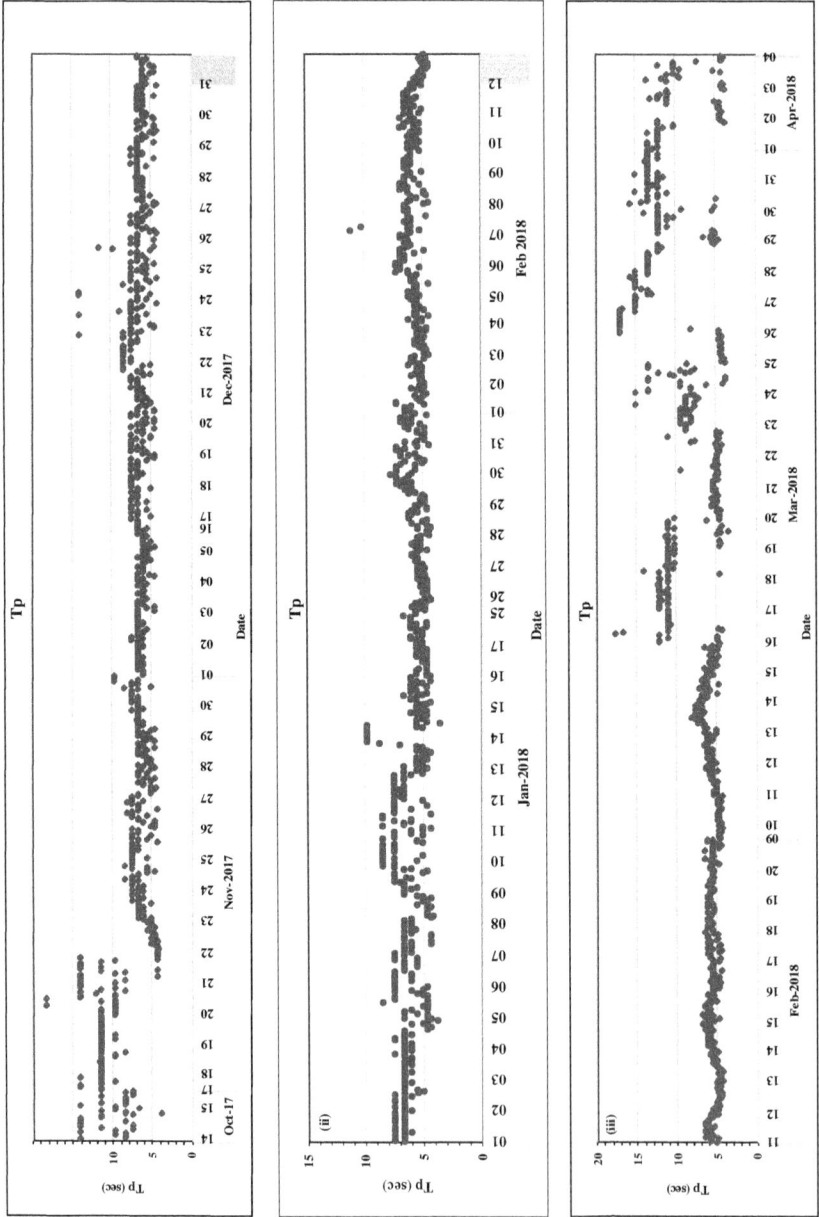

Fig. 1.1(e). Karaikal DWR Data — T_p vs date and month.

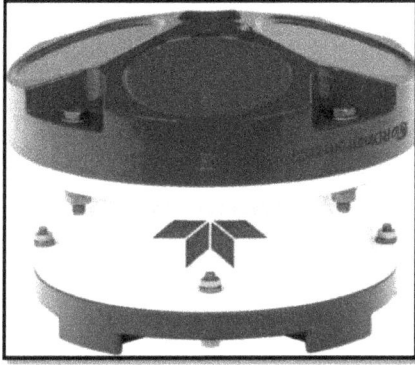

Fig. 1.2.   Acoustic Doppler Current Profiler (ADCP).

1.4.2.3.   *Recommended specification of the instrument*

The recommended technical specifications of ADCP are briefed in **Table 1.3**.

1.4.2.4.   *Deployment of the instrument*

ADCP can be mounted in different orientations, and although its recording frequency range can be between 38 kHz and few MHz, it is usually set with a frequency of about 600 kHz as this corresponds to a ringing period of the equipment at 1.5 m water depth.

ADCP is usually deployed to measure the ocean currents in water depths less than 8 m. The first ocean current measurements started at the Karaikal site on 14 October 2017. The ADCP instrument is shown in the **Fig. 1.3(a)** and was powered up by batteries shown in **Fig. 1.3(b)**. The ADCP instrument was to be deployed at the port side of the boat using three aluminium pipes provided with the instruments (shown in**Fig 1.3(c)**). After configuring the ADCP, it was connected to an external battery source and fixed on the boat with the sensors facing downward into the sea, as shown in **Fig. 1.3(d)**.

1.4.2.5.   *Typical results of ocean current observation by ADCP*

The ADCP was deployed in the ocean at a water depth of 7 m off the Karaikal coast during the period from 14 October 2017 to 3 April 2018.

Table 1.3.   Specifications of ADCP.

Description	Specifications
1. Sensor	
Stream Velocity Accuracy	1% of measured value ± 0.005 m/s
Resolution	<0.001 m/s
Sampling interval	1 sec to 10 minutes
Configuration	Minimum 4 beams
Beam angle	$>20°-30°$
Acoustic frequency	1000 kHz or higher possible frequency for adequate bottom tracking at depths of minimum 30 m in fast flowing, sediment laden waters
Number of depth cells	Programmable, 1–200 (Minimum)
Depth cell (bin) size	Programmable, minimum of 0.25 m
2. Bottom tracking	
Accuracy	1 cm/s @ 5 m/s
Stream velocity range	Same as of ADCP
Depth range	**Minimum 1 m up to 20 m or more**
3. Tilt Sensor	
Range	$-20°$ to $+20°$, both X and Y axes
Accuracy	$\pm 2°$
4. Compass	
Type	In-built flux gate
Accuracy	$\pm 5°$
Repeatability	$0.2°$C
Resolution	$0.1°$
Permissible tilt	$\pm 15°$
5. Auxiliary	
Internal memory	Two PCMCIA card slots, one memory card included
Communication interface	Serial RS 232C at PC end. The communication between ADCP and PC shall have suitable cable lengths. Bluetooth/Wi-Fi connectivity is preferable.
Baud rate	9600 BPS or more
Power supply	220 VAC $\pm 25\%$; 47–53 Hz: 10–15 VDC or 20–30 VDC
Housing	Corrosion proof, sturdy and robust
Ingress protection	Waterproof, compliant with IP66 or more, 20 m
Operating temperature	$5°-45°$C (The operating temperature range specification applies to all components of the ADCP, like sensor head, cable, interfaces etc.)
Humidity	Up to 100%
Software	Software required for data processing
OGC Compliance	**Certification required for Downloading Software**

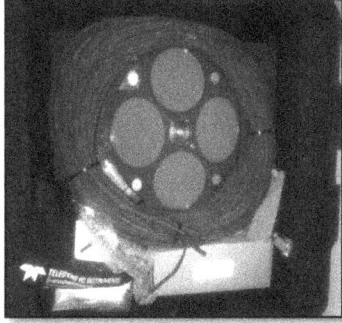

Fig. 1.3(a).   ADCP instrument sensors.

Fig. 1.3(b).   Batteries of ADCP.

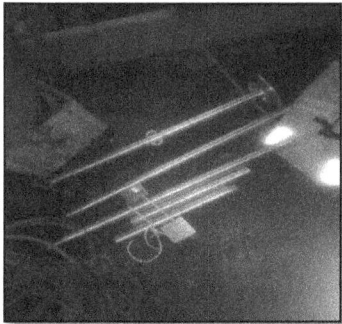

Fig. 1.3(c).   ADCP mounting poles.

Fig. 1.3(d). ADCP mounted on the side of boat.

Figure 1.4 shows the directional distribution of the percentage of occurrence of magnitude of current at different depths (bins).

1.4.2.6. *Karaikal current data*

125 days' data — 14 October 2017 to 3 April 2018.

1.4.3. *Tide*

1.4.3.1. *General*

The gravitational attraction between the sun, moon and the earth cause the level in ocean waters to rise and fall over a mean sea level (MSL), which is defined as tide. The vertical distance between the rise and fall in water levels — i.e., highest level defined as high tide line and the lowest level defined as the low tide line — is the tidal range. The average tidal range varies along the Indian coast from about 0.5 m to a maximum of about 12 m in the Gulf of Khambhat, whereas the maximum around the globe is of the order of 17 m off the Bay of Fundy, Canada. The tides are classified as semi-diurnal, diurnal and mixed types. The semi-diurnal is characterized with an amplitude of half the tidal range, whereas the diurnal tide is characterized by one high tide and one low tide within a day, and the mixed type features two uneven tides or one high and one low tide a day. The classification of tides and its distribution around the world ocean are shown in **Figs. 1.5 and 1.6**, respectively.

The occurrence of the highest tide at a specific location is defined as Highest Astronomical Tide (HAT). The average of the two high tides on the days of spring tides is defined as Mean High-Water Springs (MHWS), and

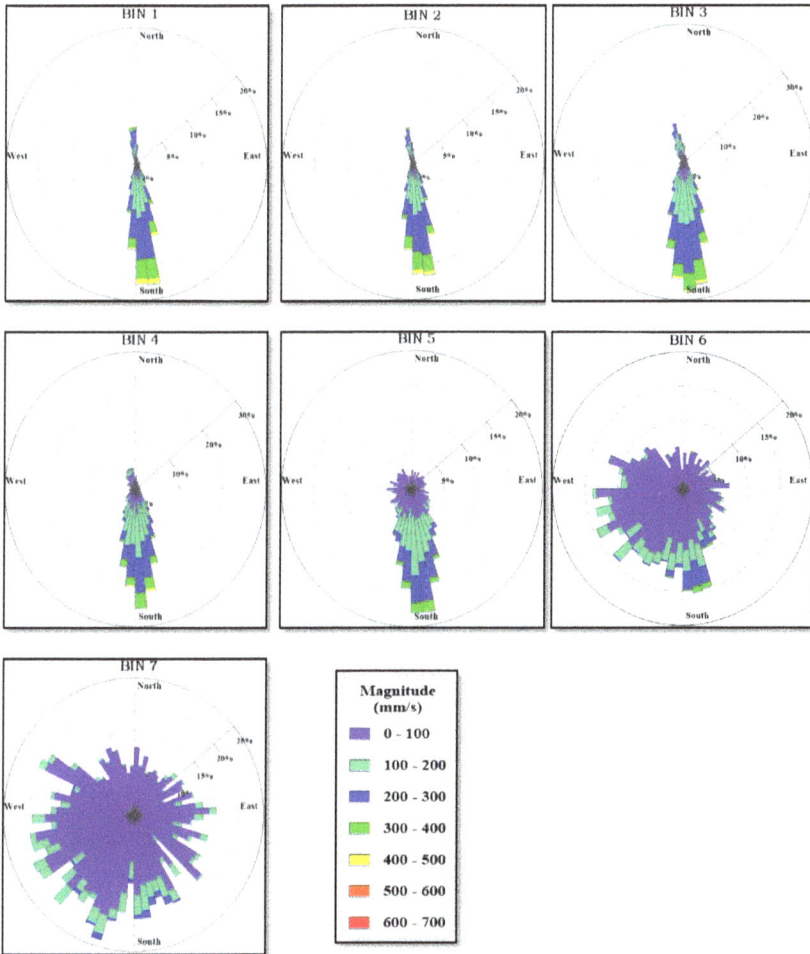

Fig. 1.4. Karaikal current rose diagram from the measurements of ADCP.

the average of the two high tides on the days of neap tides being defined as Mean High-Water Neaps (MHWN). The average of the two low tides on the days of neap tides is defined as Mean Low-Water Neaps (MLWN), whereas the average of the two low tides on the days of spring tides is the Mean Low-Water Springs (MLWS). The above description is shown in **Fig. 1.7**. The occurrence of the lowest tide at a said location is defined as the Lowest Astronomical Tide (LAT) and Chart Datum (CD).

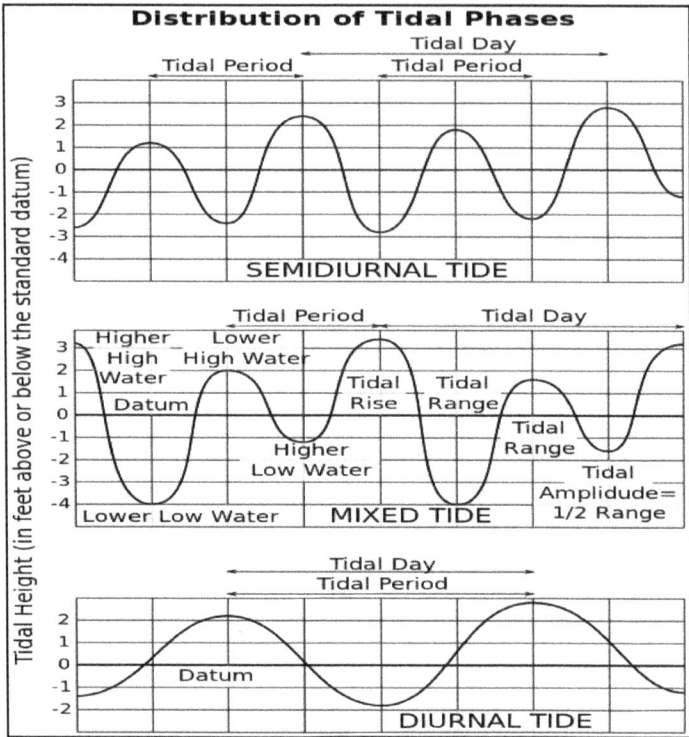

Fig. 1.5.   Distribution of tidal phases.

Fig. 1.6.   Types of tides around the world ocean.

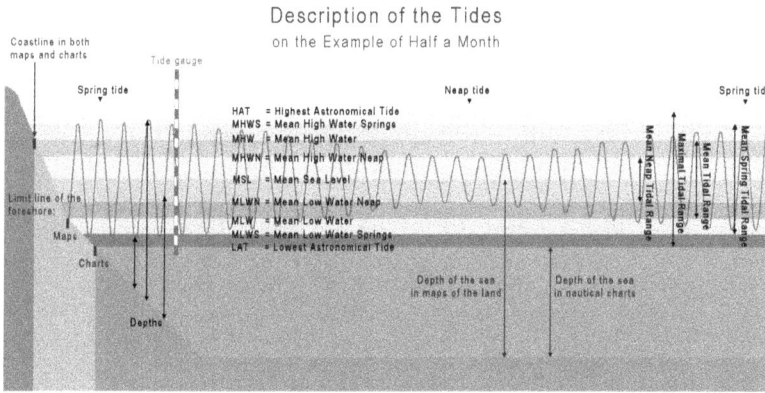

Fig. 1.7. Description of the tides.

1.4.3.2. *Importance of tidal data*

The importance of tidal data is listed below.

- The tidal elevations are essential for safe navigation in oceans and large water bodies.
- They aid in forecasting, monitoring and mitigating coastal hazards and floods.
- Tidal elevations influence sea-level rise and its effect on the coast.
- They are important to establish levels of structures along the coast as well as offshore.
- They aid in the planning and development of harbours to establish safe manoeuvring and berthing of vessels.
- They help in establishing the weather window for carrying operations in the offshore.

1.4.3.3. *Description of instrument*

Tide gauge is used for measuring the height (rise and fall) of the tide automatically producing a continuous graphic record of its variation as a function of time. New technologies such as ultrasonic sensors facilitate the transmission of real-time, remote information on the tidal level variations.

1.4.3.4. *Methodology of data collection based on pressure sensors*

The displacement of water surface at a location is measured through pressure sensors working on strain gauges or ceramic technology, which converts

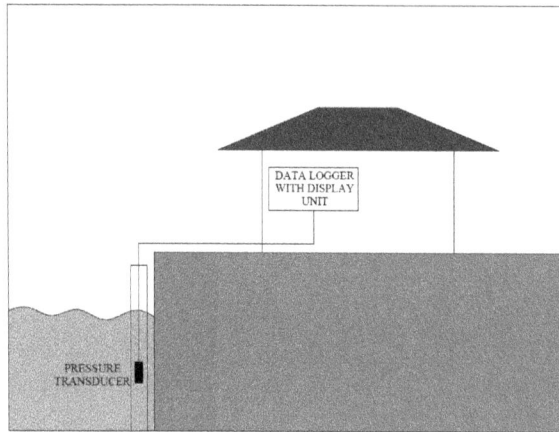

Fig. 1.8. Measurement setup.

the recorded pressures to water levels by applying field corrections. This is termed as Direct Reading Systems. The working principle would be either the change in resistance or capacitance that are recorded with changes in the water surface elevation.

The signal is amplified and may be displayed and stored in shore-based data logging equipment, as indicated in **Fig. 1.8**. Utmost care is needed while calibrating and converting the measured units to actual water elevation.

1.4.3.5.   *Recommended specifications of the instruments*

The details of water level meters incorporating pressure gauge are indicated in Table 1.4.

1.4.3.6.   *Deployment of instrument*

A view of the TideMaster, which is a small, cost-effective Water Level Recorder, is shown in **Fig. 1.9(a)**. This instrument can be continuously deployed for about a year, provided the sampling rate of data collection and retrieval are suitably fixed. A typical setup is shown in **Fig. 1.9(b)** and the views of the set-up at the measuring site under a bridge along the Arasalar river (10°54'40''N; 79°50'10''E) in Karaikal are projected in **Fig. 1.9(c)**. Typical measured tidal variation is shown in **Fig. 1.10**.

Table 1.4. Specifications for water level meters (incorporating pressure gauge).

Description	Specification
Type of equipment	Water Level Meters (Incorporating pressure gauge)
Measuring Range	0–20 m
Resolution	Better than $\pm$ 5 mm
Averaging Period	Selectable from 5 to 60 s
Working Temperature	−5°C to +50°C
Relative Humidity	0–100%.
Internal Memory	Recording should be done in an internal memory up to about two calendar years of autonomous operation at 5-minute recording intervals.
Recording Type	Self-recording
Frequency Range	20–30 GHz for wave radar
Power Supply	220V AC (50/60Hz) and 10–24 V DC

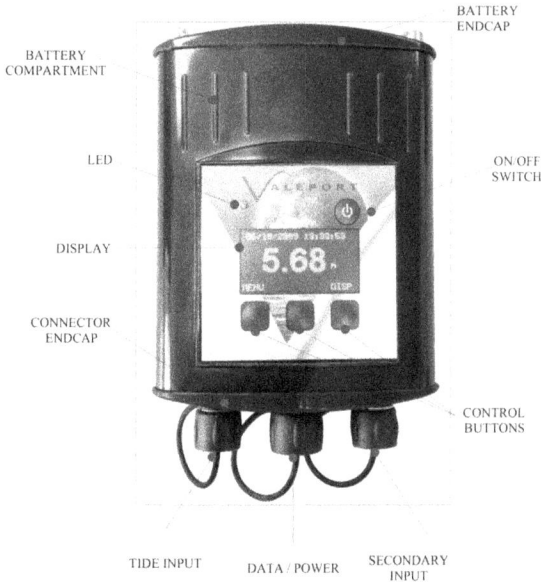

Fig. 1.9(a). TideMaster display unit.

1.4.4. *River current and discharge (near to river mouth)*

1.4.4.1. *General*

Discharge is the volume of water moving down a stream or river per unit of time, commonly expressed in cubic meter per second or gallons per day.

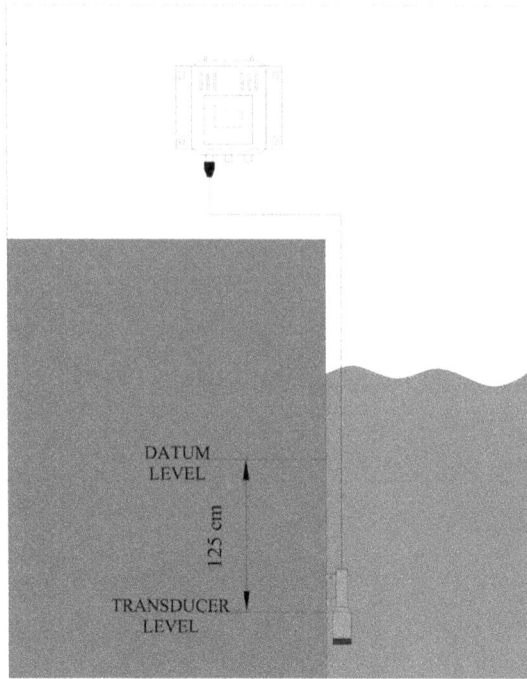

Fig. 1.9(b).   TideMaster sample setup.

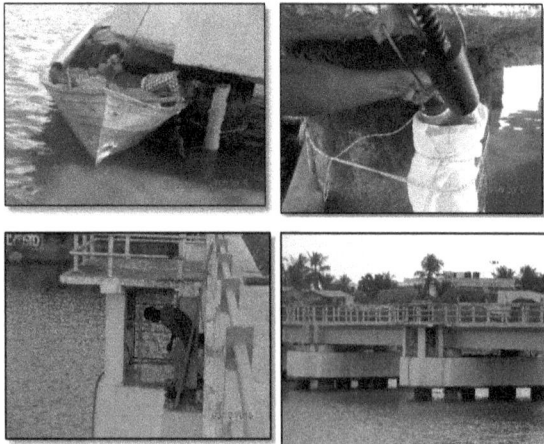

Fig. 1.9(c).   Karaikal tide gauge.

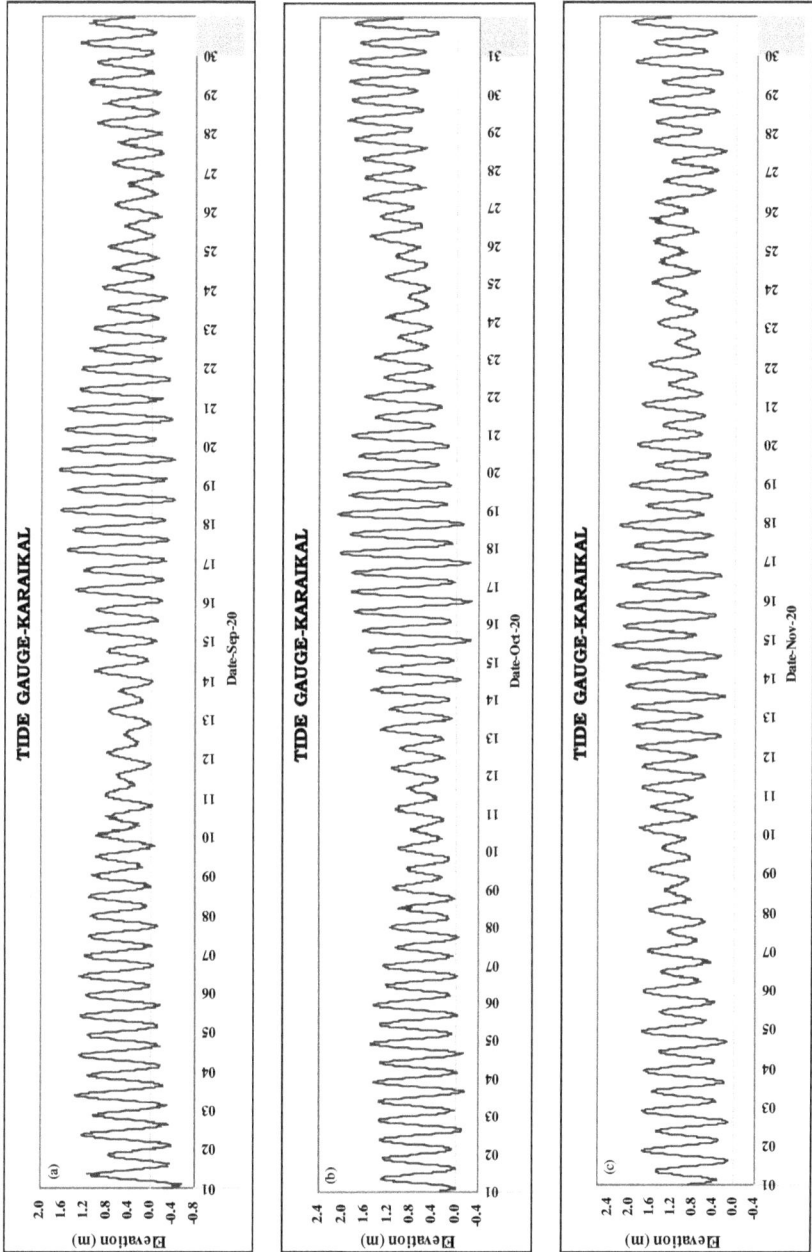

Fig. 1.10. Karaikal tide gauge data: (a) Sep. 2020, (b) Oct. 2020 and (c) Nov. 2020.

In general, river discharge is computed by multiplying the area of water in a channel cross section by the average velocity of the water in that cross section. Subsection width is generally measured using a cable, steel tape or similar piece of equipment. Subsection depth is measured using a wading rod, if conditions permit, or by suspending a sounding weight from a calibrated cable and reel system off a bridge, cableway or boat or through a hole drilled in ice.

Reliable estimation of freshwater inflow to the ocean from large tidal rivers is vital for water resources management and climate analyses. Information on time-varying water levels and magnitude and direction of tidal currents is quite critical in coastal waterways and estuarine environment management. Real-time tide is measured by tide gauge nearer to the current meter location with respect to Chart Datum. The current meter records flow velocity with direction; the water depth variation in profile and flow velocity will quantify the water discharge rate.

1.4.4.2. *Description of current meter*

The most common method used for measuring discharge is the mechanical current-meter method. In this method, the stream channel cross-section is divided into numerous vertical subsections. In each subsection, the area is obtained by measuring the width and depth of the subsection, and the water velocity is determined using a current meter (**Fig. 1.11**). The discharge in each subsection is computed by multiplying the subsection area, which was corrected with respect to real-time tide by the measured velocity. The total discharge is then computed by summing the discharge of each subsection.

The instrument model 106 that was adopted is an impeller-based meter, measuring speed and direction with optional temperature, and depth with respect to pressure at a rate of 1-second cycle. The 106 model can be operated in self-recording mode or in direct recording mode via a PC.

Self-recording mode: Instrument set and Data extraction can be carried out using the supplied lead from PC to external 10 ways SubConn connector. Direct recording mode: Over short cable lengths (up to 100 m) RS232 communication is possible via the external 10 Way SubConn and PC. Power to the model 106 may be taken from their internal battery, from a surface battery or power supply.

When power is taken from the internal battery (1.5V alkaline D cell), battery life is approximately 30 days at 10 second sampling rate, or 56 days at 5-minute sample rate. Using a 3.6V Lithium D cell, life is approximately 90 days at 10-second sample rate, or 180 days at 5-minute sample rate.

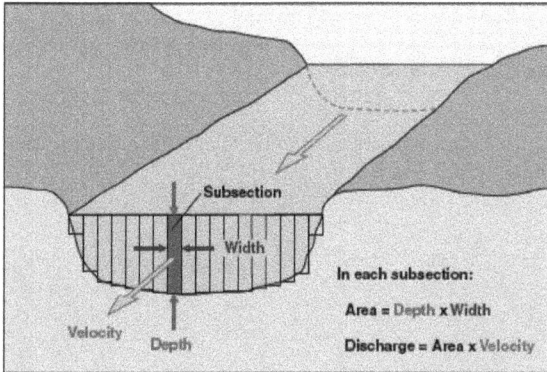

Fig. 1.11.   River discharge calculation.

The units are fitted with a 512 kb memory. This equates to storage of over 8000 speed and direction records for the 106 (over 4000 if temperature and depth are also fitted).

1.4.4.3.   *Recommended specifications of the instruments*

The specifications of the current meter are projected in Table 1.5.

1.4.4.4.   *Methodology of data collection*

Current meter observations are aimed at recording the data to analyze the magnitude and direction of flood/ebb stream. The tidal stream observations are to be carried out as per the tidal cycle. The magnitude of flood/ebb flow needs to be correlated with the timings of high water (HW). To record and analyze the minimum and maximum rates, the observations should be carried out during spring tides. The data may be collected at an hourly interval, six hours prior and after HW during spring tides.

When the meter is lowered in water and faces the current of water in the channel, the wheel rotates. A tail is attached to keep the meter facing the direction of flow. This tail aligns the meter in the direction of flow. The meter is also fitted with a streamlined weight (fish weight), which keeps the meter in a vertical position. The rate of rotation of the wheel depends on the velocity of flow. A dry battery is kept inside the current meter and an electric current is passed to the wheel from it. A commutator is fixed to the shaft of the revolving wheel.

Table 1.5. Specifications of current meter.

Description	Specifications
Speed	
Type	Any type, but it should follow the required specifications
Velocity range	0.03–5 m/s
Accuracy	±1.5% of reading above 0.15 m/s
Direction	
Range	0–360°
Accuracy	±2.5°
Resolution	0.5°
Temperature	
Type	Thermistor
Range	−5 to 35°C
Accuracy	±0.2°C
Resolution	0.01°C
Pressure	
Type	Strain gauge transducer
Range	1–100 dB
Accuracy	±0.2% Range
Resolution	0.025% Range
Power	
Internal	Minimum 30 days required
External	For external supply, 12–20 V DC is required
Software	Software required for data Processing

It makes and breaks the contact in an electrical circuit at each revolution. An automatic revolution counter is kept in the current meter with the battery, which registers the revolutions. The time taken for a required number of revolutions may be noted. The velocity of flow can be read from a rating table. The rating table is always provided with the meter.

1.4.4.5. *Methods of measuring velocity*

When the velocity is measured by the current meter at 0.5 times the depth and then the mean velocity at the section is obtained, the observed velocity should be multiplied with a coefficient 0.96. Alternatively, to obtain mean velocity of flow at a particular section, velocities may be observed at 0.2 and 0.8 times the depth from water surface.

The mean of these two readings gives the mean velocity of flow at that section. Generally, the mean velocity of flow at a section is obtained by keeping the meter depth at 0.6 times of B (breadth). In the observed velocity, corrections are made for any obliquity of the current with the cross-section line and for drift. It is necessary to conduct vertical velocity distribution experiment at each site to decide the point of average velocity.

1.4.4.6. *Typical results of river current*

The Valeport Rotan type current meter was deployed in the Arasalar River near Railway Bridge Pier at Karaikal, and typical measured current data is shown in **Fig. 1.12**.

1.4.5. **Seabed profile (near shore and offshore)**

1.4.5.1. *Bathymetry: An introduction*

An important type of hydrographic surveys is the bathymetric survey. The primary purpose of hydrographic surveying is water depth estimation, which is achieved by bathymetric surveys. Some of the other most common uses of hydrographic surveying include waterway planning, dredging analysis and wreck location. Another application of surveying is in the construction and planning for docks, harbours and dams. It is important to ensure that the water depth in and around ports are sufficient to allow for ships to safely enter and berth. Also, the portion of the seabed that supports floating structures must have a strong foundation. Construction of dams also require adequate knowledge of the surrounding terrain to ensure structural strength.

1.4.5.2. *Equipment (old technique)*

For most of the engineering works, soundings are taken from a small boat. The equipment needed for soundings are a sounding boat, sounding pole (or) lead lines (or) sounding machine.

A boat for soundings
- is used for taking soundings.
- should be sufficiently spacious and stable.
- for calm waters, a flat-bottomed boat is more suitable, but for rough waters, round-bottomed boat is more suitable.
- for regular soundings, a rowboat may be provided with a well through which soundings are taken.

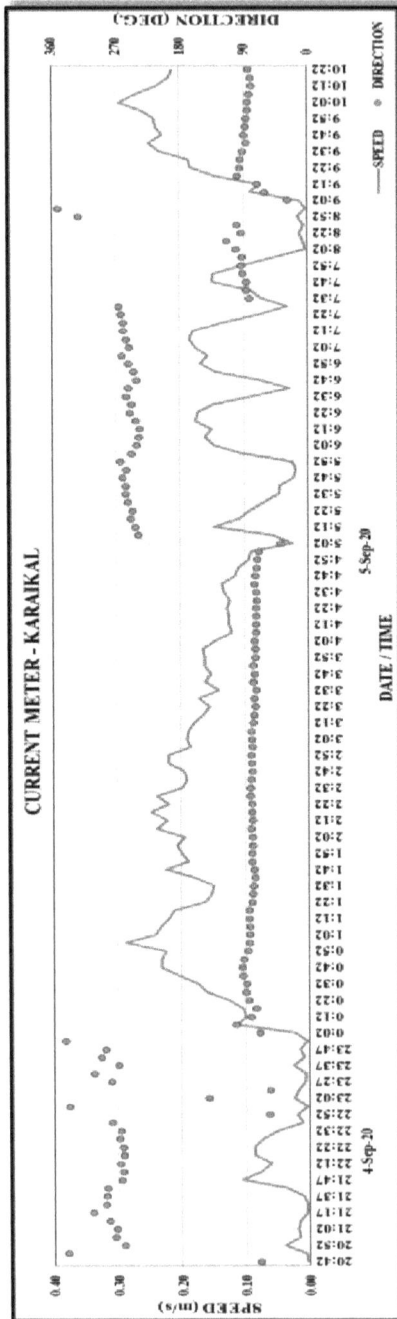

Fig. 1.12.   Typical measured current meter data.

- should be extended far enough over the side to prevent the line from striking the boat.

Sounding poles/Rods
- A sounding pole is made of a sound, straight-grained, well-seasoned tough timber usually 5–8 cm in diameter and 5–8 m in length.
- Suitable for shallow and quiet waters.
- An arrow or lead shoe of sufficient weight is fitted with end.
- Helps in holing them upright in water.
- Lead or weight should be of sufficient area so that it may not sink in mud or sand.
- Pole of length 6 m can be used to depths up to 4 m.

Lead lines
- Lead lines or a sounding line is usually a length of cord or a brass chain with sounding lead attached to the end.
- Due to prolonged use, a line of hemp of cotton is liable to get stretched.
- It should be soaked in water for about one hour before it is used for taking sounding.
- Length of the line should be tested frequently with tape.
- For regular sounding, a chain of brass, steel or iron is preferred.
- Usually used for depths over about 6 m.

Sounding machine
- Where extensive sounding measurements need to be taken, a sounding machine is very useful.
- May either be hand-driven or automatic.
- Lead weight is carried at the end of flexible wire cord attached to barrel and can be lowered at any desired rate, with the speed of barrel being controlled by means of brake.
- Readings are indicated in dials.

Typical views of sounding pole and lead line survey are shown in **Figs. 1.13** and **1.14**, respectively.

1.4.5.3.   *Types of survey*

The types of surveys can broadly be classified as offshore and nearshore surveys. For the offshore survey, a mechanized boat used for bathymetry

SOUNDING POLE

Fig. 1.13.   Sounding pole.

survey at sea depths more than 5 m cannot obtain the soundings for depths less than 5 m as the draft of the boat may jam the seabed having depth less than 3 m. In the case of nearshore surveys, up to 1 m water depth, the beach profiling survey is carried out by employing RTK-GPS at the lowest low water spring. For water depths ranging between 1 and 3 m, the age-old sounding rope technique can be adopted under calm weather conditions with the aid of RTK positioning system.

1.4.5.4.   *Description of instrument (echo-sounder)*

In areas where a detailed bathymetry is required, a precise echo-sounder may be used for hydrography. Most hydrographic echo-sounders are of dual frequency, meaning that a low-frequency pulse (typically around 20–24 kHz) can be transmitted at the same time as a high-frequency pulse (typically around 200–250 kHz). Echo-sounding is a more rapid method of measuring depth than the previous technique of lowering a sounding line until it

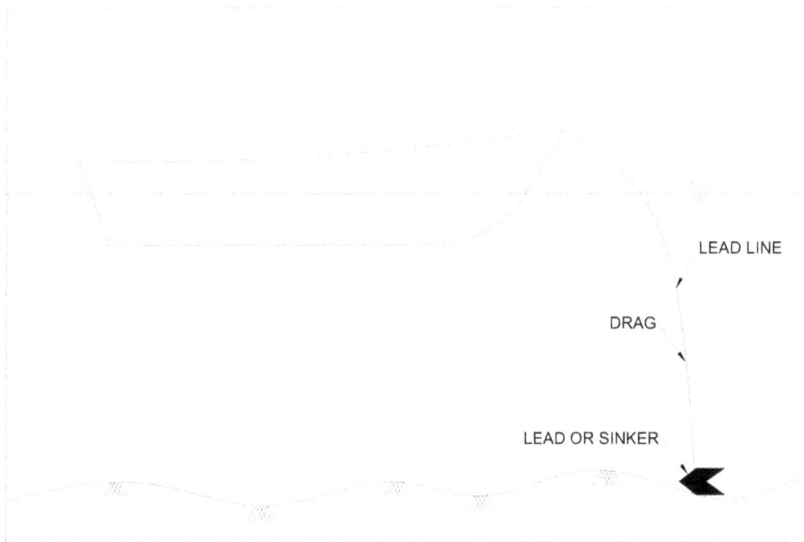

Fig. 1.14.   Depiction of lead line survey.

touches the bottom. The two types of echo-sounders used are single beam echo-sounder and multi-beam echo-sounder. A single-beam echo-sounder measures the double-way transit time of an acoustic signal reflected on the seabed, in which case the water depth under the echo-sounder base is computed knowing the sound velocity in water. In the case of multi-beam echo-sounder, the acoustic signal is generated through a wide angular lateral aperture transducer. Reflections of the lateral echoes of the seabed are received from multiples narrow beams. Water depth is extrapolated along a wide band call swath. The swath width varies from three to seven times the water depth.

1.4.5.5.   *Principle*

An echo-sounder measures the sea floor depth below the sea surface. It measures the depth of water by registering the time for a pulse of energy to travel to the seabed and back and work on the principle of reflection of acoustic energy as shown **Fig. 1.15**.

A short pulse of sound energy is transmitted vertically downward from the ship. This pulse, having been reflected from the sea bottom, returns to the ship in the form of an echo.

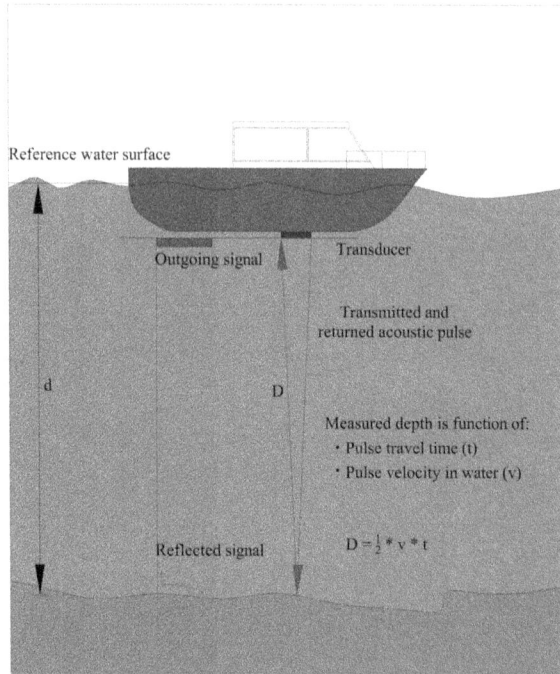

Fig. 1.15. Basic operation of an echo-sounder.

The corrected depth is obtained through the relationship

$$d = 0.5(vxt) + k + d_r$$

where d is the depth from reference water surface, v is the average velocity of sound in the water column, t is the measured elapsed time from transducer to bottom and back to transducer and d_r is the distance from reference water surface to transducer.

1.4.5.6. *Recommended specification of the instrument*

The recommended specifications for an echo-sounder are provided in **Table 1.6**.

1.4.5.7. *Setting up of the instrument*

To conduct a bathymetric survey effectively using an echo-sounder, it is necessary to obtain and load some map layers or background data. The

Table 1.6.   Specifications of echo-sounder.

Description:	Procuring a hydrographic survey system based on single beam technology, dual frequency for the purpose of bathymetry survey.
Technical specification:	
GPS	Equipment integrating a GPS receiver, echo-sounder(s) with water temperature velocity correction, and optional real time correction data.
Resolution	1 cm
Power consumption	Minimum 7.2 W
Internal/External battery	High capacity rechargeable (Minimum 10 hours)
Echo-sounder specifications:	
Depth range	200 kHz up to 100 m
	30 kHz up to 100 m
Maximum ping rate	6 Hz
Accuracy	1 cm ± 0.1% of depth
Transducer	Dual frequency 200/30 kHz, less than 10° beam width @−3 dB
OGC Compliance (Optional)	Certification required for Downloading Software

line files should be created in HYPACK Software in line file format. The operation of the Valeport MIDAS Surveyor GPS Echo-sounder, and its associated PC software Survey Log is to be ensured. The MIDAS Surveyor is a hydrographic survey system, logging a time series of position and depth data. Real-time data is displayed on the integral graphics LCD screen and output in different industry standard formats, and all logged data may subsequently be uploaded to PC. More advanced features of the MIDAS Surveyor include the ability to take corrective data from external instruments, including Valeport tide gauges and sound velocity sensors, together with heave sensors and gyrocompasses. Additionally, a second echo-sounder transducer may also be used. The system logs data asynchronously. It will log all incoming data regardless of the update rate or exact timing of each data channel and will log each data input in sequence. An on-board microprocessor applies an accurate time stamp to each data point, ensuring that a complete time series record of each data input can be generated, together with interpolated position and depth information. Any corrective data received from a heave sensor or tide gauge can be applied in real time to the depth data, and the user may choose whether to log or output either the corrected or uncorrected data. A typical display from the echo-sounder is shown in **Fig. 1.16**. Views of installing an echo-sounder and a typical bathymetry from the echo-sounder are projected in **Figs. 1.17 and 1.18**, respectively.

1.4.6.   *Wind*

1.4.6.1.   *General*

Wind speed, or wind flow velocity, is a fundamental atmospheric quantity caused by air moving from high to low pressure, usually due to changes in temperature. Note that wind direction is usually almost parallel to isobars. Wind speed is now commonly measured with an anemometer but can also be classified using the older Beaufort scale, which is based on personal observation of specifically defined wind effects. An anemometer is used to measure wind speed and consists of a vertical pillar and three or four concave cups, which capture the horizontal movement of air particles (wind speed). Another tool used to measure wind velocity includes a GPS combined with pitot. A fluid flow velocity tool, the pitot tube is primarily used to determine the air velocity of an aircraft.

1.4.6.2.   *Description of instrument*

In the present investigation, a Wind Sonic Ultrasonic Wind Sensor was installed to measure wind parameters. Wind Sonic is a robust, low-cost ultrasonic wind sensor with no moving parts. This two-axis ultrasonic wind sensor offers maintenance-free wind speed and direction monitoring for true "fit and forget" wind sensing. Automatic Weather Station (AWS) is a facility, either on land or sea, with instruments and equipment for measuring atmospheric conditions to provide information on weather forecasts and to study the weather and climate.

An AWS will typically consist of a weather-proof enclosure containing the data logger, rechargeable battery, telemetry (optional) and meteorological sensors with an attached solar panel or wind turbine and mounted upon a mast. The specific configuration may vary based on the purpose of the system. The system may report in near real time via the Argos System and the Global Telecommunications System or save the data for later recovery.

AWSs include thermometer for measuring temperature, ultrasonic wind sensor for measuring wind speed and wind direction, hygrometer for measuring humidity and rain gauge for measuring rain.

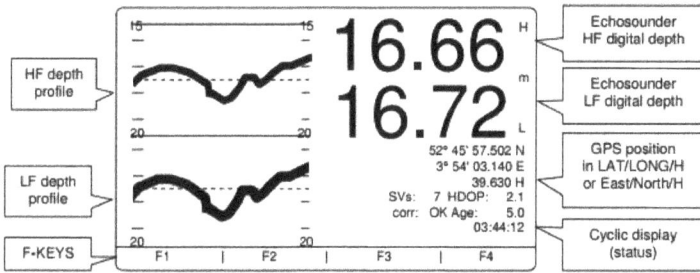

Fig. 1.16.   Echo-sounder display.

Fig. 1.17.   Setting up of instrument.

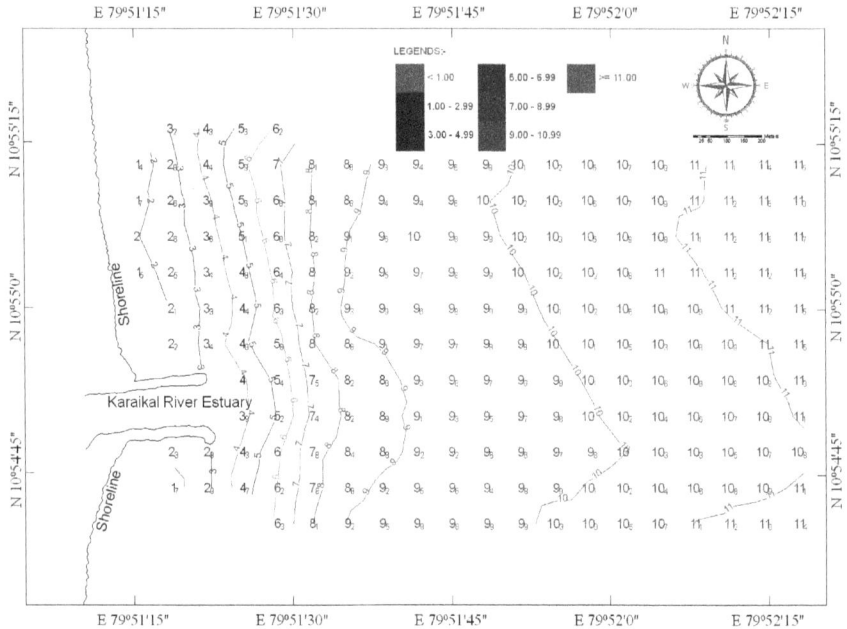

Fig. 1.18. Typical bathymetry of Karaikal.

1.4.6.3. *Recommended specification of the instrument*

The recommended specifications of an AWS are listed in **Table 1.7.**

1.4.6.4. *Deployment of the instrument*

The deployment of AWS at Karaikal is shown in **Fig. 1.19.**

1.4.6.5. *Typical results of wind observation by AWS*

Typical data acquired through the AWS deployment in Karaikal in the form of wind rose diagrams are shown in **Fig. 1.20**, which display the percentage of occurrence of wind speed with respect to the direction for the individual months.

1.4.7. *Beach profile and shoreline changes*

1.4.7.1. *General*

The beach profiles and shoreline survey provide useful information for coastal monitoring studies and management processes. Beach profile

Table 1.7. Specifications for AWS.

Description	Specifications
Wind Speed	
Range	0–60 m/s
Accuracy	±2% @ 12 m/s
Resolution	0.01 m/s (0.02 knots)
Response Time	0.25 s
Threshold	0.01 m/s
Wind Direction	
Range	0–359° (no dead band)
Accuracy	±2° @ 12 m/s
Resolution	1°
Response Time	0.25 s
Measurement	
Ultrasonic Output Rate	0.25, 0.5, 1, 2 or 4 Hz
Parameters	Wind speed and direction or U and V (vectors)
Units of Measure	m/s, knots, mph, kph, ft/min
Outputs	
Option 1	RS232
Option 2	RS232 + RS422 + RS485 + NMEA 0183
Baud Rate	2,400–38,400
Power Requirement	
Current Drain	Dependent on option selected, e.g., <2 mA @ 12V (SDI-12) to 44 mA @ 12 V (4–20 mA)
Start-up time	<5 s
Mechanical	
External Construction	LURAN S KR 2861/1C ASA/PC6
Size	142 mm × 163 mm
Weight	0.5 kg
Environmental	
Protection Class	IP66
Operating Temperature	−35°C to +70°C
Storage Temperature	−40°C to +80°C
Operating Humidity	<5% to 100% RH

surveys at suitable geographical locations repeated at appropriate time intervals will provide excellent results of beach changes. Beach morphology and volume changes can be assessed by comparing surveys taken along the same profile line at different months continuously. The study of the rate of change in shoreline is necessary for coastal morphodynamics studies that is vital for any coastal development. Beach profiling measurements are conducted along profile transects. The profile transect is a straight line running from the crest of the dune or other high point on the beach (such

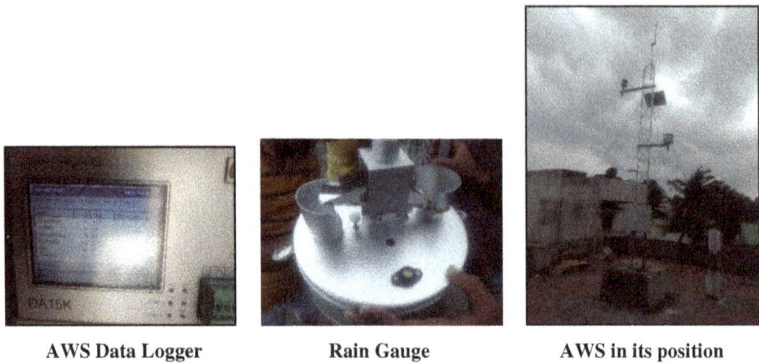

AWS Data Logger   **Rain Gauge**   **AWS in its position**

Fig. 1.19. AWS deployment.

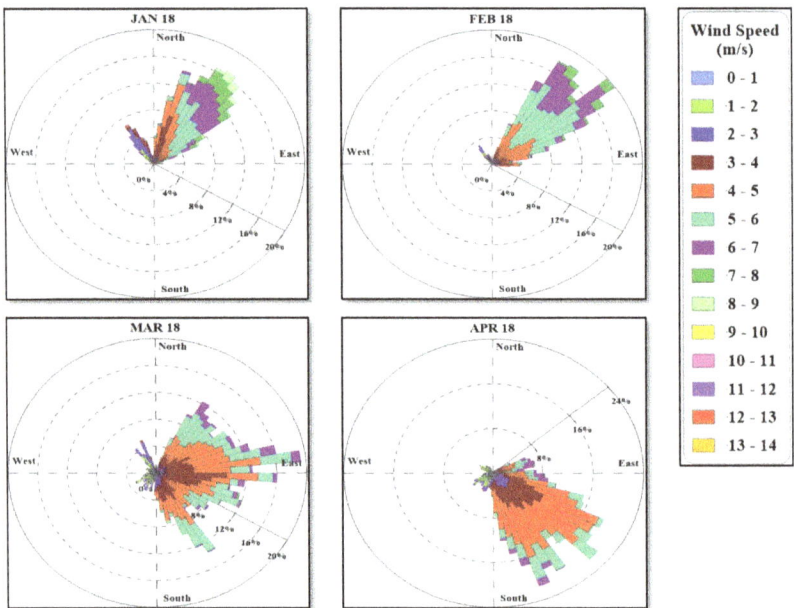

Fig. 1.20. Typical wind rose diagram acquired by AWS.

as a seawall) to the waterline. Along each transect, measurements are taken every 5 m, allowing observations of the beach to be made. Each monitoring coast has between 2 and 10 profile transects, depending upon length of the coast, and each transect has a different number so that comparisons can be made between these various lines.

In the case of shoreline mapping, the conventional techniques for assessing the rate of change of shoreline position include field measurement of present high water level and low water level, shoreline tracing from aerial photographs and topographic sheets and comparison with the historical data. Changes in shoreline through the processes of accretion and erosion can be analyzed in a geographic information system (GIS) by measuring differences in past and present shoreline locations.

1.4.7.2. *Description of instruments*

The survey instruments used to conduct the beach profile survey and shoreline mapping are a total station for beach profile survey and Real-Time Kinematic Global Positioning System (RTK-GPS) for beach profile and shoreline survey.

A total station is an electronic/optical instrument used in modern surveying. The total station is an electronic theodolite (transit) integrated with an electronic distance meter (EDM), plus internal data storage. It is designed for measuring of slant distances, horizontal and vertical angles and elevations in topographic and geodetic works, tachometric surveys as well as application of geodetic tasks. The measurement results can be recorded into the internal memory and transferred to a personal computer interface. The angles and distances are measured from the total station to points under survey, and the coordinates (X, Y and Z, or northing, easting and elevation) of surveyed points relative to the total station position are calculated using trigonometry and triangulation.

1.4.7.3. *Working principle*

Distance measurement
The EDM instrument is a major component of a total station. Its range varies from 1.5 m to 3.5 km. The accuracy of measurement varies from 1 mm + 1.5 ppm. They are used in conjunction with an automatic target recognizer. The distance measured is always sloping distance from instrument to the object with Distance = Velocity × Time.

Angle measurement
The electronic theodolite part of the total station is used for measuring the vertical and horizontal angles. For measurement of horizontal angles, any convenient direction may be taken as the reference direction. For vertical angle measurement, the vertical upward (zenith) direction is taken as the reference direction. The accuracy of angle measurement varies from $1''$ to $5''$.

Data processing

This instrument is provided with an inbuilt microprocessor. The microprocessor averages multiple observations. With the help of slope distance and
vertical and horizontal angles measured, when the height of axis of instrument and targets are supplied, the microprocessor computes the horizontal
distance and X, Y, Z coordinates. The processor can apply temperature
and pressure corrections to the measurements, if atmospheric temperature
and pressures are supplied.

1.4.7.4. *RTK-GPS*

Background

RTK-GPS is a satellite navigation technique used to enhance the precision
of position data derived from satellite-based positioning systems (global
navigation satellite systems, GNSS) such as GPS, GLONASS, Galileo and
BeiDou. It uses measurements of the phase of the signal's carrier wave in
addition to the information content of the signal and relies on a single reference station or interpolated virtual station to provide real-time corrections,
providing up to centimetre-level accuracy. With reference to GPS in particular, the system is commonly referred to as carrier-phase enhancement
or CPGPS. It has applications in land survey, hydrographic survey and
unmanned aerial vehicle navigation. The base station is at a known point,
whether fixed on a building permanently or a tripod-mounted base station.
The fact that it is in a known position allows the base station to produce
corrections. The constellation notifies the base station that it is at a slightly
different place, so corrections can be generated to send to the rover at the
unknown point. The corrections are applied in real time.

Recommended specification of the instrument

The recommended specifications of a total station and RTK-GPS are provided in **Tables 1.8** and **1.9**, respectively.

1.4.7.5. *Working with the instruments*

Total station

All the functions of a total station are controlled by its microprocessor,
which is accessed through a keyboard and display. To use the total station,
it is set over one end of the line to be measured and a reflector is positioned
at the other end such that the line of sight between the instrument and the

Table 1.8.   Specification of the total station.

Description	Technical Specifications
Angle Measurement	
Angle Display	0.2″
Angle Accuracy	1″ (In Manual, Robotic and Motorised Mode)
Telescopic Magnification	30X
Field in view	1°30″ (2.7 m at 100 m)
Minimum Focusing Distance	2.0 m
Distance Measurement Mode	With Prism and Without Prism Measurement
Using Long-Range EDM	3 km or better, Long range: up to 10 km with an accuracy of 5 + 2ppm
Accuracy	1+ 1.5 ppm up to 3 km, Time: 3–5 s
Least Count	0.1 mm
Without prism range (Reflector less)	Up to 1000 m
Accuracy Reflector less	+/−2 mm + 2 ppm up to 500 m, 4 + 2 ppm up to 1000 m
	Measuring time: 3–6 s
Measuring interval	1″ or less in tracking
Angle Measurement Method	Absolute Encoders
Motorized	Motorized movements to stake out points.
Rotation Speed	45°/s
360° prism, Target aiming/locking	Up to 1000 m/1000 m
Circular prism, target aiming/locking	Up to 1500 m/1000 m
Prism Search	Have facility to automatically search the prism up to a distance of 300 m in typically 5–10 s
Display & keyboard	Identical display of size 5 inches, WVGA, colour, touch screen with 30 or more keys with Windows EC7
Internal memory, processor and weight	2 GB, 1 GB and 5–6 Kg. External storage expandable up to 8 GB
Battery	5–8 h for a single battery. Charging time: 2–3 h. Standby battery to be given with separate charger per battery.
Interface	RS232, USB, Bluetooth and WLAN, IP Protection: IP65
	Temperature range: 10–50°C.
	Humidity: 95% non-condensing.

reflector is unobstructed. The reflector is a prism attached to a detail pole (shown in **Fig. 1.21**). The telescope is aligned and pointed at the prism. The measuring sequence is initiated, and a signal is sent to the reflector and a part of this signal is returned to the total station. This signal is then analyzed to calculate the slope distance together with the horizontal

Table 1.9. Specifications of the RTK-GPS.

Description	Technical Specifications
Satellite tracking	Tracking of most of the presently available frequencies of globally existing GNSS constellations like GPS (LI, L2, L2C, L5), GLONASS (L l, L2, L3), BeiDou (B1, B2), Galileo (El, E5A, E5B), QZSS, SBAS (WAAS/EGNOS/ MSAS/ GAGAN) and IRNSS (L5)
No of Channels	GNSS receiver system should have 300 or better channels
Measuring Mode	Static, DGPS, RTK-GPS & GPRS Module
	Technology for optimal GNSS performance
	Supported data formats: NMEA 0183, RTCM (2.2, 2.3, 3.1, 3.2, etc), CMR, CMR+, etc
	Capable to work GSM & low power UHF radio, both built into the same GNSS sensor unit
	Provision for SIM Card for GSM RTK in GNSS Sensor.
	RTK Accuracy
Horizontal	8 mm + 1 ppm
Vertical	15 mm + 1 ppm
RTK initialization	RTK initialization range in the GSM/GPRS mode: 30 km or better
Static Accuracy	
Horizontal	3 mm + 0.1 ppm
Vertical	3.5 mm + 0.4 ppm
GNSS interface	1. Both cable and Bluetooth communication between Receiver and Controller
	2. RS232, USB & Web UI for Receiver Status, Settings and Data Transfer
Power Characteristics	Battery for a minimum of 8 h operation. Standby battery to be given for each GPS and controller.
Charger	Each battery to be given a separate charger unit.
GNSS **Internal Memory/ Update rate**	Removable memory card of up to 8 GB in the GNSS sensor and update rate: 20 Hz.
Protection	Waterproof, Shockproof, Dustproof, Humidity-proof (100%) and condensation-proof as per IP68 standards.
Drop survival	2 m pole drop onto concrete.

Fig. 1.21(a).   Total station.

Fig. 1.21(b).   TS reflector prism.

and vertical angles. Some instruments have motorized drivers and can use automatic target recognition to search and lock into a prism; this is a fully automated process and does not require an operator, whereas total stations can also be controlled from the detail pole, enabling surveys to be conducted by a single person.

RTK-GPS

The RTK-GPS is based on at least two GPS receivers — a base receiver and one or more rover receivers. The base receiver take measurements from satellites in view and then broadcasts them, together with its location coordinates, to the rover receiver(s). The rover receiver also collects measurements and transmits to the satellites in view and processes them with the base station data. The rover then estimates its location relative to the base. The key to achieving centimetre-level precision with the RTK is the use of

Fig. 1.22(a).   RTK-GPS base setup.

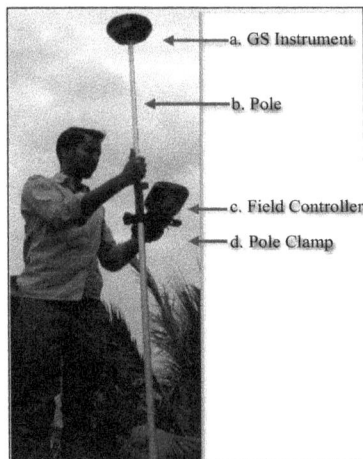

Fig. 1.22(b).   RTK-GPS rover setup.

the GPS carrier phase signals. In the receiver, carrier phase measurements are made with millimetre-level precision. Although carrier phase measurements are highly precise, they contain an unknown bias, termed the integer cycle ambiguity, or carrier phase ambiguity. The rover must resolve the carrier phase ambiguities during measurements (**Fig. 1.22**).

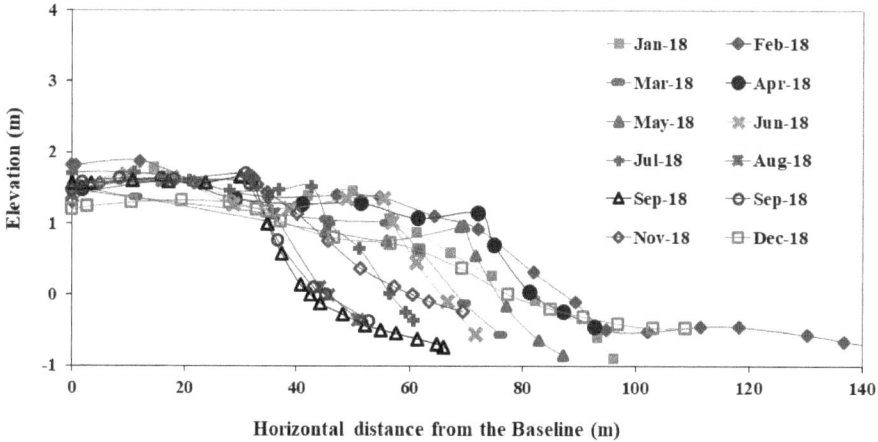

Fig. 1.23.   Typical variations of measured beach profiles.

- Profile transects are cast manually with 100–150 m spacing.
- The baseline to be fixed which is an arbitrary line that can be drawn either shoreward or landward of shorelines to serve as the origination point of transects used for calculations.
- In the study area, Local Benchmark needs to be established on the ground (often in the sand dunes and safe from erosion).
- The vertical change in elevation from the beginning to the end of the profile line is measured in meter.
- The horizontal distance from the beginning to the end of the profile.
- The survey profiles are registered back to local Benchmark.
- Comparing profile lines recorded at different months makes it possible to measure changes in the distribution of sand on the beach.
- A graph needs to be plotted with horizontal distance vs vertical elevation for each transect for each month.
- From the graph, the volume of sand (for 1 m width) from the berm to low water line can be calculated. The slope gradient of beach profile can also be calculated.

The superimposition of monthly beach profile along the Karaikal site for a typical transect and shoreline variations are shown in **Figs. 1.23** and **1.24**, respectively.

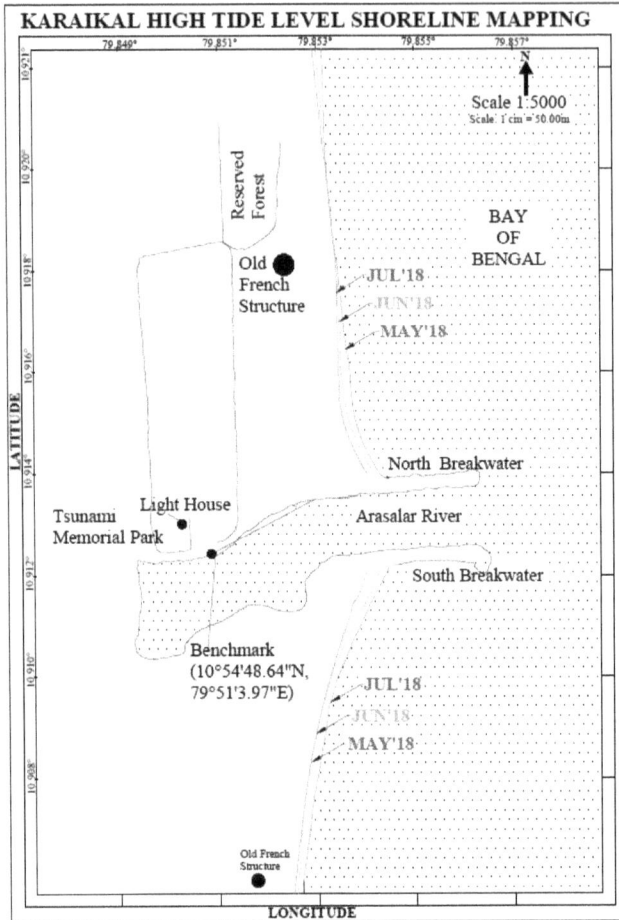

Fig. 1.24. Typical measured shoreline evolutions.

1.4.8. *Riverine data — Conductivity, temperature and depth*

1.4.8.1. *Background*

Conductivity is measured to determine the salinity of the water body. The other oceanographic parameters like Conductivity, Temperature and Depth (CTD) are measured, which is used for different kind applications. CTD is a cluster of sensors fitted to probe that is attached to the cylindrical rosette which is immersed into sea water.

1.4.8.2.   *Description of instrument*

The sensors are fitted to the instrument and measure different parameters that are in general processed through a computer software to create another set of parameters. For example, CTD will measure conductivity, temperature and pressure, and these parameters are processed for salinity. Temperature is measured by a device called thermistor, which transmits data through the thermal compression or expansion of liquid. Pressure is measured by pressure gauge, which operates on the principle of compression of a coil or tube filled with liquid, which changes shape depending on the external pressure exerted on the gauge.

1.4.8.3.   *Recommended specification of the instrument*

The detailed specifications of CTD are depicted in **Table 1.10.**

1.4.8.4.   *Instrument deployment*

The CTD instrument can be deployed from the deck or jetty structure or even from the research vessel. The instrument is fully submerged into the water, which is called the downcast, to a determined depth or to a few meters above the ocean floor, generally at a rate of about 0.5 m/sec. In general, a conducting wire cable is attached to the CTD instrument to allow instantaneous uploading and real-time visualization of the collected data on the computer screen. The interval between observations is usually 6 hours.

The order of synoptic hour is 5.30, 11.30, 17.30 and 23.30. At this time, the CTD instrument is deployed in the marine body in the vertical profile to collect the parameters. The deployment of a CTD is projected in **Fig. 1.25.**

1.4.8.5.   *Typical results of wave observation by DWR*

Typical records obtained from the deployed CTD off Karaikal are shown in **Fig. 1.26.**

1.4.9.   **Coastal sediments**

Coastal sediments involve seabed sampling and nearshore sampling to determine the particle size distribution and its specific gravity. The sediment samples (onshore and offshore) at different locations along the stretch of the coast of the study area are collected monthly/seasonally. The seabed

Table 1.10.   Specification of the CTD.

Description	Specifications
Logger	
Power Supply	Internal/external battery
Communication	USB
Housing Material	Up to 350 m
Storage Memory	more than 100 MB
Sampling Period	Variable sampling period (1 s to 24 h)
Sampling rate	4
Software	Windows-compatible software to view and present all parameters data output shall be supplied. Provision to export the data to MAT LAB for further analysis should be available.
Conductivity	
Measurement range	0 to 70 mS/cm
Accuracy	±0.01 mS/cm at 35 psu 15°C
Resolution	~1 µS/cm (marine)
Time constant	<150 ms set by flow through cell
Temperature	
Measurement range	−5°C to 35°C
Accuracy	±0.01°
Resolution	<0.001°C
Time Constant	<150 ms
Depth	
Measurement Range	1–50 m depth
Accuracy	±0.1% full scale
Resolution	<0.002% full scale
Time constant	<0.01 s
Dissolved Oxygen (Optional Sensors)	
Measurement Range	0–120%
Accuracy	±3% O_2 saturation (5–25°C)
Resolution	0.5% of saturation
PH (Optional Sensors)	
Measurement Range	2–13 pH
Accuracy	±0.03
Turbidity (Optional Sensors)	
Range	0–1000 FTU (NTU)
Accuracy	±2%
OGC Compliance	**Certification required for Downloading Software**

Fig. 1.25.   Deployment of CTD.

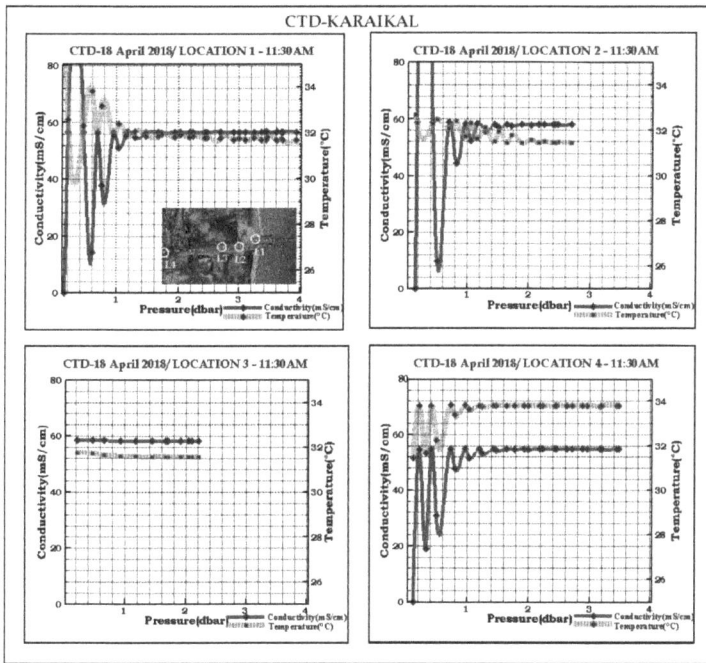

Fig. 1.26.   Karaikal-CTD data for four locations — 18 April 2018.

sediment samples usually are collected at every 2.5 m depth intervals from
the shore to about −20 m depth (i.e., 2.5 m, 5.0 m, 7.5 m, 10.0 m, 12.5 m,
15.0 m, and 20.0 m) for a particular season. Shore sediments were collected
on a monthly basis across the 5 transects (three transects to the north and
two transects to the south of Arasalar river inlet); 2 samples were collected

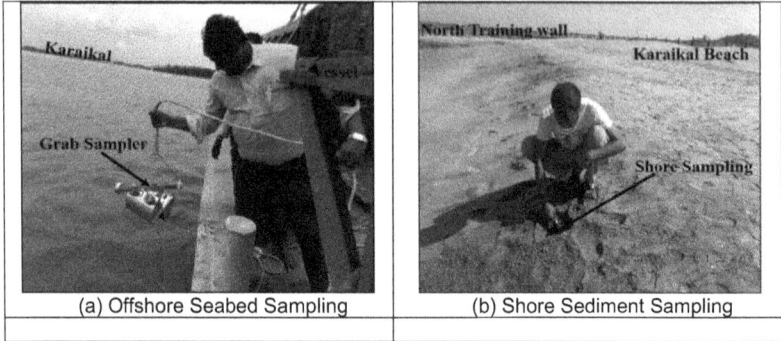

Fig. 1.27. Sediment sampling and analysis.

per transect resulting in a total of 10 location points for sample collection. The shore samples were taken along the shoreline (spring low tide) and on the berm (spring high tide) at every month and analyzed for the particle size distribution. Views of collection of seabed sample in the offshore and along the shore are shown in **Fig. 1.27**. These details are discussed in Chapter 9.

1.5. Summary

This chapter presented in detail the measurement of the different parameters that are needed to plan any type of development along the coast or mitigation measures against erosion. The instruments needed to perform a comprehensive measurement campaign along with typical data collected along the Karaikal coastline situated along the southeast coast of India, which is considered as a case study, are presented in this chapter. The chapter also considers the parameters that need to be measured to understand the morphological changes within an estuary.

Chapter 2

Sustainable Hard and Soft Measures for Coastal Protection

Abstract
Coastal developments across the globe have been rapid, in keeping pace with the growing infrastructure for ports, inland waterways, trade and commerce, tourism, fisheries, a hub for power plants, etc. This has resulted in an increase in population density along the coastal stretches. Such developments are bound to lead to instability in the shoreline, apart from the natural causes of coastal hazards. The process of erosion as well as accretion in the vicinity of approach channels of harbours, confluence tidal inlets or seas can either be natural, man-made or a combination of both. Excessive erosion or accretion problems need to be handled carefully without causing adverse effects along the adjoining shoreline, although it is a challenging task. Several such issues and the measures to combat the problems are well understood through case studies. To counter or minimize coastal erosion, the protection measures are broadly classified as hard and soft measures. A few case studies based on both approaches are considered, their performance characteristics are presented, and the merits and demerits of both methods are highlighted in this chapter.

2.1. Introduction

2.1.1. *Basic approaches to counter coastal erosion*

Developmental activities across the shore mandates the construction of facilities that promote industrial growth, such as fishing harbours, ports, jetties, training walls, quays and wharfs as well as facilities for tourism development such as surfing friendly sea-bed slopes, added beach width for recreational activities and establishment of beach resorts. The stability of coastlines is governed by a number of environmental parameters, knowledge on which is essential for planners to evaluate its status and devise protection measures for shorelines in distress facing erosion. In addition to the usual driving forces due to waves, tides, wind and currents that leads to perineal instability of the shorelines, forces induced during extreme coastal hazards such as cyclones, storm surges, depressions, tsunami and tidal bores continue to pose a challenge to the coastal engineers. The effect of extreme events on bay-like formations may be insignificant compared to open coasts. The structures constructed along the coast play a significant role in dictating the stability of shorelines adjoining them.

Any stretch of a coastline warrants protection measure when it experiences an erosion rate beyond a threshold value, and when the seasonal sediment transport/littoral drift changes fail to restore equilibrium. The predominant coastal hazards such as coastal flooding, coastal erosion, sea-level rise, storm surge and sea-water inundation are common threats to sensitive coastal communities, alongside the prevailing natural causes of melting glaciers due to global warming and extreme events such as storms and cyclones. Man-made artificial structures in the nearshore regions or in between shallow and intermediate water depths can potentially destabilize the shoreline and lead to excessive coastal erosion if not addressed properly. The behaviour of the shoreline is mostly site-specific, calling for a critical examination of a host of parameters. Necessary precautionary measures need to be taken to ensure the activities along the shore are ecologically engineered, especially along the urbanized zones. To counter or minimize coastal erosion, the protection measures are broadly classified as hard and soft measures. This chapter details the vast experience gained through several case studies.

To regulate and stabilize the coastline, shore protection measures are in place to reclaim the lost sediments due to erosion. In most scenarios, where there is a short-term effect in loss of sediments, the "do nothing" approach would naturally restore equilibrium. If the causes resulting in coastal erosion are adverse, "to protect", "to accommodate" or "to retreat" approaches (**Fig. 2.1**) can be attempted.

2.1.2. *Setback zone*

For carrying out any developmental work along the coast, a setback zone is established, i.e., the stretch from the shoreline up to a permanent vegetation line or the highest high tide line wherein either certain or all types of development are prohibited or significantly restricted. The main functions of the coastal setback zone are to provide protection along the coast for human settlements against flooding and erosion, preserve the beaches and support the enhancement of the livelihood of coastal communities, which includes fishing, tourism and other developmental activities. Coastal management

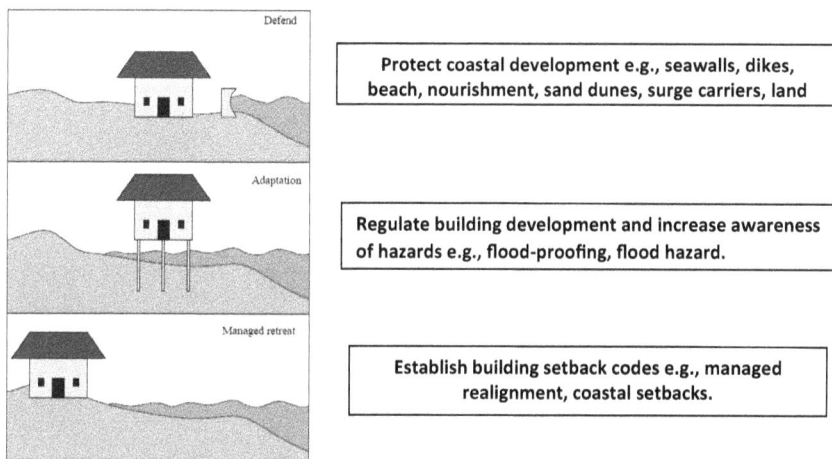

Fig. 2.1.   Basic approaches to handle coastal erosion.

involves determination of the extent of inter-tidal area to be demarcated to ensure a healthy, functional coastal system that reduces risks and adjusts to changing conditions. Where should this line run to give a sustainable coast? The setback zone is usually a specific uniform distance from the shoreline; the said distance varies based on the vulnerabilities of the coast to natural hazards. The setback zone varies with countries as well as within regions as projected in **Table 2.1.**

2.1.3.   *Technical Guidelines for setback zone*

The guidelines for fixing the setback zone are as follows.

- To map the vulnerability of the coast, four parameters are taken into account, viz., elevation, geomorphology, sea-level trends and horizontal shoreline displacement.
- The elevation data shall be obtained from the available coastal toposheets and satellite data surveys.
- The landforms will be identified on the maps based on the available toposheet and remote sensing data. The bathymetry to be derived from naval Hydrographic Charts on location-specific surveys.
- The sea-level trend data shall be based on primary data published by the Survey of India. The median estimate of mean sea-level rise in the next 100 years in terms of the Fourth Assessment Report of the Inter-Governmental Panel on Climate Change (IPCC) shall be considered.

Table 2.1. Setback line distance for different countries.

Countries	Setback zone	Countries	Setback zone
Australia NSW	200 m	New Zealand	22 m
Bahamas	5–15 m	Norway	100 m
Barbados	30 m	Oregon, USA	permanent vegetation line (variable)
Belize	20 m	Panama	20 m
Brazil	33 m	Philippines (Mangrove greenbelt)	20 m
Canada NB	30 m	Poland	200 m
Canada BC	15 m	Sweden	100 m
Chile	80 m	Spain	100–200 m
Colombia	50 m	Turkey	50 m
Costa Rica (Public Zone)	50 m	Uruguay	250 m
Costa Rica (Restricted Zone)	50–200 m	USSR Coast of the Black Sea	3000 m
Chile	80 m	Venezuela	50 m
Dominican Republic	60 m	Greece	500 m
Denmark	300 m	Hawaii	13 m
Ecuador	8 m	India	500 m
France	100 m	Indonesia	50–400 m
Germany	100–200 m	Jamaica	30 m
Nicaragua	20 m	Mexico	20 m

- The erosion/accretion data of horizontal shoreline displacement shall be obtained from long-term information derived from Survey of India Topographic maps (1967) and the latest satellite data. Horizontal shoreline displacement will be estimated (median estimate) over the next 100 years.
- The level of protection to be provided by the Setback Line will correspond to protection from coastal hazards as per the median estimates of mean sea-level rise and horizontal shoreline displacement.

2.2. Evaluation of Erosion Rate

The steps to assess the erosion rate in general are: (i) verify if the erosion is temporary due to seasonal effect or short-term extreme event; (ii) understand the importance of the coastal stretch with priority being given if densely inhabited; (iii) gauge the effect of proposed protection measures on the adjoining shoreline; (iv) identify the method of protection, including the availability of materials and labour and (v) work out the cost of different alternatives. The costs should include not only maintenance, construction, etc. but also take into account the loss of cultural values, impact on safety and the needs of the local public. Furthermore, only after ascertaining the above basic resources and needs have been met, i.e., if fund/resources are available, proceed to combat erosion permanently by proper planning. If enough funds or resources are unavailable, careful planning of temporary measures is essential.

2.3. Preliminary Studies

The primary step is to identify the causes for erosion, i.e., whether it is temporary or permanent. This can be identified through field studies or satellite imaging. Secondly, a detailed field study and data collection regarding the nature of the coast, soil properties, sediment concentration, physical features, bathymetry, tide cycles, wave climate and frequency of extreme events need to be assessed. Lastly, the socio-economic factors, including the aesthetic appeal of the coast, safety evaluation, public convenience, etc., are to be considered, and alternative temporary and more economically viable options are to be investigated and modelled with the aid of numerical or physical model studies. In the case study of a long coastal stretch, for instance, that of the entire maritime state, a preliminary survey followed by identification of vulnerable stretches and evolving with conceptual designs need to be prepared. From the conceptual designs, detailed design needs to be done with the recent shoreline positions and bathymetry. If the necessary funds are available to execute the planned coastal protection measure, it can be commissioned under suitable weather conditions to arrive at optimum results. As the different seasons dictate the direction of the sediment transport or littoral currents, the construction sequence must be planned accordingly. This is crucial at locations where groin field is planned as a coastal protection measure.

2.4. Monitoring Coastal Zones

One of the problems that coastal managers, planners and engineers face is the uncertainty in the data available on the processes that cause erosion of beaches. Several reliable and novel monitoring systems that function at a low cost are now available to analyze shoreline data over a wide range of intervals. Shoreline mapping is one of the most important coastal monitoring systems, where the shoreline has to be considered in a temporal sense whereby scale is dependent in context of the investigation. Owing to the dynamic nature of the shoreline, researchers employ various coastline indicators to represent the true shoreline position. The reliability of a monitoring technique vastly depends upon the source of data. Historical shoreline data across any given coastal stretch may often be limited, thus a choice of selection of data source does not exist. The main sources include historical maps, aerial photographs, beach profiling surveys, remote sensing and video analysis (i.e., cost-effective, continuous and long-term monitoring of beaches).

2.5. Coastal Protection Measures

The shore protection measures adopted can either be classical methods such as the construction of groin field, seawall and revetments or novel methods such as employing vegetation cover, geosynthetic materials, beach nourishment or dune stabilization. The choice of adopting an established shore protection measure for a specific field problem needs to be made after scrutinizing the local field conditions and ensuring that the proposed measure is sustainable up to or beyond its design life. The coastal protection measures, in general, as projected in **Fig. 2.2** is broadly classified as hard measures (groins, seawalls, offshore detached breakwaters) and soft measures (artificial beach nourishment, bio-shields/vegetation, dune stabilization, application of geosynthetics). Erstwhile a combination of both can also be implemented.

The hard measures are conventional methods of coastal protection that consist of gravity-type structures usually made up of rubble-mound or precast concrete units. Implementing these measures yield immediate results, yet they are expensive, unaesthetic, leave behind an ecological footprint and are sometimes difficult to construct, however, its effects might be adverse if not properly planned and executed and are irreversible. However, they are well-established and the most reliable method to attain shoreline stability within a short time-span. A few examples of successful implementation and performance of hard engineering measures are discussed in brief in the following sections. Adopting natural rocks for the armour layers for the different hard measures has resulted in its depletion, forcing the application of precast concrete blocks. Several types of concrete armour blocks (such as tetrapods, accropodes, dolos, x-blocks, etc.) are available, which can be laid as a single or double layer, the details of which are available in the Coastal Engineering Manual (2006).

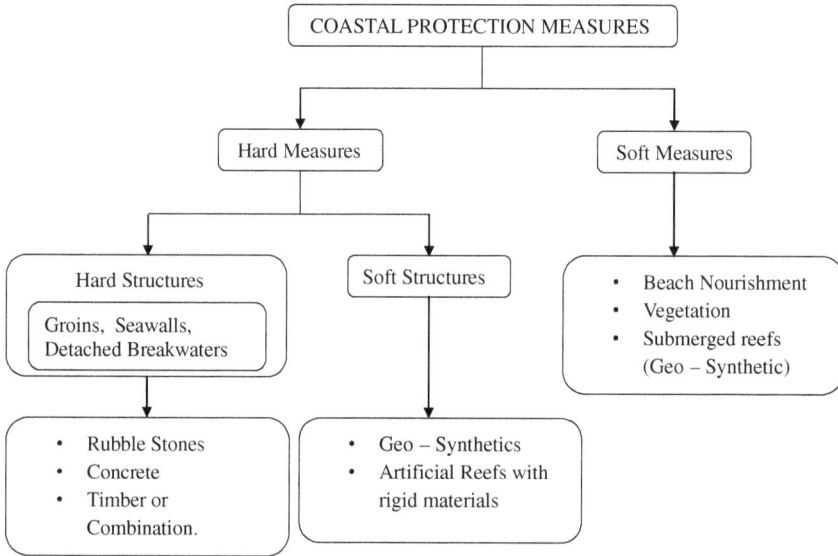

Fig. 2.2. Broad classification of Coastal Protection Measures.
Source: Sundar (2020).

The soft engineering measures form the other contrast spectrum of coastal protection techniques, which leaves behind a minimum or almost no ecological footprint. In comparison with the hard structures, the shoreline response to these measures is rather a slow process. The popular soft engineering measures are artificial beach nourishment, dune stabilization and vegetal cover protection. These methods require a higher degree of expertise for planning and execution in the field. The recent developments in the application of geosynthetics in coastal engineering problems can be termed as soft engineering structures. Pilarczyk (2000) and Koerner (2012) have discussed in detail the application and design guidelines of geosynthetic materials across various civil, hydraulic and coastal engineering perspectives. Although geotextiles were previously employed for roadway construction, it was Recio and Oumeraci (2008, 2009) who had pointed out the usefulness of the application geosynthetic materials for coastal and marine works. The conventional structures, such as seawalls, submerged breakwaters, reefs etc., can be entirely or partially made from geosynthetic materials. This is a viable option for remote locations where the transport of heavy boulders and concrete is difficult.

2.6. Soft Measures

2.6.1. *Application of geosynthetics in coastal engineering*

Geosynthetics are polymeric products that are used in combination with soil, rock or other soil-like materials for various engineering applications. Geosynthetics, as described by the International Geosynthetics Society (2009), are primarily used as:

- Barrier, to prevent the migration of liquids or gases.
- Containment, to contain soil or sediments to a specific geometry and prevent their loss. The contained fill takes the shape of the inflated at-rest geometry of the geosynthetic container.
- Drainage layers, to collect and transport fluids.
- Filter layers, to allow the passage of fluids from soil while preventing the uncontrolled passage of smaller soil particles.
- Protection layers, to prevent or reduce a localized stress development in the reduction layer such as to minimize damage to a given surface.
- Reinforcement, to resist stresses or contain deformations in geotechnical structures.
- Separation layer, to separate two dissimilar geotechnical materials or to prevent intermixing.
- Surficial erosion controller, to prevent the surface erosion of soil particles due to surface water run-off and/or wind forces.
- Frictional interlayer, which is a layer introduced within an interface to increase or reduce friction across the interface.

Geotextiles are commonly used as filter layers in most of the coastal and hydraulic structures to enhance bed protection and seepage. They are often used as composite construction elements alongside sand, sand slurry, concrete etc. Although several case studies exist only a few case studies with geosynthetics applications are discussed in this chapter.

2.6.1.1. *Stabilizing the scour embankments at the Eider barrage, Germany*

The Eider Barrage (54.264605°N; 8.844999°E) is located at the mouth of the river Eider on Germany's North Sea coast, with a primary intent to protect the coast from storm surges from the sea. The tidal barrage is one of the largest coastal protection structures in Germany, commissioned in 1973. When the Eider River storm tide barrage was constructed, bed protections

Fig. 2.3.   Landward aerial view of the storm tide barrage at the Eider River. *Source*: Ulf Jungjohann, Heide, Germany.

were arranged both on the sea and the land sides. As expected, scours developed next to the rigid bed stabilization on both sides (Heibaum, 1995). Due to the given scour geometry with steep embankments, the common construction methods could not be used. Therefore, non-woven geotextile containers with gravel filter material were adopted to offer scour protection. The landward aerial view of the storm tide barrage at the Eider River is shown in **Fig. 2.3** (Müller and Saathoff, 2015).

2.6.1.2.   *Cliff erosion protection with wrapped geotextile sand cushions at Sylt Island, Germany*

The historic monument, called the Kliffende House (54.964483° N; 8.333632° E) is situated along the western coast of the island Sylt in Germany, which experienced a series of severe storm tide incidents along the coast in the year 1990. As a result, the coast witnessed severe erosion problems, and further tidal storms could potentially risk the existence of the said monument along the coast. Attributable to the fragile site conditions, the construction of hard engineering structures such as concrete revetment

Fig. 2.4. Application of geotextiles as protection measure for Kliffende, Island of Sylt, Germany.
Source: Nickels and Heertan (2000).

or rubble-mound seawall sections were eliminated. A novel system was developed using geotextiles, where geotextile sand cushions/sandbags were employed. The geotextiles function as filter-effective protection against sediment wash-out, and the terraced layered geotextiles also acts as reinforcement for stabilization of the dune embankment. The aforesaid measure was executed in combination with artificial beach nourishment, such that even after a severe storm attack, the recovered sand can be used for renourishment. The sandbags were installed on top of each other to form a stabilized beach section. Sand trap fences made of bushes were installed and beach grass further boosted stabilization. The final construction resembled a natural dune as shown in **Fig. 2.4** (Nickels and Heertan, 2000).

2.6.1.3. *Submerged geotextile tube (dyke) at Teluk Kalong, Kemaman, Terengganu, Malaysia*

Teluk Kalong beach (4.285817°N; 103.478118°E), situated along the southeast coast of Malaysia, primarily consists of granular sandy materials, easily

(a)   (b)

Fig. 2.5. Shoreline along Teluk Kalong beach (a) prior to protection in 2006 (b) after completion of the installation of submerged geotextile tubes.

erodible due to wave attacks. A 500-m-long stretch of the beach shoreline had experienced severe erosion, causing instability of an existing seawall composed of precast concrete units. The loss of sediments behind the concrete panels had resulted in large and uneven settlements of the same. The resultant beachfront was of narrow width with a steep foreshore slope, which made it unsuitable for recreational activities. In the year 2006, a proposal was made to construct a submerged dyke using geotextile tubes. The proposed cross-section consists of geotextile tubes sitting over a layer of apron mattress. The installation extended for about 500 m in length parallel to the coast, with a design requirement of a minimum 1 m freeboard even at low tide conditions. The images of the beachfront before and after installation of the submerged dyke is shown in **Fig. 2.5** (Shin and Kim, 2018), where visible beach formation can be observed.

2.6.2.   *Coastal vegetation*

The coastal vegetation and plant ecosystems are subjected to special conditions and severe stresses from wind, waves, tides and currents. The predominant mangrove and coastal plantations are species with a high degree of adaption capabilities, which has drastically evolved to offer stability to the strong dynamic shoreline. Communities along estuaries and saltmarshes of flooded areas have a plant composition that varies according to factors such as substrate type, the intensity of flooding, level of the water table and degree of salinity. The different biotopes have specialized flora that is threatened by the proliferation of invasive species, urban development and industrial discharges (Asensi and Garretas, 2017).

Many national and international agencies have been undertaking projects to ensure reforestation in coastal belts under the Emergency Tsunami Rehabilitation Project funded by the World Bank post the 2004 Indian Ocean Tsunami. The Forest Department of Tamil Nadu state situated along the southeast coast of the Indian peninsula has initiated large-scale ($\sim$20 km^2) planting of *Casuarina* along the Karaikal and Nagapattinam coast (Mukherjee *et al.*, 2009), taking up to about 40% of the coastline in the area (Rodriguez *et al.*, 2008). Yet, the Tamil Nadu Forest Department records show that the policy of raising *Casuarina* plantations has been a consistent practice on the coast, promoted since the late 1960s, if not earlier. Thus, while what we are witnessing on the east coast of India today is a continuation of a several decades-old trend, the scale of planting exotic trees is likely greater than at any time in the past and is now facilitated by the inputs of international funds (Mukherjee *et al.*, 2009).

Bio-shield or plantations located adjacent to or behind coastal communities are often the primary disaster management strategy along the coast, possibly giving a misleading feeling of security to policymakers. The design principles derived through a comprehensive study on the effectiveness of coastal vegetation in reducing the inundation height and distance along the coastal region was proved through field surveys post the Indian Ocean Tsunami of 2004. Structures having direct contact with the ingress of the tsunami suffered great damage while the damage of those fronted by vegetation was less. The results of examining the wave-induced pressures and forces exerted on a wall in the absence and presence of different configurations of vegetation cover through a comprehensive experimental program are reported by Sundar *et al.* (2011).

2.6.3. *Sand dunes*

2.6.3.1. *General*

The sediment retention capacity over sand dunes is vastly dependent on the nature of coastal vegetation. Coastal vegetation serves as a "bio-shield" against extreme events. Native vegetation within the first kilometre of the coast is typically adapted to a dynamic environment, including among other features, episodic conditions of saltwater inundation or salt spray, mass sediment movement and relatively rapid succession or spatial migration after disturbance. Sand dunes, considered important both ecologically and recreationally, are wind-driven and formed with huge sediment deposits just landward of normal high tides, as discussed by French (2001). They are also formed artificially with carefully planned stability measures. The construction, maintenance and rehabilitation of artificial dunes are important as they serve as protection against erosion and coastal flooding. The simplest method of artificially constructing dunes is placing the dredged sediments in eroding areas, which can also fall under the category of artificial beach nourishment to maintain or enhance the stability of such formed dunes. There are several methods of dune rehabilitation. One such method is to build fences on the seaward side of an existing dune to trap sand and help stabilize any bare sand surfaces (USACE, 2003), which can also be used to promote dune growth after a structure has been created. Natural materials, such as branches or reed stakes, are commonly used for fence construction because they tend to break down once they have accomplished their sand-trapping objective (Nordstrom and Arens, 1998).

2.6.3.2. *Rehabilitation of Leirosa sand dunes in Portugal*

The reconstruction of Leirosa (40.069565°N; −8.884645°E) sand dunes in the central west Atlantic coast of Portugal that involved increasing its stability through vegetation was carried out in 2001 (Reis and Freitas, 2002), which was destroyed overnight due to its exposure to a storm event. Soon after the damage, a continuous monitoring plan for about a year was undertaken and it was disclosed to be stable with vegetation of about 25 plant species that lead to a strong green belt cover. The images of the sand dune slopes with and without vegetation, prior and post the storm event is projected in **Figs. 2.6(a)** and **2.6(b)**, respectively.

The damaged sand dune was then stabilized by placing geotextile containers filled with sand placed in layers as shown in **Fig. 2.7**. The sand containers protective barrier was covered by a 1.0 m layer of sand over which a vegetation cover was planted. This completed coastal defence barrier system resulted in the formation of an excellent greenery coastal dune, the details of which are discussed by Reis *et al.* (2005).

2.6.3.3. *Sand dunes in Gopalpur, India*

The geomorphology, flora and hydrogeology of the coastal sand dunes and anthropogenic impact on sand dunes from the region of Gopalpur (19.244633°N; 84.894405°E), along the southern coast of Odisha (**Fig. 2.8**) are discussed in detail by Kumar and Hota (2014). The composition and distribution from coastal sand dunes examined by Sridhar and Bhagya (2007) revealed a total of 338 species along the coast of India. Coastal

(a) May 2000

(b) February 2001

Fig. 2.6. Reconstructed sand dune system, (a) in May 2000 and (b) its destruction due to a winter storm in early 2001.

Source: Reis and Freitas (2002).

Fig. 2.7. Geotextile containers for dune stability at Leirosa, Portugal.

Fig. 2.8. Sand dunes in Gopalpur, Odisha.

sand dunes serve as barriers against inundation particularly during extreme wave characteristics and their stability is enhanced by promoting vegetation cover.

2.6.3.4. *Coastal dune stabilization at Central Italy*

The coastal stretch along the Circeo National Park (41.356497°N; 12.957816°E) is a micro-tidal sandy beach lagoon system characterized by a single continuous dune ridge, approximately 22 km long, running parallel to the shoreline. The beach dune system is the remnant of a barrier island that separated the coastal plain from the sea. The dunes are mostly colonized by specialized coastal vegetation typical of the Mediterranean (Acosta *et al.*, 2005; 2007).

Analyzing the morphological parameters along the coastal stretch, it is observed that the elevation and foredune width in the southern part are almost double of those in the northern area. The elevation of the emerged beach and dune crest increases southward, whereas the foredune foot has

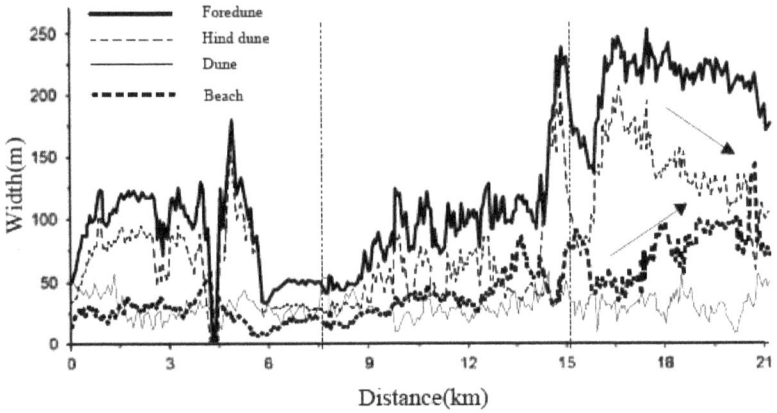

(a)

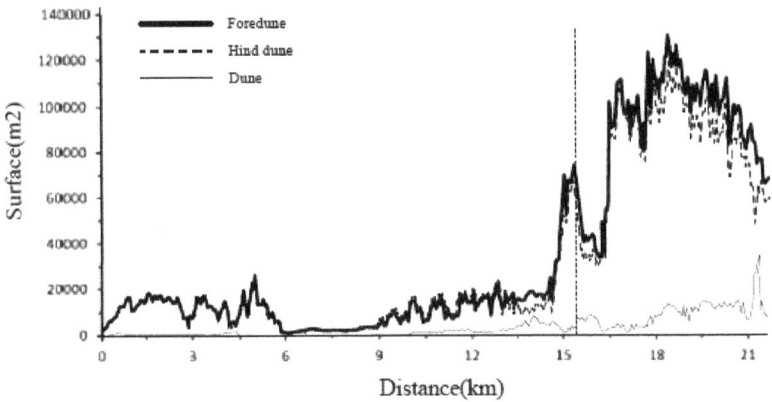

(b)

Fig. 2.9. Morphological behaviour from North to South: (a) width vs distance and (b) profile underlying the surface vs distance.

an elevation of about 2 m. The observed changes in the parameters are projected in **Fig. 2.9** (Valentini *et al.*, 2020).

2.6.4. *Artificial beach nourishment*

Artificial nourishments refer to a process when a coastal system is fed with sediments from a borrowed source (which could be either transported from a remote location or locally reclaimed). Artificial nourishments are widely adopted to replenish the apparent sediment losses owing to severe erosion out of any stretch of the coastal environment such as sandy beaches, dunes, lagoons, etc.

| (a) | (b) |

Fig. 2.10. Shoreline retention. The dotted line indicates the shoreline position in August 2006 (picture taken in 1999).
Source: Güler *et al.* (2008).

2.6.4.1. *Shore stabilization by artificial nourishment at Side, Turkey*

A coastal erosion problem that is observed at Side (36.801440°N; 31.351043°E) located in Turkey is discussed herewith as a case study for the application of beach nourishment, a soft measure approach to address an erosion problem. The problem area of the Side beach is shown in **Fig. 2.10**.

Prior to 1999, the beach width was found to be approximately 50 m. The beach was protected against the impact of waves by a submerged rubble mound breakwater, located at 1.5 m water depth and 100 m away from the shoreline. The rock formations parallel to the western end of the beach were removed post 1999, which resulted in a significant erosion along the western end, as can be seen in the above figure. The changes brought about owing to the removal of rock formations in the local bathymetry has in turn drastically affected the sediment transport mechanism at the beach and efforts to solve the erosion problem by artificial beach nourishment alone were unsuccessful. Thus, a numerical simulation including a submerged breakwater, a groin at the western end of the beach, an artificial beach nourishment and a gabion berm system in front of the artificial nourishment was carried out. The resultant shoreline changes within one year of its implementation through the simulation is projected in **Fig. 2.11**, where a stabilized beach is achieved (Güler *et al.*, 2008).

2.6.4.2. *Beach nourishment with coastal protection structures at Norderney, Germany*

Severe dune erosion jeopardized the health resort settlement on the west end of Norderney (53.720001°N; 7.263515°E), one of the East Frisian Barrier

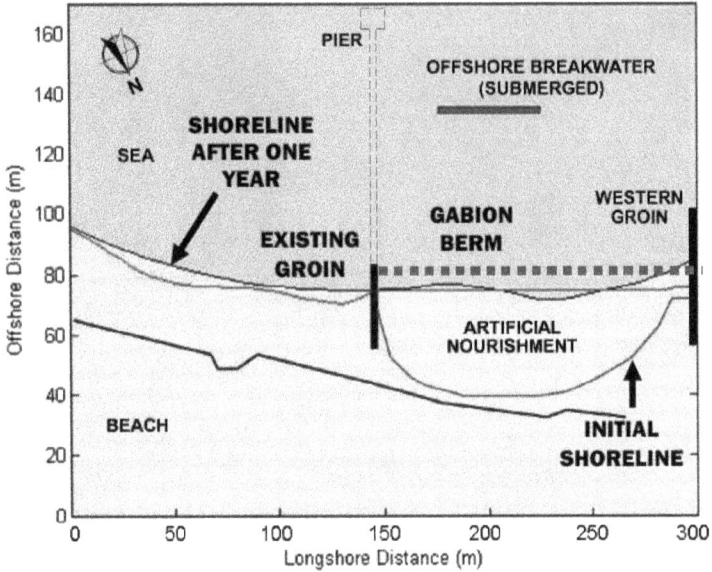

Fig. 2.11. Shoreline changes over a year after the application of the solution.

Islands that extends along the western part of the German North Sea coastline, and necessitated the construction of solid coastal protection structures
since 1857. These structures were successful in stopping the migration of the
inlet and preventing further dune erosion, but they were not able to maintain stable beaches. Artificial beach nourishment was initiated in the year
1951. It has been proved that Norderney beach nourishment is an appropriate solution to protect the existing structures from failure during severe
storm-floods. If a lower extent of sediment loss is anticipated, the beach
needs to be nourished not higher than necessary to achieve the protection
goals. Above a "critical beach profile" in certain areas, losses of nourished
material increase considerably with its height. The short parallel structures
and groynes constructed as early as 1857 were insufficient to stabilize the
coast and prevent shifting of the inlet mouth. In addition, these measures
required regular maintenance works and often, the extension of the length
of these structures was also mandated.

    To achieve the protection goals, it must be ensured that the beach
area above MLW (mean low water) level is considerably higher than
what the "natural beach" would be without beach restoration. Calculations of the beach volumes above the dynamic natural equilibrium profiles

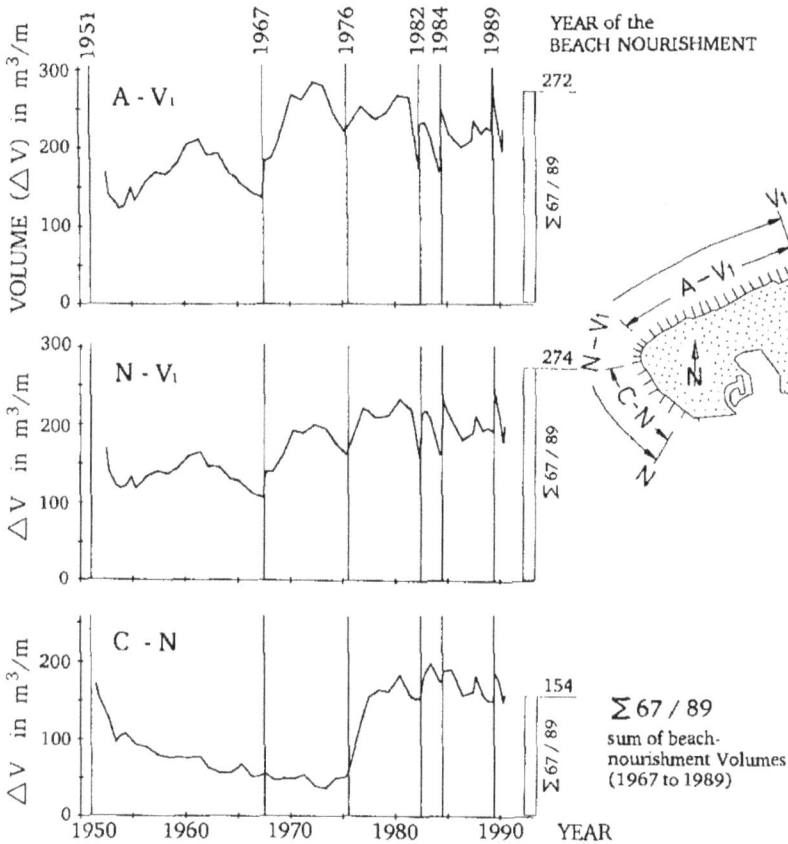

Fig. 2.12. Volume of beach nourishment along Nordeney.
Source: Kunz (1990).

(approximated by "reference profiles") over a long-term period (1951–1990) led to relationships between the losses of nourished sand and beach volume, as illustrated in **Fig. 2.12** (Kunz, 1990). Above a "critical beach profile", the losses of fulfilled material increase considerably with height (divergence area).

2.6.4.3. *Beach nourishment along Visakhapatnam and Paradeep ports, India*

One of the perineal problems hindering the efficient functioning of harbours is siltation, which can be minimized through regular maintenance

Fig. 2.13. Beach nourishment by Rainbow technique along Visakhapatnam coast.

dredging. The dredged quantity of sediments for all the major ports in India in maintaining the approach channels together is an average of about 50 million m^3 per year (Sarma, 2015). The dredged sediments are partly utilized for direct nourishment on the downdrift shoreline and partly dumped offshore on the assumption that the dredged soil will be transported by waves. The direct nourishment of the shoreline is usually through pumping of the dredged sand through floating pipelines or by dispersal of sand along the coast through the well-known rainbow technique as shown in **Fig. 2.13.** The maintenance dredging quantities used for direct nourishment of downdrift coastline and dumped offshore to maintain the approach channel of Visakhapatnam (17.681191°N; 83.295279°E) and Paradeep (20.255606°N; 86.667551°E) ports from the available records have shown that an average about 0.2–0.3 million m^3 of sand used to pump on the down drift shoreline at Visakhapatnam port whereas about 0.5–0.6 million cubic meters of sand pumping takes place along eroded coastline of Paradeep port. The balance quantity from maintenance dredging is used to dump at the earmarked offshore disposal grounds. From the estimates of littoral drift, it is known that the net drift blocked by the southern breakwaters of Visakhapatnam and Paradeep port are 0.55 and 1.68 million cubic meters per year, respectively (Sarma, 2015). It means that that the quantities pumped for the nourishment of eroded beaches are inadequate.

Fig. 2.14. Coast of Cancun, Mexico, after Hurricane Wilma in 2005.

2.6.4.4. *Nourishment along the coast of Cancun, Mexico*

Cancun beach (21.259321°N; −86.747168°E) is located on the Caribbean coast of the Yucatan Peninsula, in Mexico. The barrier island between the Caribbean Sea and a system of coastal lagoons has been exploited since the 1960s, resulting in a virtual reduction in neighbouring mangrove forests. A dense construction of tourism-related infrastructure (such as hotels, resorts, parks, etc.) along the Cancun coast has effectively hindered the exchange of fresh water and seawater across the lagoons. An average sediment loss of about 1.8 m/year recession in the beach width between 1967 and 2005 was reported by Martell *et al.* (2010). Hurricane Wilma hit the Cancun coast in October 2005, eroding about 8 million m^3 of sand, leaving the coast with virtually no sediments, as can be seen in **Fig. 2.14**.

The initial beach nourishment, carried out in the year 2006, restored 2.7 million cubic meters of sand on the shores of Cancun, borrowed and dredged from various neighbouring resources. This nourishment yielded only short-term benefits, and the coast continued to experience erosion. A reduction in the beach width to an extent of about 30 m was experienced between September 2006 and 2007. The coast was further damaged by Hurricane Dean in August 2007, which removed the previously nourished sediments.

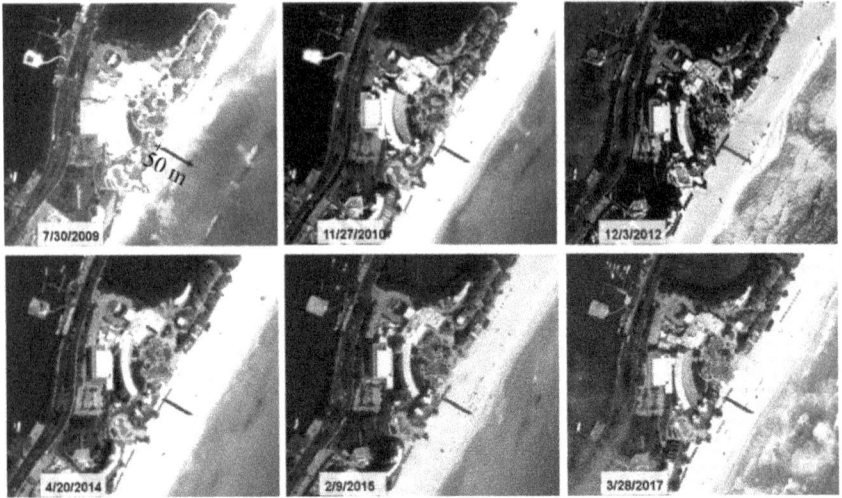

Fig. 2.15. Satellite imagery of Cancun beach between 2009 and 2017.
Source: Martell *et al.* (2020).

The 60 m beach width originally restored after beach nourishment in 2006, decreased to as low as 14 m in 2009 (González *et al.*, 2013). The sediment balance/budget is negative, which prevents natural self-regulation of the coastal system, causes very low sediment input into the system, and leads to large amounts of sediment being transported out of the system by hurricane-induced high energy waves. Since the beach had not reached a dynamic equilibrium, a secondary beach nourishment was performed in 2010, through which about 5.2 million m^3 of sediments up to a beach width of 80 m were deposited alongside the construction of a 305-m-long breakwater with a crest elevation 2.5 m above MSL. From satellite imagery, it is evident that even 10 years after the second nourishment, the beach at Cancun has managed to conserve a nearly stable width of approximately 30 m, as seen in **Fig. 2.15**.

2.7. Hard Measures

2.7.1. *Offshore breakwaters*

These are non-shore connected structures, constructed seaward of and parallel to the shoreline. The primary function of such structures is to protect an erosion-affected stretch of the shoreline by diffracting and reducing the incident energy, which would further modify the local sediment dynamics and induce settlement of sediments at locations of less wave energy, more so on the lee side of the breakwaters.

Submerged breakwaters are mostly considered to reduce the amount of energy reaching the shore by forcing the waves to break and by extending the delay time of sediment motion in a sheltered region (Basiński *et al.*, 1993; Creter *et al.*, 1994). In other words, they function like a reef system, where the deployed units facilitate premature breaking of incident waves. They are usually built on sandy shores where the increased abrasion and the associated sediment deficit are the main problems. Based on the 2D physical model studies conducted by Lorenzoni *et al.* (2012) it is concluded that wave transmission co-efficient are higher for submerged type compared to the emerging type offshore breakwaters. Moreover, the submerged configuration of breakwaters exhibits larger shoreline retreats than emerging ones for the same layout and configuration subjected to identical wave climate, thus concluding the emerged type offshore breakwaters to possess higher defence efficiency.

2.7.1.1. *Nagahama beach, Japan*

Japan has widely adopted offshore breakwaters as a coastal protection measure to protect its shoreline (Isobe, 2001). Owing to the severe existing wave climate along the Japanese coast, construction of such offshore structures is expensive, and the impact of offshore breakwater along Nagahama coast (37.163905°N; 138.175066°E) of Joetsu district are briefly discussed herein as a case study. The layout of offshore detached breakwaters at Nagahama beach along with the corresponding shoreline variation between 1994 and 2004 is projected in **Fig. 2.16**. Construction of these structures has promoted accretion on the lee side of the breakwaters and resulted in the formation of salient and tombolo as seen in **Fig. 2.17**.

2.7.1.2. *Along the Spanish coast*

The effects of a detached breakwater as coastal protection measure along the Spanish coastline were analyzed by Bricio *et al.* (2008) to check the validity of the empirical relations proposed by several researchers. All the formulas studied are based on the ratio of a detached breakwater length, B, to its distance from the affected shoreline, X, and predicted that tombolo formation takes place if B/X about 1.3, whereas the salient formation occur when B/X is between 0.5 and 1.3.and shore response being limited if B/X is about 0.5. The study has summarized the most significant analytical

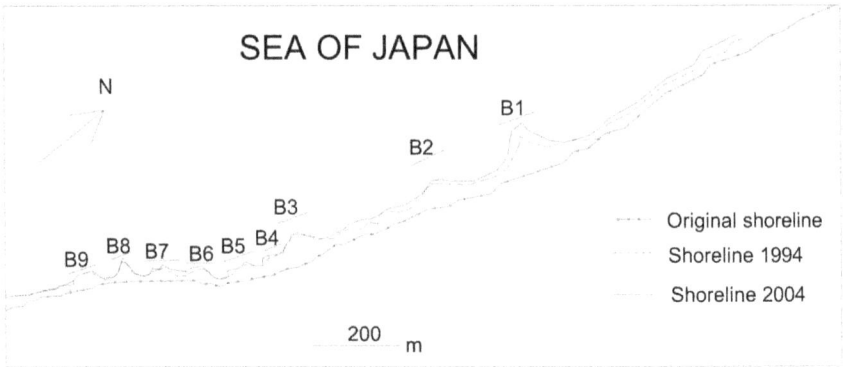

Fig. 2.16. Shoreline changes at Nagahama beach due to offshore breakwaters. *Source*: Isobe (2001).

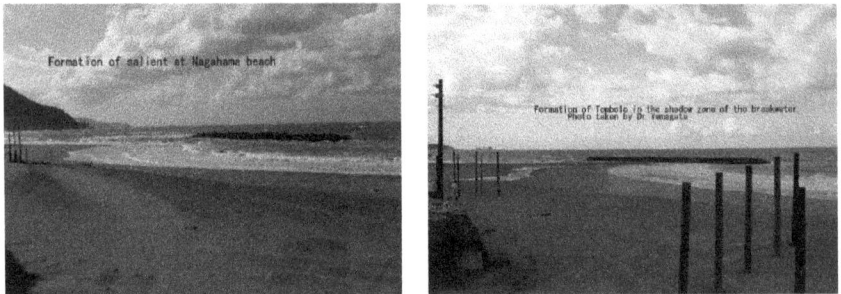

Fig. 2.17. Salient and tombolo formation at Nagahama beach. *Source*: Sane *et al.* (2007).

models for investigating the relationship between the characteristics of a detached breakwater and its effect as a coastal protection measure through field studies.

2.7.1.3. *Marawila Beach, Sri Lanka*

Erosion to an extent of about 10 m per year along the sandy beach of Marawila (7.409082°N; 79.816805°E) situated in the north-western coast of Sri Lanka has been reported (CC&CRMD, 2006). A comprehensive study on the historical changes of about 50 coastal stretches focusing on Marawila Beach is reported by Samarasekara (2019). The study focuses on the 58

Fig. 2.18.   An aerial view of the offshore breakwaters Marawila Beach in 2021. *Source*: Google Earth.

historical shoreline changes and adopted management for approximately 40 years for one of the most vulnerable beaches in Sri Lanka.

Through the said comprehensive study, it was concluded that continuous beach nourishment, along with the deployment of detached breakwaters (**Fig. 2.18**), is an acceptable solution for combating erosion that has resulted in shoreline instability over the last four decades.

2.7.2.   *Groins*

Groins are shore-connected, narrow, perpendicular hard structures designed to intercept the longshore sediment transport by trapping sediments carried by the longshore/littoral currents that would otherwise be carried forward alongshore. Groins are usually built on exposed and moderately exposed sedimentary coastlines to address persistent erosion problems. They can be constructed using rock armour, concrete armour units, geosynthetics, steel piling or timber. There exist multiple case studies to ascertain the effectiveness of the construction of groin field along an eroding coastline.

2.7.2.1.   *Southwest coast of Tamil Nadu, India*

The coast of Simon Colony, Vaniyakudi and Kurumbanai villages, was experiencing the problem of erosion due to high wave action in the Arabian Sea. The site conditions prior to implementation of the measure is shown in **Fig. 2.19**. Based on the site conditions, six groins were proposed for this

Fig. 2.19. Status of coast prior to the implementation of groin field as coastal protection.

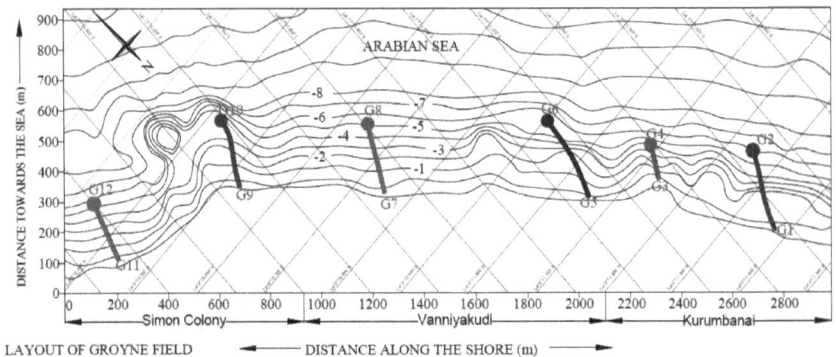

LAYOUT OF GROYNE FIELD   ◄——— DISTANCE ALONG THE SHORE (m) ———►

Fig. 2.20. Layout of Groin field — coast of Tamil Nadu (southeast coast of India).

study area, in which two groins at Kurumbanai, one groin at Vaniyakudi, two groins at Kodimunai and one groin at Simon colony are as projected in **Fig. 2.20**. These groins constructed from mid to end of 2002 were oriented at about 15° inclination to shore normal from the shore towards the off-shore direction, and the end of the groin is curved to take care of the waves coming during Southwest monsoon. The complete details, including the numerical model studies, design of the layout, and considering the advantages of outcrops as head sections for the groins, are discussed by Sundar *et al.* (2004). Since then, the groin field has been serving well in trapping the longshore sediments. The groin field has also been serving not only as coastal protections measure for the villages but also as a fishing harbour for anchoring vessels. The different views of the above-stated aspects are depicted in **Fig. 2.21**.

2.7.2.2. *Tip of Indian peninsula*

The stretch of the coast along the Ratchagan street (8.087992°N and 77.555309E) along the tip of the Indian peninsula has been protected by a groin field consisting of seven groins constructed in 1999. A bird's eye view of the groin field is projected in **Fig. 2.22(a)**. As this stretch does not

(a)

(b)

Fig. 2.21.   (a) Construction of Groin G1–G2. (b) Simon Colony (outcrop acts as head of the groin).

(c)

(d)

Fig. 2.21. (*Continued*) (c) Beach in between Groins G7–G8 and G5–G6. (d) Status of the groin field after almost two decades.

experience littoral drift to the extent it could be trapped, the groin field was ineffective in serving as a coastal protection measure.

A comprehensive study by Sundar and Sankarbabu (2005) and Sundar (2006) post the 2004 Indian Ocean Tsunami and the probable mitigation measures carried out considered the stretch of the coast being discussed herein. The ingress during the tsunami claimed several lives, and it was clear that the presence of groins had neither served as coastal protection measure nor it had any effect on reducing the inundation. As there was growing demand for a fishing harbour, it was decided to extend the southernmost groin such that it can serve as a wave attenuator during extreme coastal hazards. A view of the implemented proposal and a view of harbouring small vessels are shown in **Figs. 2.22(b)** and **2.22(c)**.

2.7.2.3. *Kadakkarapally coast of Kerala, India*

A case study of Kadakkarapally coast of Kerala (9.688515°N; 76.289794°E), southwest of the Indian peninsula, is discussed herein. Sundar and Murali (2007) proposed the construction of a series of T-groins, which were successfully implemented in the year 2015. A remote sensing analysis was carried out to study the effectiveness of the groin field and reported the coast to experience high and moderate accretion over a period of five years (2015–2020) as shown in **Fig. 2.23**.

2.7.3. *Seawalls and bulkheads*

Seawalls and bulkheads are structures placed parallel or nearly parallel to the shoreline to separate the land from water area. The primary purpose of a bulkhead is to retain or prevent sliding of the land, with a secondary purpose of affording protection to the back shore against damage by wave action. The most common and widespread coastal engineering tool for the protection of shorelines is the construction of a seawall as a hard measure. Seawall protection is, for all practical purposes, an irreversible act because the beach in front of it is often removed. The seawall will eventually have to be rehabilitated at constant intervals with bigger-sized stones. In locations, where, large-size natural rocks are scanty, gabions (wire net filled with stones of smaller size) can be adopted. At locations of abundance of sand (near river mouths, which may be dredged as a part of river training works) geo-bags or geo-tubes can be adopted.

(a)

(b)

(c)

Fig. 2.22. (a) View of the six groins from groin 1. (b) Extension of groin1 to serve as a mini harbour as well as reduce inundation. (c) Extension of groin 1 serving as a mini harbour.

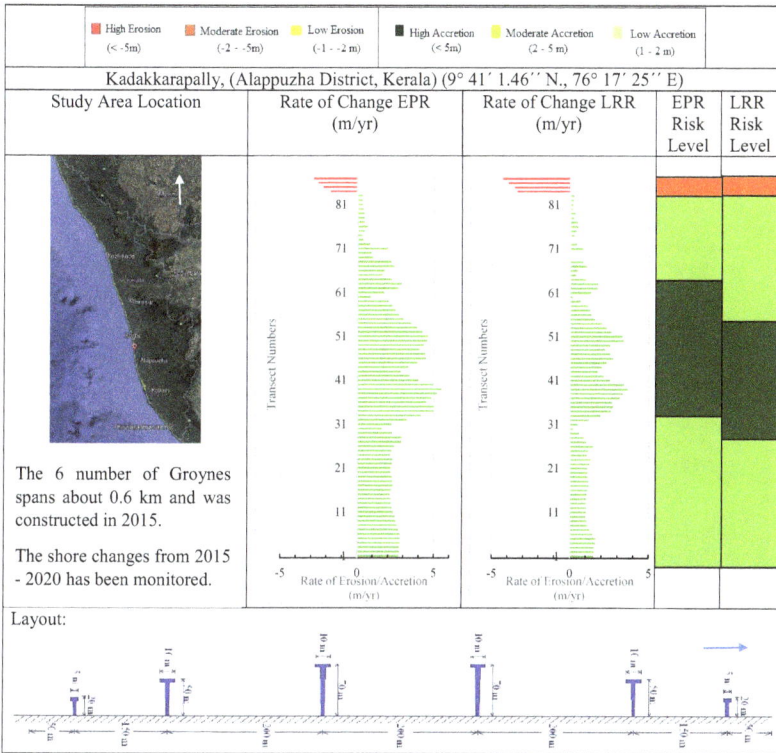

Fig. 2.23. Performance of T-groins at Kadakkarapally, Kerala.
Note: EPR: Endpoint rate; LRR: Linear regression rate.

2.7.3.1. *Thazhampally, Kerala, India*

The coastal stretch of Thazhampally (8.636995°N; 76.783762°E) about 300 m north of the northern breakwater of the Muthalapozhy harbour is an unprotected virgin beach (**Fig. 2.24**), prone to severe wave attack. A proposal was made by Sundar and Murali (2007) to construct a seawall with a seaward slope of 1:3 as shown in **Fig. 2.25**. The seawall section along the beachfront has successfully protected the Thazhampally coast against erosion and seawater inundation, especially during storm surges.

2.8. Training Walls

Sandbar formation and chocking of river mouths are one of the most commonly occurring phenomena along a riverine/tidal inlet. Sandbars, spits and

Fig. 2.24. Beachfront north of Muthalapozhy harbour, Kerala.

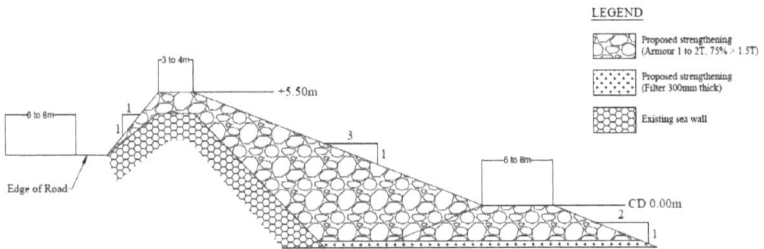

Fig. 2.25. Typical seawall cross section at Thazhampally.

shoal formations are characteristic of any coast with predominant longshore sediment transport. The heavy sediment-laden longshore currents tend to deposit the sediments in sinks such as an inlet or an estuary.

2.8.1. *Along the Kerala coast, India*

The nearshore sediment processes at the vicinity of inlets in Kerala, along the southwest coast of India, was discussed by Sundar and Sannasiraj (2016). The construction of training walls proved to be beneficial in maintaining the river mouth to be open across all seasons; moreover, it did

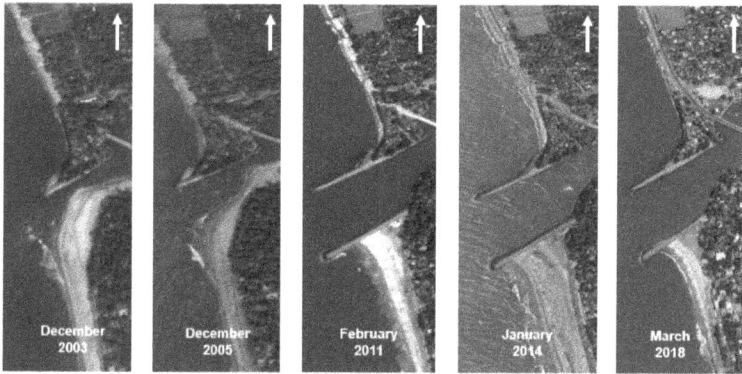

(a)

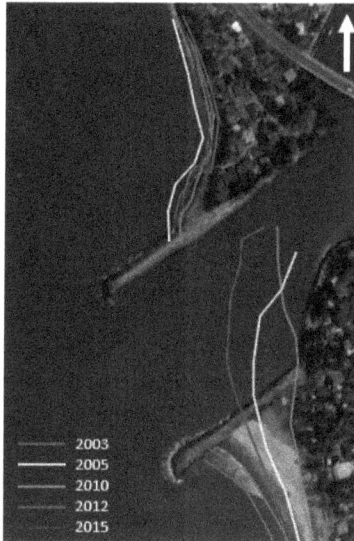

(b)

Fig. 2.26. (a) Timeline variation of Korapuzha inlet. (b) Shoreline overlay at Korapuzha inlet.

Source: Google Earth Pro.

not impair the shoreline dynamics, and no visible erosion on the downdrift side were witnessed even years after the construction of the well-planned and designed training wall section. The Google Earth satellite imagery of Korapuzha inlet (11.227133°E; 75.779221°N) across varying timelines is projected in **Figs. 2.26(a)** and **2.26(b)**.

Fig. 2.27. Tweed River inlet.
Source: Ware and Banhalmi-Zakar (2017).

2.8.2. *Tweed River Entrance, Australia*

The Tweed River Entrance ($-28.169849°$E; $153.552560°$N) is situated in the Northern Rivers district of New South Wales, Australia. A Sand Bypass Project is a large system, composed of a jetty and pumping station (on Letitia Spit just south of the Tweed River in Northern NSW) and a series of pipes, which transfer sand to an outlet at Snapper Rocks located at the southern end of the Gold Coast in Queensland (**Fig. 2.27**). The extension of the Tweed training walls in 1964 trapped the longshore transport of sand into Queensland, which reduced the width of beaches and rendered coastal infrastructure and property vulnerable to erosion. The sand bypassing system at Tweed River mouth maintains a navigable depth of water of at least 3.5 metres in the approach and within the entrance channel. A continuous supply of sand to the Southern Gold Coast beaches at a rate consistent with the natural littoral drift rates. The restoration of beach width on both the updrift and downdrift side with the supply of sand through dredged sediments facilitates to promote recreational infrastructure and maintain the same. The total volume of sand bypassed between 2000 and 2015 is estimated to be about 81,03,065 m^3 (Ware and Banhalmi-Zakar, 2017) of sediments.

Table 2.2.   Comparison between hard and soft measures.

Hard measures	Soft measures
Constitutes large, gravity structures extending either parallel or normal or inclined to the shoreline, which might often affect the aesthetics of a coast.	Sandy beaches, vegetal green cover, sand dunes etc., are ideal tourism and eco-friendly in addition, to serving the purpose of shore protection.
Hard measures are ideal to immediately restore lost beach or instantaneous prevention of coastal flooding or overtopping.	Instantaneous results cannot be expected after implementation, although, these measures can stabilize the shore in a due course of time.
The capital cost for procuring materials, employing heavy machinery, labour cost and duration of construction are high. Even after an initial investment certain rehabilitation towards restoration and maintenance are mandatory that involves recurring expenditure.	Installation of slurry pumps and procurement of geosynthetic materials are expensive.
Widely applicable for all environmental and climatic conditions. These measures leave a lasting and irreversible ecological impact of the coast.	These measures are ideal only for certain environmental and climatic conditions and not universally applicable.
Construction work can be carried out with transportation or fabrication of heavy quarry stones or in-situ precast concrete units with the aid of heavy machinery and unskilled labour.	Skilled labour is mandatory.
Hard structures tend to hinder free public access to the beachfront and could potentially increase the risk of initiating rip currents.	The aesthetic soft measures can be potentially destroyed due to vandalism, especially geosynthetics.
Scouring beneath structures built across different water depths is a common problem, which needs to be duly addressed failing which the structural integrity of the protection measure is questionable.	The problems due to scour are relatively insignificant.
Many years of expertise, abundance of well-established and published research articles along with guidance manuals such as coastal engineering manual, shore protection manual make these more desirable.	Lack of well-established design manuals and guidelines for geosynthetic applications in coastal engineering.
These measures have proved to have a long survival life even up to 100 years.	The design life of soft measures cannot be directly deciphered.

2.9.   Summary

This chapter has discussed a number of case studies across a few global locations and their unique site conditions. It is to be remembered by the readers that an optimum coastal protection measure for a given site vastly depends on the site-specific parameters and thus generalization of merits/demerits of a system is not valid. However, a few common pointers can be mentioned for the hard and soft measures as listed in **Table 2.2**. A thorough understanding of the nearshore physical processes and an in-depth knowledge of the behaviour of proposed protection measures are inevitable to combat coastal erosion problems.

References

Acosta, A., Carranza, M.L. and Izzi, C.F. (2005). Combining land cover mapping of coastal dunes with vegetation analysis. *Appl. Veg. Sci.*, 8, 133–138.

Acosta, A., Ercole, S., Stanisci, A., Pillar, V.D.P. and Blasi, C. (2007). Coastal vegetation zonation and dune morphology in some Mediterranean ecosystems. *J. Coast. Res.*, 23(6), 1518–1524.

Asensi, A., and Diez-Garretas, B. (2017). Coastal vegetation. In *The Vegetation of the Iberian Peninsula* (pp. 397–432). Springer, Cham.

Coastal Engineering Manual (CEM) (2006). U.S. Army Corps of Engineers, Manual 1110-2-1100, January.

Basiński, T., Pruszak, Z., Tarnowska, M. and Zeidler, R. (1993). *Seashores Protection*. Publisher Institute of Hydroengineering PAS, Gdańsk.

Bricio, L., Negro. V. and Diez. J.J. (2008). Geometric detached breakwater indicators on the Spanish northeast coastline, *J. Coast. Res.*, 24, 5.

CC&CRMD (2006). Coastal Zone Management Plan (CZMP) 2004 [WWW Document]. Gaz. 618 Extraordinary Part I Sec Gaz. Extraordinary Democr. Social. Repub. Sri Lanka 2006. 619. Available at: http://www.coastal.gov.lk/downloads/pdf/CZMP English.pdf [Accessed 10 September 2017].

Creter, R.E., Garaffa, T.D. and Schmidt, C.J., (1994). Enhancement of beach fill performance by combination with an artificial submerged reef system. In L.S. Tate (Ed.), *Proc. 7th National Conference on Beach Preservation Technology*. Florida Shore and Beach Preservation Association, Tallahassee, FL, pp. 69–89.

French, P.W. (2001). *Coastal Defences: Processes, Problems and Solutions*. London: Routledge.

González-Leija, M., Mariño-Tapia, I., Silva, R., Enriquez, C., Mendoza, E., Escalante-Mancera, E., Ruíz-Rentería, F. and Uc-Sánchez, E. (2013). Morphodynamic evolution and sediment transport processes of Cancun Beach. *J. Coast. Res.*, 29, 1146–1157.

Güler, I., Baykal, C. and Ergin, A. (2008). Shore stabilization by artificial nourishment, a case study: A coastal erosion problem in Side, Turkey. In Proc. of the 7th International Conference on Coastal and Port Engineering in Developing Countries (COPEDEC), Paper (No. 90).

Heibaum, M. (1995). Sanierung der Kolke am Eidersperrwerk- Geotechnische Stabilität von Deckwerk und Untergrund Mitteilungsblatt Bundesanstalt für Wasserbau. 73, 111–122.

International Geosynthetics Society (IGS) Secretariat (2009). Recommended Descriptions of Geosynthetics, Functions, Geosynthetics Terminology. Mathematical and Graphical Symbols, Easley, SC, p. 1.

Isobe, M., (2001). A theory of integrated coastal zone management in Japan. Department of Civil Engineering, University of Tokyo, Japan, 18 pp.

Koerner, R.M. (2012). *Designing with Geosynthetics*, 6th Edition, Xlibris Publishing Co., New York.

Kumar, N.P. and Hota, R.N. (2014). Geomorphological study of sand dunes with special reference to their hydrogeology in southern coast of Odisha, India. *International Research Journal of Earth Sciences*, 2(9), 15–21.

Kunz, H. (1990). Artificial beach nourishment on Norderney: A case study. *Proc. Coastal Eng.*, July 2–6, Delft, The Netherlands.

Lorenzoni, C., Postacchini, M., Mancinelli, A. and Brocchini, M. (2012). The morphological response of beaches protected by different breakwater configurations. *Coast. Eng. Proc.*, 1. 10.9753/icce.v33.sediment.52.

Martell, R., Mariño, I., Mendoza, E. and Silva, R. (2010). Variaciones morfológicas a largo plazo del perfil de playa en Cancún, México. *Proc. XXI Congreso Nacional De Hidráulica*, Jalisco, México, 25 November, p. 8.

Martell, R., Mendoza, E., Mariño-Tapia, I., Odériz, I. and Silva, R. (2020). How effective were the beach nourishments at Cancun? *J. Mar. Sci. Eng.*, 8, 388. DOI: https://doi.org/10.3390/jmse8060388.

Mukherjee, N., Balakrishnan M. and Shanker K. (2009). Bioshields and ecological restoration in tsunami-affected areas in India. In E. Dahl, E. Moskness, J. Stottrup (Eds.), *Integrated Coastal Zone Management*. Wiley Blackwell Publishing, Hoboken, New Jersey, pp. 131–134.

Müller, W.W. and Saathoff, F. (2015). Geosynthetics in geoenvironmental engineering. *Science and Technology of Advanced Materials*, 16(3), 034605. DOI: 10.1088/1468-6996/16/3/034605.

Nickels, H. and Heerten, G. (2000). Objektschutz Haus Kliffende. HANSA – Schiffahrt, *Schiffbau*, 137(3), 72–75.

Nordstrom, K.F. and Arens, S.M. (1998). The role of human actions in evolution and management of foredunes in The Netherlands and New Jersey, USA. *J. Coast. Conserv.*, 4, 169–180.

Pilarczyk., K.W. (2000). *Geosynthetics and Geosystems in Hydraulic and Coastal Engineering*. A.A. Balkema Publishers, Brookfield, USA.

Recio, J. and Oumeraci, H. (2008). Hydraulic permeability of structures made of geotextile sand containers: Laboratory tests and conceptual model. *Geotext. Geomembr.*, 26, 473–487.

Recio, J. and Oumeraci, H. (2009). Process based stability formulae for coastal structures made of geotextile sand containers. *Coast. Eng.*, 56, 632–658.

Reis, C.S. and Freitas, H. (2002). Rehabilitation of the Leirosa sand dunes. In EuroCoast-Portugal Association (Ed.), *Littoral 2002*, Porto, 22–26 September. Porto, Portugal. III, pp. 381–384.

Reis, C.S, Freitas, H. and Antunes do Carmo, J.S. (2005). Leirosa sand dunes: A case study on coastal prototection. *Proc. IMAM — Maritime Transportation and Exploitation of Ocean and Coastal Resources*, Lisboa, 2–30 September, pp. 1469–1474. Taylor & Francis/Balkema.

Rodriguez, S., Balasubramanian G., Peter S.M., Duraiswamy M. and Jaiprakash P. (2008). Beyond the tsunami: Community perceptions of resources, policy and development, post-tsunami interventions and community institutions in Tamil Nadu, India. UNDP/UNTRS, Chennai and ATREE, Bangalore.

Samarasekara, R.S.M., Sasaki, J., Jayaratne, R., Suzuki, T., Ranawaka, R.A.S. and Pathmsiri, S.D. (2019). Historical changes in the shoreline and management of Marawila Beach, Sri Lanka, from 1980 to 2017, *J. Ocean Coast. Manage.*, DOI: 10.1016/j.ocecoaman.2018.09.012.

Sane, M., Yamagishi, H., Tateishi, M. and Yamagishi, T. (2007). Environmental impacts of shore-parallel breakwaters along Nagahama and Ohgata, District of Joetsu, Japan. *J. Env. Manage.*, 82. 399–409. DOI: 10.1016/j.jenvman.2005.01.030.

Sarma, K.G.S. (2015). Siltation and coastal erosion at shoreline harbours. *Proc. 8th Intl Conf. on Asia and Pacific Coasta*, APAC 2015, *Procedia Eng.*, 116, 12–19.

Shin, E.C. and Kim, S.H. (2018). Case study of application geotextile tube in the construction of sea dike and shore protection. *Int. Conf. Disaster Manage.* (ICDM 2018). DOI: https://doi.org/10.1051/matecconf/201822904021.

Sridhar, K.R. and Bhagya, B. (2007). Coastal sand dune vegetation: A potential source of food, fodder, and pharmaceuticals. *Livestock Res. Rural Dev.* 19(84). Available at: http://www.lrrd.org/lrrd19/6/srid19084 [Accesfsed 1 September 2011].

Sundar, V., Sundaravadivelu., R. and Nagabhushan, N. (2004). Groin Field as coastal protection measure for the coast of Kanyakumari, Tamil Nadu. *Proc. Intl. Seminar on Coastal Area Construction Management*, 1–2 Nov, Mumbai, pp. 49–56.

Sundar, V. and Murali, K. (2007). Planning of Coastal Protection Measures along Kerala Coast. Report submitted to the Govt. of Kerala, by the Dept of Ocean Engineering, IIT Madras.

Sundar, V., Murali, K. and Narayanan, L. (2011). Resistance of flexible emergent vegetation and their effects on the forces and run-up due to waves. Chapter 8. In N.-A. Moerner (Ed.), *Tsunami, Research and Technologies*, InTech, Japan, 978-953-307-552-5.

Sundar, V. and Sannasiraj, S.A. (2016). Training of a few river mouths of Kerala coast in India. In B. Crookston and B. Tullis (Eds.), *Hydraulic Structures and Water System Management, 6th IAHR International Symposium on Hydraulic Structures*, Portland, OR, 27–30 June, pp. 178–187. DOI:10.15142/T3570628160853 (ISBN 978-1-884575-75-4).

Sundar, V., Sannasiraj, S.A. and Sukanya, R. (2021). Sustainable hard and soft measures for coastal protection – Case studies along the Indian coast. *Marine Geo-resource. Geotechnol.*, DOI: 10.1080/1064119X.2021.1920650.

Sundar, V. (2022). Sustainable hard and soft measures for coastal protection. *ISS-MGE TC 213 workshop on Scour and Erosion* 16 December, Indian Geotechnical Society and Andhra University in Visakhapatnam, pp. 41–58. DOI: 978-981-16-4783-3_4, © 2022.

Sundar, V. (2006). "Protection Measures Against Tsunami-type Hazards for the Coast of Tamil Nadu, India", in T. S. Murty, U. Aswathanarayana and N. Nirupama, (Eds.), *The Indian Ocean Tsunami*, Taylor & Francis/Balkema, The Netherlands, pp. 411–421.

Sundar, V. and Sankar Babu, K. (2005). Planning of coastal protection considering the effects of tsunami along the coast. Presented in *1st Int. Conf. Coastal Zone Manage. Eng.*, Dubai, November 27–29.

Valentini, E., Taramelli, A., Cappucci, S., Filipponi, F., and Nguyen Xuan, A. (2020). Exploring the Dunes: The correlations between vegetation cover pattern and morphology for sediment retention assessment using airborne multisensor acquisition. *Remote Sensing*, 12(8), 1229. DOI: https://doi.org/10.3390/rs12081229.

Ware, D.M. and Banhalmi-Zakar, Z. (2017). Funding coastal protection in a changing climate: Lessons from three projects in Australia. Report. National Climate Change Adaptation Research Facility, Sydney, NSW, Australia. https://researchonline.jcu.edu.au/50228/1/Funding%20Coastal%20Protection_ACCARNSI_Discussion_Paper_1_Final.pdf.

Chapter 3

Coastal and Marine Data Information System for Maritime Spatial Planning

Abstract

This chapter describes a coastal and marine data information system that can be used for Maritime Spatial Planning. The system takes advantage of Internet mapping and web services technology to publish data as maps. These maps are created dynamically and enable the visualization of a diverse and growing collection of environmental data. Capabilities of the system include (i) visualization of environmental parameters on 2D maps; (ii) visualization of location-specific depth profiles of selected parameters; (iii) time animation of grids of environmental parameters and (iv) preparation of digital elevation and cultural object data sets for 3D terrain visualization (Durairaju *et al.*, 2003, 2010). The function and interaction between key components of the system are described to illustrate the working of the system. The use of a mature map publishing engine greatly eased the development of these capabilities into a web-based system.

3.1. Introduction

The management and planning of coastal and marine areas are complex processes that are increasingly gaining importance to effectively support the coordinated development of socio-economic activities while preserving the environment using ecosystem-based approaches (European Union, 2014; Center for Ocean Solutions, 2011; Douvere, 2008). Practical tools to support the implementation of the various steps of Maritime Spatial Planning (MSP) have been developed in various contexts and analyzed to evaluate their usability for different purposes (Stelzenmuller *et al.*, 2013a). Considering the management of conflicts between marine applications as a central point of MSP, the project COEXIST (Stelzenmuller *et al.*, 2013b) developed a tool to analyze the level of coexistence which in-turn enables the provision of important information for decision makers in spatial management processes. Various authors proposed methodologies to create cumulative impact maps to reconnect the effects of human intervention of the coast on environmental components, beginning with the methodology first introduced by Halpern *et al.* (2008) at a global scale and then implemented in several marine regions (Mediterranean Sea by Micheli *et al.* (2013), Baltic Sea by Korpinen *et al.* (2013), and North Sea by Andersen *et al.* (2013)). In particular, Stock (2016) developed an open-source software for mapping human impact on marine ecosystems.

The MSP process tends to involve several kinds of users, from data producers (e.g., domain experts such as ecologists and modellers) to planners, who combine data and information to create a more comprehensive picture of a marine area and reallocation of the human activities. Therefore, the availability of high-quality geospatial data and information from numerous fields (e.g., environmental sciences, tourism, human activities and infrastructure) is a key issue of the MSP process (Menegon *et al.*, 2018). In this regard, the continuous development of Spatial Data Infrastructures (SDI) provides a favourable context for environmental management and planning (Georis-Creuseveau *et al.*, 2017) while the importance of the integration of geoportals in the context of SDIs has been highlighted by various authors (Maguire and Longley, 2005) and the role of a user-driven and community-based development is considered as fundamental for effective and efficient use of the resource (De Longueville, 2010; Georis-Creuseveau *et al.*, 2016).

This chapter explains the development of the Coastal and Marine Data Information System (CMDIS) that integrates diverse information collected on coastal and marine water ecosystems, such as physical, environmental and biological, as well as data from predicted models. By integrating these diverse datasets using data management and GIS techniques, one can search based on date and location. The spatially attributed data can be converted as maps based on a diverse set of data sources collected through different field techniques. The development of the system enables MSP through the integration of a diverse and growing collection of environmental data sets, as well as increasing the visibility of key data sets, and data sharing. The application with the web mapping functionality, can address research challenges and the development of innovative tools for marine spatial planning and management.

3.2.   CMDIS

CMDIS is an interactive marine environmental mapping system developed to visualize digital maps over the World Wide Web. The system design allows for ingesting a diverse set of data sources collected over the years by researchers, institutions and practitioners. The development of the system enables the organization to integrate a diverse and growing collection of environmental data sets, as well as increases the visibility of key data sets (Durairaju *et al.*, 2003).

The data sources in CMDIS include those on physical characteristics such as tides, currents and bathymetry, coastal habitats, human activity in the coastal zone, environmental quality and sources of hazards collected from coastal, intertidal zones, surface and subsurface marine environments. The data collected was organized spatially to create an integrated marine database, creating a valuable resource for the marine science research community.

3.2.1.   *Design of CMDIS*

CMDIS comprises a set of services to perform the following:

- Ingestion of data sets in different formats.
- Metadata capture, cataloguing and search of data sets.
- Interactive visualization as 2D or 3D graphics, maps or reports.
- Data export after conversion to other application formats.

3.2.1.1.   *System architecture*

In the view of the integration of available software and development of new tools, the architecture described in **Fig. 3.1** has been developed. The schema also highlights the final users specifying their interaction with the system. The system is designed using a client-server architecture and comprises a web mapping server, database systems and applications that can help visualize data as tables and charts. The system allows users to put forward interactive spatial queries through linkage of spatial and non-spatial data. This utility aids the users in location-specific decision-making through innovative and interactive tools.

3.2.2.   *Services and applications*

The functionality of the system is organized around the major areas of metadata management, viz., mapping and visualization, data conversion, data analysis and 3D visualization. The system is based on a web-based

Data Flow / Integration

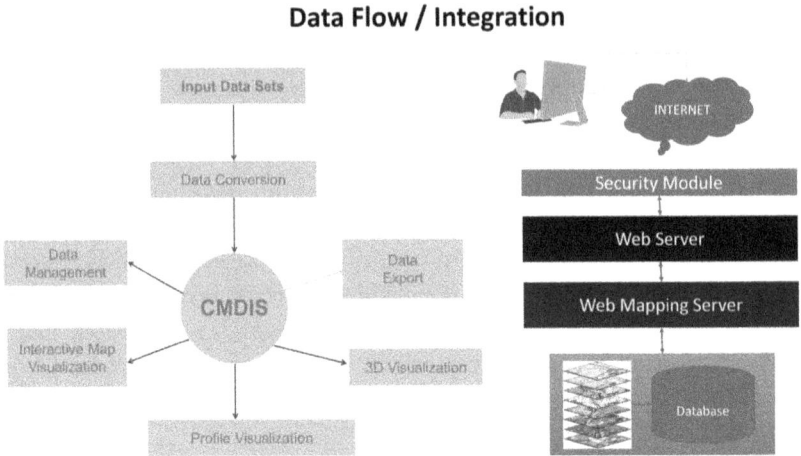

Fig. 3.1. System architecture for data management and analysis.

client-server architecture. Future development effort is directed at evolving spatial database infrastructure and advanced ocean analytics that can interoperate with other databases using open GIS standards.

3.2.2.1. *Metadata management*

Metadata management consists of cataloguing the data sets so as to allow users to locate data sets that are related by spatial location, theme or time. Cataloguing in CMDIS uses the Extensible Markup Language (XML) approach of the Dublin Core and the Federal Geographic Data Committee (FGDC) standard to create flexible, extensible metadata that will be easily accessible and searchable. The catalogue not only allows free text search but also supports search on specific attributes such as location, parameter, file type, etc. The retrieved catalogue information includes citation information, data set description, time, spatial domain, a preview map of the dataset and an URL link to the data set itself. Additionally, users can edit metadata information for the data sets they had collected through a menu-driven interface.

3.2.2.2. *Data conversion*

The data conversion process is a server-based module for the user to import data from different data sources into the database. These data sources include data from various agencies, historical and real-time observations, satellite data and outputs from models. For data sets like bathymetry, temperature and salinity profiles, part of the data conversion process is a determination of the uniqueness of the data point to be added by checking its metadata attributes against a set of validation rules. A screen status report allows the user to decide whether to add the new data point or use it to replace older data points.

3.2.2.3. *Interactive map visualization*

This refers to an interactive functionality to retrieve and view depth profiles of environmental parameters. A map display of locations with depth profiles allows the user to select profiles. Once a selection is made, the metadata of profiles identified in the selection will be returned to the user to refine the selection, and the updated selection is reflected on the map. Finally, the user can choose to view the depth profiles of the points selected and export the associated data to different file formats. 2D vector graphics was used to visualize the parameter profiles. It was chosen because the profile data was stored in the database as XML files, and it was easy to convert these to SVG graphics (**Fig. 3.2**).

3.2.3. *Data analysis*

Multi-disciplinary and Cross-Cutting Thematic Data and Information were used to analyze the different multi-scale data and facilitate additional applications of the data for different scenarios and weightage-based trade-offs to enable spatial planning. This tool will enable users to obtain information from the database of scenarios that were calculated based on the time-series observations and predictions using a web mapping system. The system allows retrieving data based on time-based selection and trend plots for a specific location and selected scenarios. In addition, the customized application will allow users to interactively analyze digital maps through spatial overlay techniques through which users present their queries to databases and information stores. The cross-cutting application will make use of the web services and the web mapping services to access a time-series database and multiple-thematic maps to create outputs in the form of GIS data layers tailored to the specific domains for water resources, coastal resources, biodiversity, public health, energy demand and integrated coastal and MSP.

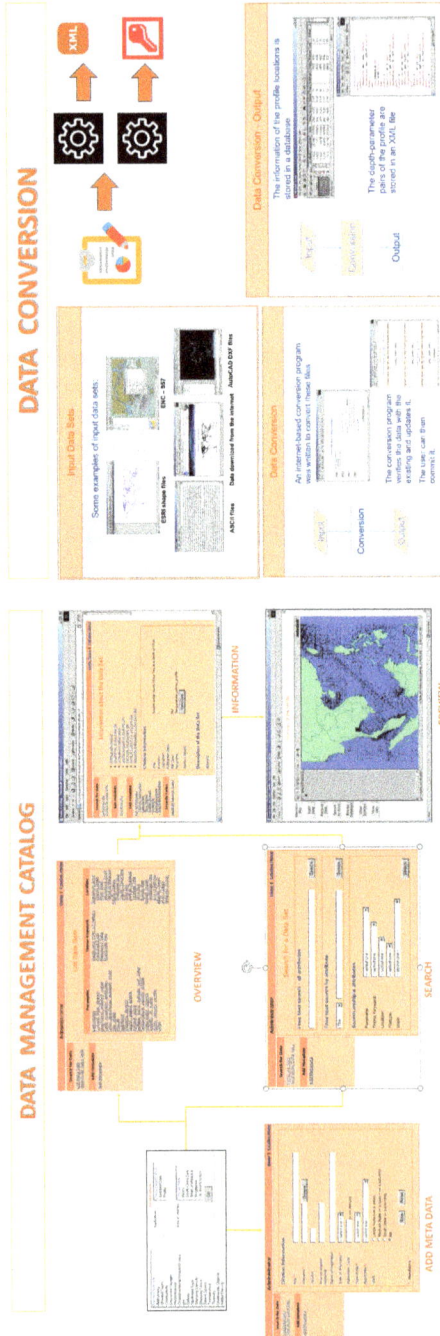

Fig. 3.2.   Interactive web-based application of CMDIS.

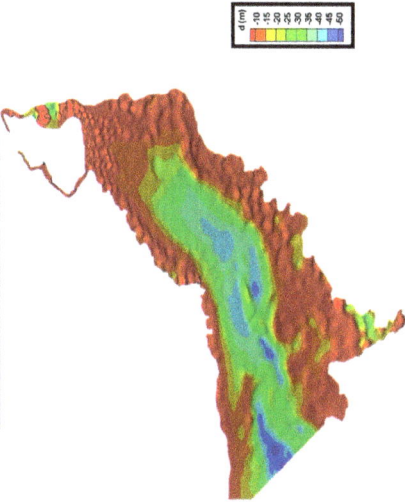

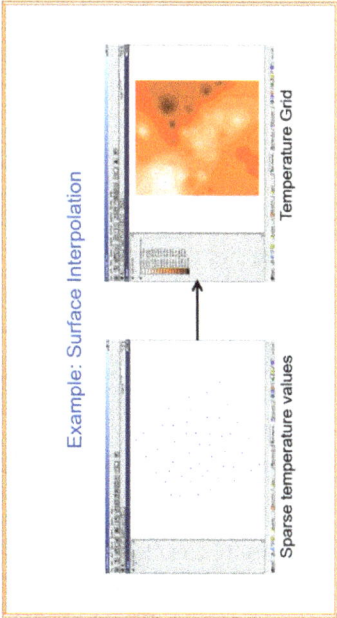

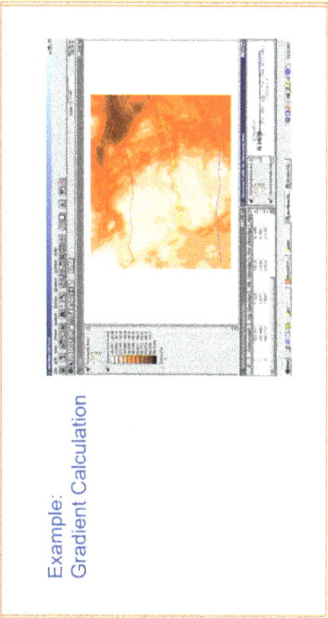

3D VISUALIZATION

DATA ANALYSIS

Example: Surface Interpolation

Sparse temperature values

Temperature Grid

Example:
Gradient Calculation

Fig. 3.2.   (*Continued*)

3.2.3.1. *3D terrain visualization*

CMDIS applies 3D terrain visualization techniques for underwater visualization. This is achieved using a specialized program in underwater seabed terrain visualization. The 3D virtual reality scenes, consisting of the seabed terrain and underwater objects on the seabed, are created from data sets selected by the user through a mapping interface. Special routines were developed to automatically construct man-made objects like underwater pipelines and other coastal infrastructure elements directly from the information in GIS map layers. A special effort was made to ensure that the auxiliary objects such as pipelines — followed the undulation in the seabed terrain as much as possible (**Fig. 3.2**). A "swim-through" can be conducted by controlling path of a submersible vehicle using mouse or joystick.

3.3.   Smart Maritime Information

"Smart port," defined broadly as a new concept, is part of modern infrastructure development that aims to use the ICT revolution, including smart devices, sensors, networks and data analytics, to improve a port's overall economy and operations (**Fig. 3.3**). With the use of data and technology, the port can be operated in a predicted way to improve the connectivity with users and stakeholders for smooth operation of ports and sustainable development.

Developing a port into a "smart port" (**Fig. 3.4**) involves a step-by-step approach where a port transforms its mode of operations through automation. The transformation method involves:

- Building the infrastructure based on the requirements.
- Employing the relevant and correct technology.
- Equipping the port with skillsets that caters to the new requirements.
- Enabling the interoperability of devices, machinery, ICT and related workflows and processes.
- Connecting port users and stakeholders.
- Ensuring connectivity of the ports to other ports globally, including linking to global trade and supply chain.
- Employing data analytics over time-based for improvement of the port.

The concept of "Digital Twins" is gaining popularity in developing the concept for smart systems that enable users to develop data-enabled intelligence in the system. The data being collected from the system and the date generated will aid in effective decision-making. Digital twin simulation

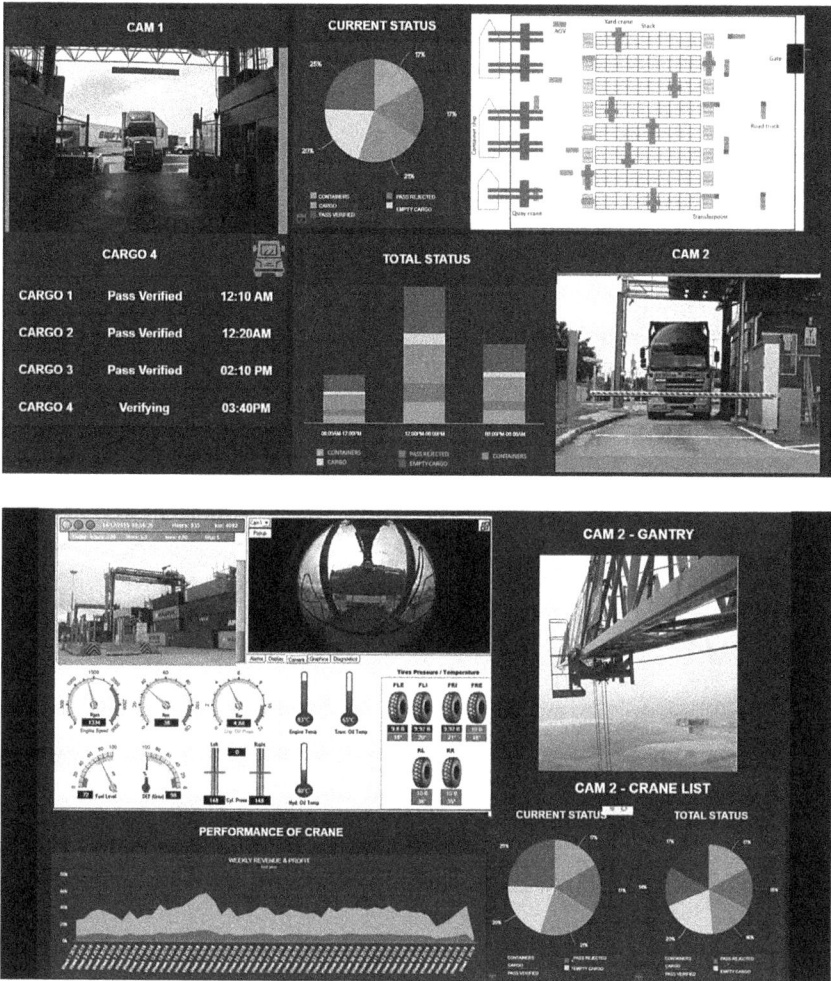

Fig. 3.3. Monitoring terminals of the port area through a dashboard system.

allows ports in the creation of testing scenarios as part of the "smart port" transformation to prepare for possible disruptions to operations, including natural disasters and extreme weather (**Fig. 3.5**).

3.4. Advanced Ocean Analytics and Internet of Things

The use of state-of-the-art advanced Internet of Things (IoT)-based marine environment helps to control and monitor devices and the transmission of

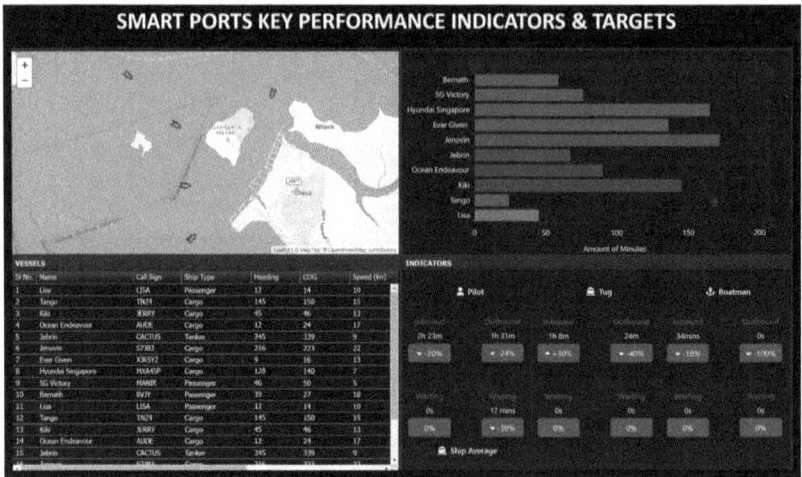

Fig. 3.4. Smart ports performance overview dashboard.

Fig. 3.5. Digital twins for ports.

real-time marine environmental data using web services (**Fig. 3.6**). The system is new and innovative and equipped with the dynamic discovery of sensors that allows interoperability and failure detection.

The web services technology enables heterogeneous systems that can work irrespective of the network to provide spontaneous information using Services-Oriented Architecture. The integration of such dynamic sensing enables the visualization of objects that are critical to safety and operations at sea. It can be further enhanced through marine IoT, artificial intelligence (AI) and machine learning (ML) for advanced ocean analytics and precise forecasting of the marine environment.

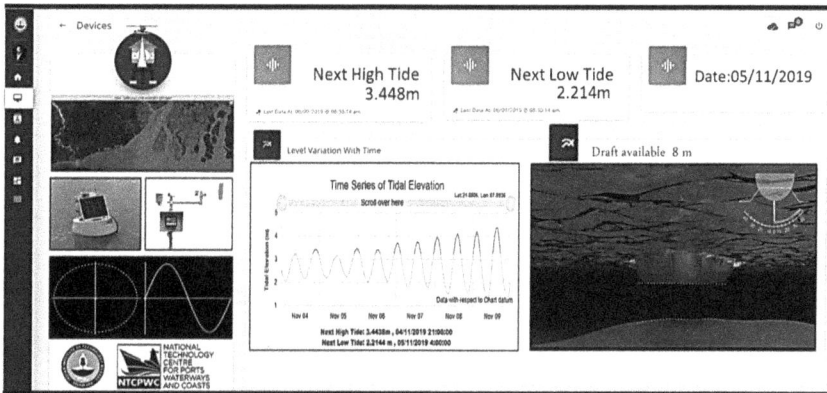

Fig. 3.6. Sensor-based observation and prediction through advanced ocean analytics.

The advanced ocean analytics framework require complete mapping, monitoring and modelling of the oceans and seabed and coordinated network operations.

The various tasks involve:

(1) Satellite- and laser imaging, detection, and ranging (LiDAR)-based nearshore bathymetry mapping.
(2) Sensor fusion and Wireless sensor network (WSN) networks through deep learning techniques.
(3) Real-time processing of payload data.
(4) High-performance and data-driven simulation modelling for ocean state prediction.
(5) Integration and analytics through IoT, ML and AI for decision-making.

3.5. MSP Tools

CMDIS was designed to perform a set of tasks for MSP tools through developing and organizing comprehensive and large-scale baseline environmental thematic layers for coastal and marine waters, such as coastal zone, near and offshore observations and modelling data sets, as part of the system development. The system allows to house, through a generic database, schema pattern to accommodate all the datasets for centralized and effective search and visualization through intuitive Graphical user interface (GUI) tools to allow users to select spatially the area of interest and to interactively visualize and to carry out spatial analysis.

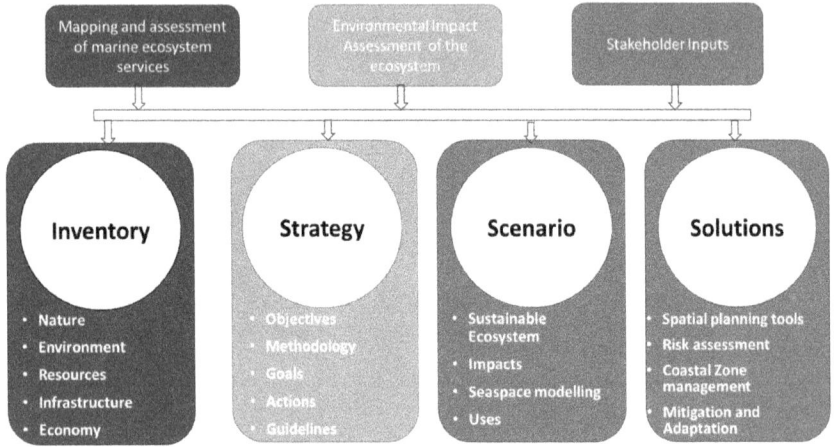

Fig. 3.7.   Framework for MSP tools.

The Decision Support System (DSS) tools allow spatial planning of coastal and marine sectors based on a sustainable approach through the spatial integration of thematic information and other environmental for capacity building of governments as part of national-level development plans. The introduction of new and innovative tools supports interactive analysis using geospatial tools to perform location-specific analysis for various decision-makers (**Fig. 3.7**).

The MSP tools provide integrated decision-making tools with special reference to planning tools that are spatially explicit with an integrated approach for the assessment of economic, environmental, socio-cultural risks and opportunities. The tool supports decision-makers to monitor and maintain the environment, given the climate change and extreme events risks. The major contribution lies in the identification of sensitive areas while spatial planning on establishing protected areas that are of biological importance. The tools help stakeholders to reduce conflicts while decision-making through identification of marine spaces and allocation on communities-based use.

3.6.   Summary

The CMDIS is developed largely from open data standards and open architecture to a variety of marine data sets and further integration with real-time observation and leverage on advanced ocean analytics that makes use of AI, ML and big data. The system needs to house a set of open-source tools to support the implementation of Coastal Zone Management and MSP to manage data over the entire workflow, from the collaborative upload of data, to the creation of metadata, portrayal styles, aggregated maps, the setup of uses cases and the elaboration through specific modules producing final maps and descriptive reports. The integration of such a suite of open-source tools allows a transparent, reproducible and highly interactive application of solid methodologies already applied with a specific focus on the analysis of conflicts between marine uses and cumulative impacts of human activities on marine environments (Menegon *et al.*, 2016). The integration of technologies such as real-time data feeds from marine environment observations and advanced ocean analytics will further improve the decision-support capabilities of the system. The outputs measured in terms of Key Performance Indicators that are being used to support the development of maritime spatial plans for further implementation process in various case study areas and marine waters in the region. Future development efforts would be directed at evolving the current system to that of a spatial data infrastructure that can interoperate with other spatial data information systems for effective and sustainable maritime planning.

References

Andersen, J., Stock, A., Heinanen, S., Mannerla, M., and Vinther, M. (2013). Human uses, pressures and impacts in the eastern north sea. Technical report, Aarhus University, DCE-Danish Centre for Environment and Energy.

Center for Ocean Solutions (2011). Decision Guide for Selecting Decision Support Tools for Marine Spatial Planning. Technical report, The Woods Institute for the Environment, Stanford University.

De Longueville, B. (2010). Community-based geoportals: The next generation? Concepts and methods for the geospatial Web 2.0. *Computers, Environment and Urban Systems*, 34(4): 299–308.

Douvere, F. (2008). The importance of marine spatial planning in advancing ecosystem-based sea use management. *Marine Policy*, 32(5):762–771.

Durairaju K. R., Li R. R., Chan W. T., and Van Winkel, I. (2003) Marine Data Information System. *GIS@ Development*, 7(8):29–31.

Durairaju K. R., Clews, E., Tigli, J. Y., Lavirotte, S., and Rey, G. (2010). Sensors networks, SOA and web based approach for fresh water environmental monitoring. *9th International Conference on Hydroinformatics (HIC)*.

European Union (2014). Directive 2014/89/EU of the European Parliament and of the Council of 23 July 2014 establishing a framework for Marine spatial planning. O.J. L 257/135.

Georis-Creuseveau, J., Claramunt, C., and Gourmelon, F. (2017). A modelling framework for the study of spatial data infrastructures applied to coastal management and planning. *International Journal of Geographical Informa-tion Science*, 31(1):122–138.

Halpern, B. S., McLeod, K. L., Rosenberg, A. A., and Crowder, L. B. (2008). Managing for cumulative impacts in ecosystem-based management through ocean zoning. *Ocean & Coastal Management*, 51(3):203–211.

Korpinen, S., Meidinger, M., and Laamanen, M. (2013). Cumulative impacts on seabed habitats: An indicator for assessments of good environmental status. *Marine Pollution Bulletin*, 74(1):311–319.

Maguire, D. J. and Longley, P. A. (2005). The emergence of geoportals and their role in spatial data infrastructures. *Computers, Environment and Urban Systems*, 29(1):3–14.

Menegon, S., Sarretta, A., Barbanti, A., Gissi, E., and Venier, C. (2016). Open source tools to support integrated coastal management and maritime spatial planning (No. e2245v2). PeerJ Preprints.

Menegon, S., Depellegrin, D., Farella, G., Gissi, E., Ghezzo, M., Sarretta, A., Venier, C., and Barbanti, A. (2018). A modelling framework for MSP-oriented cumulative effects assessment. *Ecological Indicators*, 91:171–181.

Micheli, F., Halpern, B. S., Walbridge, S., Ciriaco, S., Ferretti, F., Fraschetti, S., Lewison, R., Nykjaer, L., and Rosenberg, A. A. (2013). Cumulative human impacts on Mediterranean and Black Sea marine ecosystems: Assessing current pressures and opportunities. *PLoS ONE*, 8(12):e79889.

Stelzenmuller, V., Lee, J., South, A., Foden, J., and Rogers, S. I. (2013a). Practical tools to support marine spatial planning: a review and some prototype tools. *Marine Policy*, 38:214–227.

Stelzenmuller, V., Schulze, T., Gimpel, A., Bartelings, H., Bello, E., Bergh, Ø., Bolman, B., Caetano, M., Davaasuren, N., Fabi, G., *et al.* (2013b). Guidance on a better integration of aquaculture, fisheries, and other activities in the coastal zone: From tools to practical examples. Technical report, COEXIST project.

Stock, A. (2016). Open source software for mapping human impacts on marine ecosystems with an additive model. *Journal of Open Research Software*, 4(1):e21. DOI: http://doi.org/10.5334/jors.88.

Section B

Observation of Coastal Environment

Chapter 4

Data Analysis Methods and Significance

Abstract
The vast amount of wave data obtained from physical measurements require systematic analysis to derive useful information for design and other widely varying applications such as sediment transport. The time-domain statistical analysis and frequency-domain spectral analysis provide the necessary characteristics of data sets for various applications. Further, the temporal variation of frequency distribution of energy could be explored by different methods such as phase-time, Hilbert transform and wavelet analysis.

4.1. Analysis of Random Waves

The realistic sea surface is composed of a number of waves with different directions, frequencies, phases and amplitudes. Understanding the random wave process and analysis is important for an adequate description of the sea surface and for practical applications.

4.1.1. *Waves in the ocean as a random process*

Random waves can be idealized by the linear superposition of harmonic waves discrete in space for all practical purposes. On the assumption that a random wave field follows a stationary and ergodic process, the spatial distribution of the wave profile is idealized into a time series of sufficient length for the analysis. The stationary property implies the invariant statistical properties if the time window is moving over a random process and the probability distribution is the same for any sufficiently longer time window over the period of interest.

In a real ocean wave field, a typical measurement of 20 min record is sampled to represent a wave field of 3 h (**Fig. 4.1**). The time-averaged statistics over a 20-min record is equal to the time-averaged statistics of every 20-min

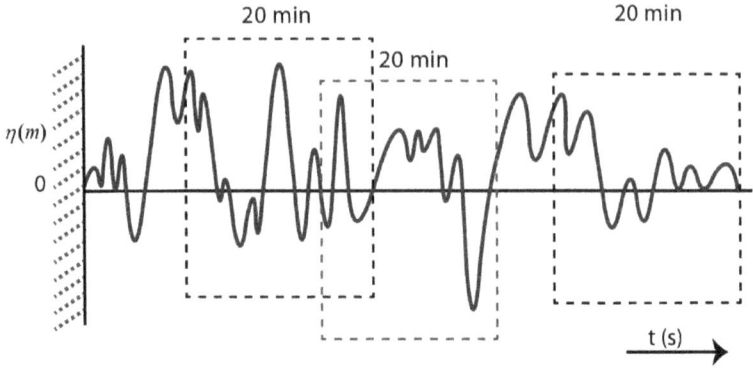

Fig. 4.1. Typical wave record.

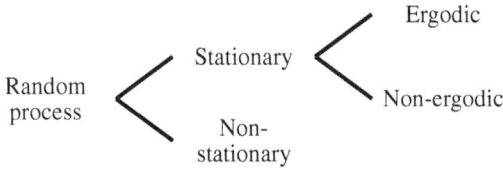

Fig. 4.2. Broad classification of random process.

record within the 3-h time interval. In addition, the event-averaged statistics (at a particular time frame across all the 20-min records) is also equal to the time-averaged statistics of one frame following the ergodicity. If the stationary ergodic process is not valid, and one has to measure for a longer period to take statistical averages to represent a wave climate for 3-h intervals. A broad classification of the random process is illustrated in **Fig. 4.2.**

Expected value of $x(t) = \overline{x(t)} = \langle k(t) \rangle$.

Average or mean value of a quantity:

$$E[x(t)] = \lim_{T \to \infty} \frac{1}{T} \int_0^T x(t)dt$$

Mean square value:

$$E[x^2(t)] = \overline{x^2} = \lim_{T \to \infty} \frac{1}{T} \int_0^T x^2 dt$$

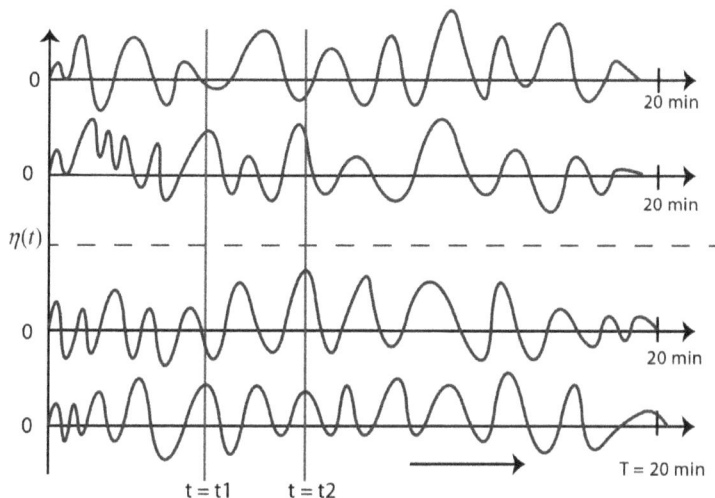

Fig. 4.3.   Discrete 20-min wave records.

Variance:

$$\sigma^2 = \text{mean square value about the mean}$$

$$= \lim_{T \to \infty} \frac{1}{T} \int_0^T (x - \bar{x})^2 dt$$

$$\sigma^2 = \overline{x^2} - (\bar{x})^2$$

where σ = Standard deviation is a measure of the spread about the mean [small $\sigma \to$ narrower probability curve, $p(x)$].

A sample discrete random wave elevation records are projected in **Fig. 4.3**. The simplest approach to represent such a random event is the concept of the spectrum of ocean waves. The distribution of wave energies across different wave frequencies/wave lengths are given in the spectrum. The statistical and spectral analyses of a random wave process are discussed below.

4.1.2.   *Statistical and spectral analyses*

The random process such as a random wave, $\eta(t)$ can be analyzed in either time domain or frequency domain. The assumption of linear superposition (and hence, the process is assumed to be linear) makes a good correlation between the two types of analysis, namely the statistical (time domain) and spectral (frequency domain).

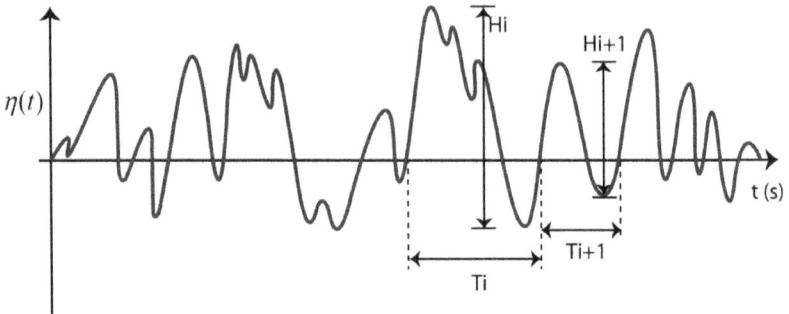

Fig. 4.4.   Representation of individual wave record within the time series.

4.1.2.1. *Statistical analysis*

Statistical analysis is a direct analysis without subjecting the time series into any conversion process. Hence, it is valuable and can be taken as primary information as far as the physical representation along the time scale and the ordinate scale (it can be wave elevation or pressure etc.).

If we measure the length scale of the time series of consecutive waves, we can have,

- along the horizontal time scale, $T_1, T_2, \ldots, T_n$ for n number of waves;
- along the vertical wave elevation scale, $H_1, H_2, \ldots, H_n$ for n number of waves.

In **Fig. 4.4**, the origin of the ith wave is defined from the point where the time series is crossing the mean value (here, zero for all the zero-mean process) and the slope of time series has positive value at the origin of ith position. This is called up-cross analysis.

On the other hand, if our starting point of the time series is such that at the zero crossing, at the mean value, the progressive slope is negative, then it is called down-cross analysis.

The choice of the appropriate type of analysis depends on the considered variable. For the wave surface elevation, if we observe the time progress of the event, $\eta(t)$, the wave propagates from right to left and our time scale is positive from left to right. Hence, in the real field (on our assumption of a conventional wave, where a crest is followed by a trough), if the trough accompanying a crest crosses the "time" step ahead of the following crest, down-cross analysis is preferred. However, if the convention is different, up-cross analysis can be adopted.

Since the random waves comprise different wave heights (H_i) and periods (T_i), an engineer or a navigator is obviously confronted with the dilemma of selecting the appropriate value from the time series. There should not be any ambiguity between two different users who want to obtain the wave field estimate at the same location. So, it is important to define a unique parameter that is also matching with a visual field observer. It is observed that a visual observer usually makes a bias in the estimation of the wave height above its mean value. The estimate of the visual observer in general is found to correlate with the average of one-third of the largest wave heights among the group. Hence, in general, the "significant wave height" is defined as the average of the highest one-third of the waves. The procedure to evaluate the significant wave height $H_{1/3}$ from a group of n values is to rearrange the H_i values in a descending order and to take the average of $n/3$ values.

$$\text{Significant wave height, } H_{1/3} = \frac{\sum_{i=1}^{n/3} H_i}{n/3} \tag{4.1}$$

where, H_i values are listed in descending order of its magnitude.

The engineer would also be concerned to find the maximum value and other statistical properties useful for design and operational purposes.

Different statistical properties can be arrived at from the list of descending order of wave heights as follows:

$$\text{Maximum wave height, } H_{\max} = H_1 \tag{4.2}$$

$$\text{Average of highest 0.2\% waves, } H_{1/500} = \frac{\sum_{i=1}^{n/500} H_i}{n/500} \tag{4.3}$$

$$\text{Average of highest one-hundredth, } H_{1/100} = \frac{\sum_{i=1}^{n/10} H_i}{n/100} \tag{4.4}$$

$$\text{Average of highest one-tenth, } H_{1/10} = \frac{\sum_{i=1}^{n/10} H_i}{n/10} \tag{4.5}$$

$$\text{Mean wave height, } \bar{H} = H_{av} = \frac{\sum_{i=1}^{n} H_i}{n} \tag{4.6}$$

The above calculations require a long time series of sufficient records, i.e., to say to calculate $H_{1/500}$, one should have more than 1,000 records in the series. What should be the optimum value to find an average? In general, the time series with a record length of 3,000 is required.

Now, if the typical record of 20 min is considered to represent a wave climate, say with an average wave period of 6 s in that location, there will be about 200 records in the series. This is not sufficient to adopt in the above form of calculations.

The distribution of wave heights, H_i, is found to follow Rayleigh probability distribution, which addresses our concern for the estimate. Following Rayleigh distribution, the various statistical parameters can be estimated from the characteristic wave height, i.e., significant wave height, $H_{1/3}$.

$$\text{Average of highest } 0.2\% \text{ waves, } H_{1/500} = 1.91\, H_{1/3} \qquad (4.7)$$

$$\text{Average of highest one-hundredth, } H_{1/100} = 1.67\, H_{1/3} \qquad (4.8)$$

$$\text{Average of highest one-tenth, } H_{1/10} = 1.27\, H_{1/3} \qquad (4.9)$$

$$\text{Mean wave height, } H_{av} = 0.63\, H_{1/3} \qquad (4.10)$$

Since, the Rayleigh distribution has no upper bound, the maximum wave height, $H_{\max}$ cannot be estimated from the characteristic estimate.

An approximate estimate, however, can be given as,

$$\text{Maximum wave height, } H_{\max} = 2\, H_{1/3} \qquad (4.11)$$

Another salient parameter a design engineer would be looking for is the root mean square wave height (H_{rms}),

$$H_{\mathrm{rms}} = \sqrt{\frac{\sum_{i=1}^{n} H_i^2}{n}} \qquad (4.12)$$

Similarly, third- and fourth-order statistical parameters such as skewness and kurtosis can be evaluated to explore the nonlinearity in the random signal. The analysis of a nonlinear signal will be dealt later.

The above process can be used for any variable of interest. However, depending on the distribution of variables, the fitting coefficients in Eqs. (4.7) to (4.11) have to be carefully chosen. For example, the wave surface elevation, $\eta(t)$ follows Gaussian distribution and extreme values of wave heights for long-term statistics follow Weibull or Gumbel distribution.

Similar to the concept in deriving statistical parameters for the wave height, the characteristic wave period parameters can also be derived. Some of the details will be dealt in the next section.

4.1.2.2. Spectral analysis

Even though statistical analysis provides a comprehensive direct data analysis, the information on the distribution of concentration of wave energy at different frequency bands is lacking. It is particularly important if the offshore system under design has natural frequency of the same order as the wave frequency at which the maximum energy concentrates. This is supplemented by the frequency domain analysis by decomposing the time series into various frequency components using Fourier Transform (FT).

In simple terms, the distribution of wave energy as a function of its frequency is known as a wave spectrum, which is the interpretation of the total energy transmitted by a wave-field. The data transformation is often executed to facilitate better understanding of the existing wave climate. It may be difficult to interpret a sine wave in time scale; therefore, the application of Fourier series representation of wave time histories can easily transform the original time signal for easier data interpretation.

Only a brief overview of FT to obtain a frequency spectrum has been provided in this chapter. The salient aspects that need attention in the analysis, particularly to extract design parameters, are dealt here.

4.2. Random Waves and Wave Spectrum

Most of the information provided earlier pertains to only monochromatic waves, which have only one frequency. In reality, the sea surface is composed of a number of waves with different directions, frequencies, phases and amplitudes. Understanding the random wave process and analysis is important for an adequate description of sea surface and for practical applications. A simple approach to represent such a random event is the concept of the spectrum of ocean waves. The distribution of wave energies across different wave frequencies/wave lengths is given in the spectrum.

4.2.1. Stationary and ergodic random process

A random process is called a stationary ergodic random process if the time average statistics and event average statistics are equal for a zero-mean process. With the random process, $\eta(x, t)$, assuming that the expected value being zero is not always possible. If the expected value equals some constant a_0, the random process can be adjusted such that the expected value is indeed zero:

$$\eta(x, t) = \chi(t, x) - a_0 \tag{4.13}$$

4.2.1.1. *Auto correlation function*

For a random process, $x(t)$, the auto correlation function is defined as the average value of the product of $x(t)$ and $x(t + \tau)$.

$$R(\tau) = E\{\eta(t, x)\eta(t + \tau, x)\} = R'(\tau) = \lim_{T \to \infty} \frac{1}{T} \int_0^T \eta_i(t)\eta_i(t + \tau)dt$$

$$(4.14)$$

The correlation properties are:

(1) $R(0) = \text{variance} = \sigma^2 = (\text{RMS})^2$
(2) $R(\tau) = R(-\tau)$
(3) $R(0) \geq |R(\tau)|$

For example, considering the waves following random process that is a summation of cosines of different frequencies.

$$\eta(x, t) = \sum_{n=1}^N a_n \cos(\omega_n t + \psi_n(x)) \qquad (4.15a)$$

where, $\psi_n(x)$ are all independent random phases in $[0, 2\pi]$ with a uniform probability density function. This random process is stationary and ergodic with an expected value of zero.

The autocorrelation (τ) is thus,

$$R(\tau) = \sum_{n=1}^N \frac{a_n^2}{2} \cos(\omega_n \tau) \qquad (4.15b)$$

If $x(t)$ is stationary, $E[x(t)] = E[x(t + \tau)] = m$

$$\sigma_{x(t)} = \sigma_{x(t+\tau)} = \sigma$$

$$\text{Correlation coefficient } \rho = \frac{R_x(\tau) - m^2}{\sigma^2}$$

4.2.2. *Spectrum*

Joseph Fourier developed the concept that any function $x(t)$ over the interval $(-T/2 < t < T/2)$ can be represented as the sum of an infinite series of sine and cosine functions with harmonic wave frequencies, based on which the concept of spectrum is arrived

$$x(t) = \frac{a_0}{2} + \sum_{n=1}^\infty (a_n \cos n\omega't + b_n \sin n\omega't) \qquad (4.16)$$

where

$$a_n = \frac{2}{T} \int_{-T/2}^{T/2} x(t) \cos n\omega' t \, dt, \quad (n = 0, 1, 2, \ldots)$$

$$b_n = \frac{2}{T} \int_{-T/2}^{T/2} x(t) \sin n\omega' t \, dt, \quad (n = 0, 1, 2, \ldots)$$

$\omega' = 2\pi f' = 2\pi/T$ is the fundamental frequency, and nf' are harmonics of the fundamental frequency. This form of $x(t)$ is called a Fourier series, and a_0 is the mean value of $x(t)$ over the interval.

The above equations can be written in a complex form:

$$\exp(i \, n\omega' t) = \cos(n\omega' t) + i \sin(n\omega' t)$$

and,

$$x(t) = \sum_{n=-\infty}^{\infty} Z_n \exp^{in\omega' t} \tag{4.17}$$

where,

$$Z_n = \frac{1}{T} \int_{-T/2}^{T/2} x(t) \exp^{-in\omega' t} dt, \quad (n = 0, 1, 2, \ldots)$$

Z_n is called the FT of $\eta(t)$.

The spectrum $S(f)$ of $x(t)$ is:

$$S_\eta(f) = Z_n Z_n^*$$

where Z^* is the complex conjugate of Z. The computation of ocean wave spectra follows from these forms for the Fourier series and spectra.

Thus, in the complex form of a random signal, $x(t)$ and its Fourier transform can be written as below.

$$X(\omega) = \int_{-\infty}^{\infty} x(t) e^{-i\omega t} dt \tag{4.18a}$$

$$x(t) = \frac{1}{2\pi} \int_{-\infty}^{\infty} X(\omega) e^{i\omega t} d\omega \tag{4.18b}$$

where $X(\omega)$ is the Fourier transform of $x(t)$
$x(t)$ is the Inverse FT of $X(\omega)$.

Note: The factor $1/2\pi$ can appear in any one of the above equations.

4.2.3. *Spectral density*

For a stationary process, $x(t)$ goes to infinity and the condition

$$\int_{-\infty}^{\infty} |x(t)|dt < \infty \text{ is not satisfied,} \qquad (4.19)$$

therefore, the theory of Fourier analysis cannot be applied to a sample function. This difficulty can be overcome by analyzing its auto correlation function $R_x(\tau)$ since, $R_x(\tau \to \infty) = 0$ for non-periodic wave with zero mean process and the condition.

$$\int_{-\infty}^{\infty} |R(\tau)|dt < \infty \text{ is satisfied}$$

For a stationary ergodic random process, with an expected value of zero and autocorrelation, $R(\tau)$, the power spectral density or spectrum of the random process is defined as the FT of the autocorrelation.

$$S(\omega) = \int_{-\infty}^{\infty} R(\tau)e^{-i\omega\tau}d\tau \qquad (4.20a)$$

Conversely, the autocorrelation, $R(\tau)$, is the inverse FT of the spectrum

$$R(\tau) = \frac{1}{2\pi} \int_{-\infty}^{\infty} S(\omega)e^{i\omega\tau}d\omega \qquad (4.20b)$$

Properties of the spectrum $S(\omega)$ of $\eta(x,t)$:

(1) $S(\omega)$ is a real and even function, since $R(\tau)$ is real and even.

(2) $$\int_{-\infty}^{\infty} R(\tau)e^{-i\omega t}d\tau = \int_{-\infty}^{\infty} R(\tau)(\cos\omega\tau - i\sin\omega\tau)d\tau \qquad (4.21a)$$

It can be shown that the sine component integrates to zero.

(3) The variance of the random process can be found from the spectrum:

$$\sigma^2 = (\text{RMS})^2 = R(0) = \frac{1}{2\pi} \int_{-\infty}^{\infty} S(\omega)d\omega \qquad (4.21b)$$

(4) The spectrum is positive always: $S(\omega) \geq 0$

(5) With some restriction it can also be established that

$$S(\omega) = \lim_{T \to \infty} \left| \int_{-T}^{T} x(t, x_k)e^{-i\omega t}dt \right| \qquad (4.21c)$$

A spectrum covers the range of frequencies from minus infinity to plus infinity $(-\infty < \omega < +\infty)$.

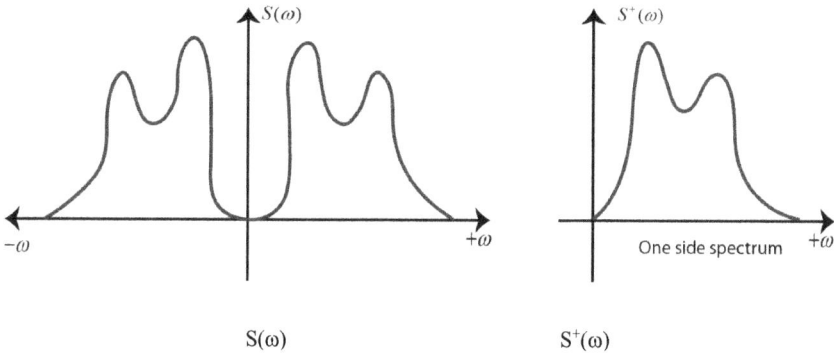

Fig. 4.5. Spectral representation.

When an entire spectrum is represented only in the positive frequency domain, it is known as a one-sided spectrum, $S^+(\omega)$. The data representation in such one-sided spectrum is rather convenient and traditional and yet it may not be strictly correct. A sample representation of a typical spectrum profile shape is shown in Fig. 4.5. By "folding," the energy over $\omega = 0$ and introducing the $\frac{1}{2\pi}$ factor, we get:

$$S^+(\omega) = \begin{cases} \dfrac{2}{2\pi} S(\omega) \omega \geq 0 \\ 0 \end{cases} \tag{4.22}$$

The variance for this one-sided spectrum, $R(0)$ is represented as,

$$R(0) = \sigma^2 = \frac{l}{2\pi} \int_{-\infty}^{\infty} S(\omega) d\omega = \frac{2}{2\pi} \int_{-\infty}^{\infty} S(\omega) d\omega \tag{4.23}$$

which can be rewritten in terms of the one-sided spectrum

$$\sigma^2 = \int_0^{\infty} S^+(\omega) d\omega \tag{4.24}$$

where $S^+(\omega) = \frac{2}{2\pi} S(\omega)$; for $\omega \geq 0$

The spectrum provides a distributed amplitude, or "probability density" of amplitudes, indicating the energy of the system. Hereafter, $S(\omega)$ represents $S^+(\omega)$, i.e., "+" sign is conventionally not added.

The surface ocean waves can be represented in terms of $\eta(x, y)$ using similar techniques used for demonstrating a Fourier series. Therefore, any surface wave variation can be established as an infinite series of sine and cosine functions oriented in all possible directions of wave orientations.

Considering the random process, $\eta(x, t)$, following the stationary and ergodic processes, it is assumed that the expected value of the random process is zero, which is not always true. If the expected value is arrived as a constant, say a_0, then the random process is to be suitably adjusted to arrive at the expected value of zero.

$$\eta(x, t) = \eta(t, x) - a_0 \qquad (4.25)$$

In Fourier series analysis, we assume that the coefficients (a_n, b_n, Z_n) are constant. For representation of ocean waves covered over a duration of about an hour and propagation distances in the order of tens of kilometres, the aforesaid assumption is valid. Moreover, the nonlinearity among wave–wave interactions is very weak; thus, a local sea surface profile can be represented by linear superposition of real, sine waves having many different frequencies and different phases travelling in many different directions. The spectrum of the wave-height gives the distribution of the variance of sea-surface height at the wave staff as a function of frequency. The spectrum is also known as the *energy spectrum* or the *wave-height spectrum* since the wave energy is proportional to its variance. Typically, three hours of wave staff data are used to compute a spectrum of wave-height.

4.2.4. *Window functions*

In the above frequency representation of a typical time series, it is assumed that the series is continuous. However, in practice, there is a definite time step between successive data. The time step is small enough such that the event is presented as a smooth functional variation over time. This discrete nature of data forces the adoption of discrete FT and in turn adds noise in the estimate. In addition, a sudden initiation of the event (represented by the time series at the initial step, say $t = 0$) and an abrupt end of the event induce higher frequency noises, which is otherwise unwanted change in the energy level. This is avoided by introducing a *windowing function*.

In signal processing, a windowing function is a mathematical function that is zero-valued outside of some chosen interval. When another function or a signal (data) is multiplied by a window function, the product is also zero-valued outside the interval: all that is left is the part where they overlap.

The following window functions, $w(n)$ are commonly adopted for filtering furious noises.

- Cosine tapered window.
- Hanning window.
- Welch window.

Cosine tapering

$$w(n) = 0.5 \left(1 - \cos \left(\frac{\pi n}{M+1} \right) \right) \qquad 0 \le n \le M$$

$$w(n) = 1 \qquad M \le n \le (N - M - 2)$$

$$w(n) = 0.5 \left(1 - \cos \frac{(N - n - 1)\pi}{M+1} \right) \qquad (N - M - 2) \le n \le N - 1$$

$$\text{where,} \quad M = INT \left[\frac{N-2}{10} \right]$$

$$(4.26a)$$

Hanning

$$w(n) = 0.5 \left(1 - \cos \left(\frac{2\pi n}{N-1} \right) \right) \qquad 0 \le n \le N - 1 \qquad (4.26b)$$

Welch

$$w(n) = \left(\frac{n - 0.5N}{0.5N} \right)^2 \qquad (4.26c)$$

where, N is the number of time steps in the time series and M is the number of time steps for windowing.

4.2.5. Spectral smoothening

Similar to the discrete time series, the frequency spectrum is derived in discrete frequency steps. This resulted in leaking of energy in between discrete steps and can be rectified by smoothening the spectral curve. A 5-point or higher-order smoothening can be carried out to perform this task.

4.2.6. Statistics from the spectral method

Now, the statistics of $\eta(t)$ given by the spectrum $S_\eta(\omega)$ needs to be established. The wave heights, H_i and wave periods of interest T_i are the random variables in this problem. As time statistics are equal to the event statistics, if $\eta(t)$ is a realization of the random process $\eta(t, x)$, then ergodicity says that H_i and T_i will provide the statistics on $\eta(t, x)$ and vice versa.

Before defining the various statistics, let us define the moments of the spectrum as follows:

Zeroth Moment:

$$m_0 = \int_0^\infty S(\omega)d\omega = \sigma^2 = VARIANCE \tag{4.27a}$$

Second Moment:

$$m_2 = \int_0^\infty S(\omega)\omega^2 d\omega \tag{4.27b}$$

Note, it can be shown that m_1, m_3, etc... are zero (for n odd).

Fourth Moment:

$$m_4 = \int_0^\infty S(\omega)\omega^4 d\omega \tag{4.27c}$$

The root mean square wave height, H_{rms} or standard deviation, σ_o is given by,

$$H_{\text{rms}} = \sigma_o = \sqrt{m_o}$$

$$H_s = 4\sqrt{m_o}$$

Mean wave height, $\bar{H} = 2.5\sqrt{m_o}$.
Average of highest $1/10^{\text{th}}$ waves, $H_1/10 = 5.09\sqrt{m_o}$.
Average of highest $1/100^{\text{th}}$ waves, $H_1/100 = 6.67\sqrt{m_o}$.
The average period, $\bar{T}$ can be found by calculating the centre of the area of the spectrum.

$$\bar{T} = 2\pi \frac{m_o}{m_1}$$

The peak period, T_p is the wave period at which the wave energy is maximum. This can be calculated either by differentiating the spectral function or interpreting the spectral values.

$$\bar{T}_p = 2\pi \sqrt{\frac{m_2}{m_4}}$$

The mean zero crossing period, $\bar{T}_z$ can be estimated as follows:

$$\bar{T}_z = 2\pi \sqrt{\frac{m_o}{m_2}}$$

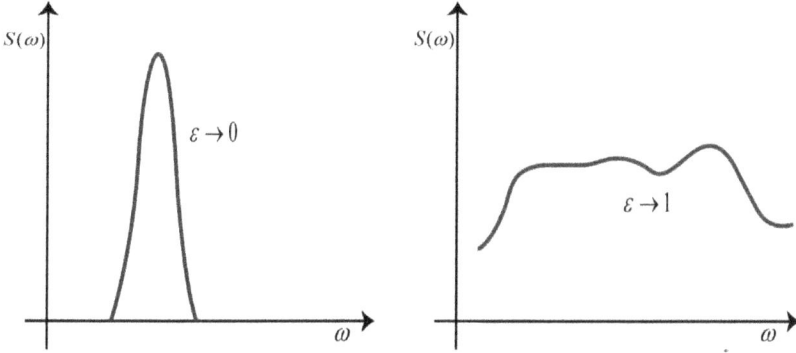

Fig. 4.6. Narrow and broad band spectrum.

Hence, $\bar{\eta}(A)$, the average frequency of up crossings past a certain level A (crossing above the threshold elevation A per time) can be estimated as,

$$\bar{\eta}(A) = \frac{1}{T_z} e^{-A^2/2m_0} \tag{4.28a}$$

$$\bar{\eta}(A) = \frac{1}{2\pi} \sqrt{\frac{m_2}{m_0}} e^{-A^2/2m_0} \tag{4.28b}$$

and, $\bar{\eta}(0)$, the average frequency of all up crossings (past a zero level).

$$\bar{\eta}(0) = \frac{1}{2\pi} \sqrt{\frac{m_2}{m_0}} \tag{4.28c}$$

4.2.7. *Bandwidth of the spectrum*

The spectral bandwidth dictates the width of the spectrum. For example, consider a simple harmonic signal with a single frequency: the bandwidth for this signal is virtually zero with a single narrow peak, although for a signal with multiple frequencies, the bandwidth increases accordingly. The bandwidth for white noise approaches to 1.

The bandwidth parameter, ε, called the spectral width parameter, defines the width of the spectrum. It can be estimated from the spread of energy over the frequencies.

$$\varepsilon^2 = 1 - \frac{\bar{T}_p}{\bar{T}_z} = 1 - \frac{m_2^2}{m_0 m_4} \tag{4.29a}$$

The bandwidth parameter is 0 for a narrow spectrum and 1 for a broad-band spectrum as seen in **Fig. 4.6.** In the ocean, a bandwidth between 0.6 and 0.8 is common. In general, $\varepsilon > 0.6$ is called broad band spectrum and

$\varepsilon < 0.6$ is called narrow band spectrum. In general, it can be said that most sea spectra are relatively narrow banded.

It is to be noted that the significant wave height, H_s is dependent on the bandwidth of the spectrum and can be estimated as below.

$$H_s = 4\sqrt{m_o\left(1 - \frac{\varepsilon^2}{2}\right)} \qquad (4.29\text{b})$$

On the assumption of wave elevation being a narrow band process, the significant wave height ($\varepsilon = 0$) is given by,

$$H_s = 4\sqrt{m_o} \qquad (4.29\text{c})$$

If the spectrum is wide band ($\varepsilon = 1$),

$$H_s = 2.83\sqrt{m_o} \qquad (4.29\text{d})$$

However, the following points have to be noted before the analysis is continued in comparison to the statistical analysis that we have seen in the earlier section. The maximum wave period obtained from the above procedure may be misleading since it would not correspond to the maximum wave energy.

Problem 4.1 Given the wave climate, evaluate the frequency spectrum and spectral characteristics.

f(Hz)	a (m)	$S_\eta(f)$ $(m^2 - s)$
0.05	0.0	0.0
0.075	0.0012	3.0E-05
0.100	0.1415189	0.4005518
0.125	0.3610530	2.607185
0.150	0.3915139	3.065662
0.175	0.3352098	2.247313
0.200	0.2684360	1.441158
0.225	0.2122135	0.9006911
0.250	0.1687059	0.5692335
0.275	0.1356967	0.3682719

Answer: $H_s = 2.15\ m, f_p = 0.15\ \text{Hz}$

4.3.   Algorithm

4.3.1.   *Statistical analysis*

The first step is to transform the given time series into a zero-mean process, for which the mean of the wave elevations is calculated and then this mean is subtracted from all the individual wave elevation values.

From this zero-mean time series, locate the time values where the wave elevation becomes zero. Let the elevation be h_1 at time t_1 and h_2 at a time t_2. Then, if the wave elevation reaches zero between t_1 and t_2 then h_1 and h_2 must be of opposite signs, assuming that the time steps are small enough. Thus the zero-crossing should lie between two time instances t_1 and t_2, such that the product of h_1 and h_2 is negative.

Hence, two points $P_1(t_1, h_1)$ and $P_2(t_2, h_2)$ are obtained between which the curve representing the time-series crosses the x-axis, say at the point P. This point P can be conveniently and accurately obtained by assuming the portion of the curve between P_1 and P_2 to be a straight line, provided the time steps are small enough.

Furthermore, if there is a down-crossing between P_1 and P_2, the slope of the line joining P_1 and P_2 will be negative and for an up-crossing, the slope will be positive.

The following chunk of MATLAB code reads the time series from start to end (N = total number of points in the time-series), finds all the zero-crossings and then stores the zero down-crossings in a variable "down cross."

```
-------------------------------------------------------------------
While (next <= N)
 if(height(current)*height(next) < 0)
    p1 = [time(current) height(current)];
    p2 = [time(next) height(next)];
    slope = (p2(2) - p1(2))/(p2(1) - p1(1));
    if(slope < 0)
       time_value = p1(1) + p1(2)*(p1(1) - p2(1))/(p2(2) - p1(2));
       downcross(index,1) = time_value;
       points(index,1) = current;
       index = index + 1;
    end
 end
 current = current + 1;
 next = current + 1;
end
-------------------------------------------------------------------
```

The time period is the difference between any two consecutive zero down-crossings. A number of time periods $T_1, T_2, T_3, \ldots, T_{n-1}$ etc., are obtained from n zero down-crossings. Also, the maximum and minimum heights between any two down-crossings are noted and then the wave-height for that time-range is $H_{\max} - H_{\min}$. A total of $n-1$ wave-heights are obtained corresponding to $n-1$ time periods. The MATLAB code that calculates this is shown below:

```
for i = 1:(length(downcross)-1)
   T_value(i,1) = downcross(i+1,1) - downcross(i,1);
end
for i = 1:(length(points)-1)
   point1 = points(i);
   point2 = points(i+1);
   temp = height(point1:point2);
   h_min = min(temp);
   h_max = max(temp);
   H_value(i,1) = h_max - h_min;
end
```

Example 4.2

Calculate various statistical averages of wave height ($H_{1/3}$, $H_{1/10}$, $H_{1/100}$, $H_{1/500}$, etc.) and wave period ($T_{1/3}$, $T_{1/10}$, $T_{1/100}$, $T_{1/500}$, etc.) for the given random wave surface elevation time history in **Fig. 4.7**.

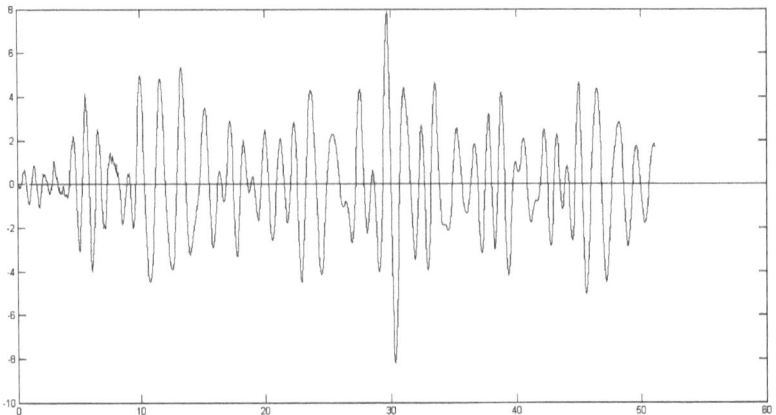

Fig. 4.7. Random wave surface elevation time history.

Table 4.1. Results from sample calculations of wave height and period.

$T_{1/3}$	1.7 s	$H_{1/3}$	8.7 m
$T_{1/10}$	1.9 s	$H_{1/10}$	10.8 m
$T_{1/100}$	2.8 s	$H_{1/100}$	14.5 m
$T_{1/500}$	3.2 s	$H_{1/500}$	16.6 m
T_{mean}	1.1 s	H_{mean}	5.5 m

Simple calculations from standard formulae follow after arranging both, the wave heights and the periods, in descending order. The results are provided in **Table 4.1**. All the values have been rounded to one decimal place.

4.3.2. *Spectral analysis*

Problem 4.3

(1) Read the given data using "LOAD" command.
(2) The given time series has first column time (t) and second column, the wave elevation (η).
(3) Transform it to zero mean process.
(4) Apply windowing function.
(5) Take "FT" of η by using *fft* command.
(6) Evaluate the spectrum using,

$$\text{ffty} = \text{fft}(\eta)$$

$$f = (0 : N/2)/(\Delta t \cdot N)$$

$$sf = 2 \times dt \times \text{abs}(\text{ffty})^\wedge 2/(N \times \Delta t \times 0.875);$$

(7) Smoothing using 5-point smoothening.
(8) Find zero-, second- and fourth-order moments of spectrum.
(9) Evaluate H_{mean}, H_{rms}, $H_{1/3}$ and $H_{1/10}$ using the moments.
(10) Determine the spectral width parameter.

Results

$$\text{Mean\_WAVE\_Height} = 5.4684 \text{ m}$$

$$\text{WAve\_Height\_rms} = 6.1726 \text{ m}$$

$$\text{H\_one\_third} = 9\,\text{m}$$
$$\text{H\_one\_tenth} = 11\,\text{m}$$
$$\text{Mean time period} = 1.2463\,\text{s}$$
$$\text{Time rms} = 1.2998\,\text{s}$$
$$\text{Spectral width parameter} = 0.78$$

It is a broad band spectrum

4.3.3.   Standard spectrum

Based on the measured spectra and theoretical results, several standard forms have been developed. This would provide a convenient platform for specifying the spectral representation to represent a particular sea-state.

4.3.3.1.   Bretschneider or International Towing Tank Conference (ITTC) spectrum

The two-parameter spectrum is defined below:

$$S_\eta(\omega) = \frac{A}{\omega^5} \exp\left(\frac{-B}{\omega^4}\right) \tag{4.30}$$

where,

$$A = 172.75\frac{H_{char}}{\bar{T}^4} \quad \text{and} \quad B = \frac{691}{\bar{T}^4}$$

The two parameters here are the characteristic wave height, H_{char} and the average wave period, $\bar{T}$.

The total energy content of the wave packet relates A and B.
i.e.,

$$m_o = \frac{A}{4B}$$

$$T_z = 0.92\bar{T}$$

and $H_{char} = 4\sqrt{m_o}$, which is the significant wave height for the case of a narrow-banded spectrum. However, the Bretschneider or ITTC two-parameter spectrum is a broad band spectrum and contains many frequencies up to higher ranges. In practice, the high-frequency ripples are neglected and the spectrum would become narrow banded and hence, the assumption of $H_{\text{char}} \approx H_{\text{s}}$ is valid.

The peak period may be found by differentiating the wave spectrum for finding the maximum (where slope is zero).

$$\text{i.e., } \frac{dS_\eta}{d\omega} = 0 \quad \Rightarrow \quad \omega_p = \sqrt[4]{\frac{4B}{5}} = \frac{4.849}{\bar{T}}$$

where, $T_p = 1.296\bar{T} = 1.41\bar{T}_z$.

Given the above representation and the relation between A and B for a fully developed sea state, it can be stated as one parameter by redefining A as the Philip's constant.

$$A = 0.00811\,\mathrm{g}^2 \quad \text{and} \quad B = \frac{3.11}{H_s^2}.$$

This is due to the fact that in a fully developed sea state, each characteristic wave height is associated with a corresponding characteristic wave period.

4.3.3.2. Joint North Sea WAve Project

The Joint North Sea WAve Project (JONSWAP) spectrum is developed based on the wave measurement in the North Sea, in which the fetch is limited. It is based on the ITTC spectrum, but the JONSWAP has a narrower peak than the ITTC.

$$S_\eta(\omega) = 0.658\,S_{ITTC}(\omega)\gamma^\chi \tag{4.31}$$

where,

$$\chi = \exp\left[\frac{-1}{2\sigma^2}\left(\frac{\omega}{\omega_p} - 1\right)^2\right]$$

With the peak enhancement factor, $\gamma = 3.3$ and,

$$\sigma = \begin{cases} 0.07 & \omega < \omega_p \\ 0.09 & \omega > \omega_p \end{cases}$$

4.3.3.3. Pierson–Moskowitz spectrum

PM spectrum was developed for representing the fully developed seas in the North Atlantic generated by local winds. The PM spectrum is defined with nominal wind speed, U_w in m/s at a height of 19.5 m above the sea surface. Hence, it was developed as one-parameter (i.e., wind speed) spectrum. However, in practice, we use this as a two-parameter (H_s and T_p) spectrum in which both H_s and T_p are defined in terms of U_w in a fully-developed sea.

The general spectral representation is same as defined for Bretschneider and the constants are defined as follows.

The Philip's constant $A = 0.00811\,g^2$
and,

$$B = \frac{0.74g^4}{U_w^4}$$

4.3.3.4. Det Norske Veritas (DNV) spectrum

DNV presented a more generalized spectral formulation with the above cases as special cases, i.e., the peak enhancement factor, $\gamma = 3.3$ for JON-SWAP and $\gamma = 1.0$ for PM spectrum, where γ is determined from the characteristics of the wave spectrum, i.e., H_s and T_p.

$$S_\eta(\omega) = \frac{\alpha}{\omega^5} \exp\left(\frac{-\beta}{\omega^4}\right) \gamma^\varphi \tag{4.32}$$

$$\gamma = \begin{cases} 5.0 & \dfrac{T_p}{\sqrt{H_s}} \le 3.6 \\[2mm] \exp\left(5.75 - \dfrac{1.15 T_p}{\sqrt{H_2}}\right) & 3.6 < \dfrac{T_p}{\sqrt{H_s}} \le 5.0 \\[2mm] 1.0 & 5.0 < \dfrac{T_p}{\sqrt{H_s}} \end{cases}$$

where α and β are generalized form of A and B defined earlier.

$$\alpha = \frac{5}{16}(1 - 0.287\ln(\gamma))H_s^2\omega_p^4$$

and

$$\beta = \frac{5}{4}\omega_p^4$$

Before proceeding with an assumption of a particular form of a standard spectrum, one has to keep in mind the following limitations on empirical spectra: fetch limitations, state of development or decay, seafloor topography, local current and effect of distant storms (swells). It has to be noted that a developing sea has a broader spectral peak and a decaying sea has a narrower peak.

4.4.   Univariate and Multivariate Spectral Analysis

4.4.1.   *General*

In studying a physical process, if one or more independent variables govern one dependent variable, then the analysis is called as univariate analysis. For example, in the process of wind-induced waves, wind is an independent variable and wave is the corresponding dependent variable. If more than one independent variable governs more than one dependent variable, then it is essential to evaluate the dependent variables simultaneously. The process is called multivariate analysis. For instance, does only wind speed dictate the cyclonic condition? There are many atmospheric variables governing the system. And the resulting wind-waves and storm-surge setup are salient features in the process. It is to be noted that, in general, we are interested in predicting the dependent variable for the given independent variable. It is essentially to find out the strength of association between the variables but not to test whether the independent variable causes the dependent variable. Module I discussed mainly univariate analysis. The following are covered in this module with particular focus on univariate nonlinear wave simulation/ analysis and multivariate analysis.

- Hilbert transforms.
- Phase-time method.
- Wavelet transforms.
- Bi-spectral analysis of nonlinear waves.
- Principal component analysis.

4.4.2.   *Hilbert transforms*

Hilbert transforms act as an essential tool in many problems, such as in understanding many modulation methods, in representing the relationship between the real and imaginary parts of a complex signal, in creating special class of casual signals called analytic which are important in simulation, in representing band pass signals as complex signals, etc. In the wave simulation, Hilbert transforms are useful in creating signals with one-sided FTs.

Coming to the simple mathematical representation of Hilbert transform, it is a simple operator which just phase shifts the signal by 90° independent of frequency, i.e.,

$$\cos t >>>> \text{ Hilbert transform, } H[\,] >>>>>> \sin t$$

which is similar to $\cos(t - 90) = \sin t$

i.e., the cosine wave is transformed to a sinusoid. On similar basis,

$$\text{Sin}\,t \ggg\gg \text{Hilbert transform} \ggg\gg\gg -\cos t$$

which is similar to $\sin(t - 90) = -\cos t$. This phase shift occurs regardless of the original phase of the signal.

Let us take a sinusoid with arbitrary phase (ϕ), $\cos(t + \phi)$. It can be decomposed into sine and cosine components.

$$\cos(t + \phi) = \cos t \cdot \cos \phi - \sin t \sin \phi$$

Now, transform the components, i.e., the individual components are subjected to phase shift.

i.e.,

$$\cos t \cos(\phi - 90) - \sin t \sin(\phi - 90) = \cos t \sin \phi + \sin t \cos \phi = \sin(t + \phi)$$
$$= \cos(t + \phi - 90)$$
$$= H[\cos(t + \phi)]$$

Now, if we sequentially transform a signal four times, the original signal will evolve.

$$\text{i.e.,}\quad \cos t \to \sin t \to -\cos t \to -\sin t \to \cos t$$

Hence, Hilbert transform can also be called a *quadrature filter*.

Following the above, it can be seen that the H [ ] is not a complex concept. However, in terms of its generic representation, it is related to convolution.

If $x(t)$ is the given signal, the Hilbert transform of $x(t)$ is represented by $\widehat{x}(t)$.

$$x(t) \ggg\gg\gg \text{Hilbert transform} \ggg\gg\gg \widehat{x}(t)$$

A real function $x(t)$ and its Hilbert transform, $\widehat{x}(t)$ are related to each other in such a way to create a strong analytic signal. The derivative of the phase of the analytic signal gives the instantaneous frequency.

4.4.2.1. *Representation in frequency domain*

To represent the Hilbert transform in the frequency domain, consider the spectral amplitudes. The cosine spectral amplitudes are both positive and lie in real plane. The sine wave has spectral components of opposite sign and lie in imaginary plane.

Now, following our earlier observation of $H[\cos\theta] \rightarrow \sin\theta$ and, $H[\cos(-\theta)] \rightarrow -\sin\theta$.

i.e., all positive frequencies of a signal shift by $-90°$ and all negative frequencies shift by $+90°$. Hilbert transform of a generic signal, $x(\omega t)$, has the following property.

$$\widehat{x}(\omega t) = \begin{cases} -i & for\ \omega > 0 \\ i & for\ \omega < 0 \end{cases} \tag{4.33}$$

Hence, the operator for the transform is, $H(\omega) = -iSgn(\omega)$.

The Signum function is defined as,

$$Sgn(\omega) = \begin{cases} 1 & \omega > 0 \\ 0 & \omega = 0 \\ -1 & \omega < 0 \end{cases} \tag{4.34}$$

4.4.2.2. Representation in time domain

In the time domain representation, $H[x(t)]$ is expressed as a convolution between the Hilbert transformer, $1/\pi t$ and $x(t)$. This is achieved by determining the inverse Fourier transform of $H(\omega)$.

$$\hat{x}(t) = \frac{1}{\pi t} \times x(t) \tag{4.35}$$

$$\text{Also,} \quad \widehat{x}(t) = \frac{1}{\pi} P \int_{-\infty}^{\infty} \frac{x(\tau)}{t - \tau} d\tau \ \text{when the integral exists.} \tag{4.36}$$

Here, P denotes Cauchy principal value, thus making the integral exist for a wide class of functions; else an improper integral would be the resultant t the pole ($\tau = t$).

The characteristic features of Hilbert transform can be summarized as below.

(1) It changes the phase of the signal but the amplitude does not change.
(2) The function acts like a transformer that converts the phase of the spectral components depending on the sign of their frequency.
(3) It can also be realized that $x(t)$ and $\widehat{x}(t)$ are orthogonal.
(4) $x(t)$ and $\widehat{x}(t)$ have identical energy since phase shifts do not change the energy.

4.4.2.3. Analytic signal

The creation of an analytic signal using Hilbert transform is useful in single- and double side-band processing and also to split the I and Q components

Table 4.2.   Analytic signals of cosine and sine functions.

$x(t)$	$\widehat{x}(t)$	$x^+(t)$
Sin t	$-$Cos t	$e^{-i\omega t}$
Cos t	Sin t	$e^{i\omega t}$

of a real signal. The analytic signal $(x^+(t))$ is a complex signal that can be formed by adding the signal $(x(t))$ with its Hilbert transform $\widehat{x}(t)$ on its quadrature.

$$x^+(t) = x(t) + i\widehat{x}(t) \qquad (4.37)$$

It is also called the pre-envelope of the real signal and is useful in the simulation of waves of required characteristics. **Table 4.2** presents the analytic signal of basic cosine and sine functions.

One may note that spectra of cosine and sine functions are two-sided (i.e., they have both positive and negative frequency components) but the complex exponential has one-sided (positive only) spectrum, i.e., the analytic signal of any given signal has only one-sided spectrum in positive frequency domain. It may be noted that the converse of analytic signal falls only in the negative frequency domain.

Now, from Eq. (4.37), FT of $x^+(t)$ leads to,

$$X^+(\omega) = X(\omega) + i[-i\text{sgn}(\omega)X(\omega)] \qquad (4.38)$$

Since, $\text{sin}(\omega) = 2u(\omega) - 1$,

$$X^+(\omega) = 2X(\omega)u(\omega) \quad \text{where,} \ u(\omega) = \begin{matrix} 1 & \omega \geq 0 \\ 0 & \omega < 0 \end{matrix} \qquad (4.39)$$

$$\text{i.e.,} \ X^+(\omega) = \begin{cases} 2X(\omega) & \textit{for } \omega > 0 \\ X(0) & \textit{for } \omega = 0 \\ 0 & \textit{for } \omega < 0 \end{cases} \qquad (4.40)$$

This is an important characteristic to create low pass and modulated (band pass) signals, the details of which are discussed in the wave group simulation procedure in one of the modules.

4.5.   Phase-Time Method

4.5.1.   *General*

The time or frequency domain investigation of the ocean wave time histories generally gives a better estimation of global data such as power spectral densities and various probability density functions. Although these results are valuable, the location of the different events within the measured time histories cannot be extracted. The knowledge of the local properties of ocean waves is also important for the ocean-related studies. For example, the properties of wave groups, interactions between waves and the local deformation of the wave profile during a breaking event require an analysis, which would reveal the localized properties of the ocean surface waves. The phase-time method (PTM), based on the Hilbert transformation, was introduced by Huang *et al.* (1992) to study the frequency modulation, structure of the wave group and wave breaking in the measured wave elevation time histories, which was later adopted by Griffin *et al.* (1996) to study the local properties of the breaking waves. The details of PTM are given below.

An analytic signal can be derived from the time histories $(x(t))$ using Eq. (4.16).

Then the phase function, $\Phi(t)$ can be obtained as follows:

$$\Phi(t) = \tan^{-1}\left(\frac{\widehat{x}(t)}{x(t)}\right) \tag{4.41}$$

$\Phi(t)$ can be decomposed into its mean value, which is the product of the mean frequency f_0, and time t and a time varying component, Griffin $\Theta(t)$,

$$\Phi(t) = f_o t + \Theta(t) \tag{4.42}$$

Huang *et al.* (1992) noted that the phase function is not usually associated with events occurring in real time, which limits the information on local properties of the wave field. The phase function is usually wrapped around $\pm\pi$ or between 0 and 2π, which would not yield an informative view of the wave elevation time histories. In the PTM, the phase is unwrapped and then de-trended by subtracting the mean from Eq. (4.42). By definition, the time derivative of the phase function is the local frequency of the time.

$$f = \frac{\partial \Phi}{\partial t} = f_o + \frac{\partial \Theta}{\partial t} \tag{4.43}$$

The above equation gives the variation of the local frequency from the mean frequency of the time history in the time domain. The number of localized frequencies equivalent to the number of data in the time history would yield important information about the presence of nonlinearities, such as breaking, in a wave train. From an experimental study, Zimmermann and Seymour (2002) concluded that the PTM is effective for the time histories with a sampling rate of 25 times the peak frequency in the wave elevations, and accuracy would greatly improve with a sampling rate of 50 times the peak frequency.

4.6.   Wavelet Transformation

FT is an important tool for the analysis and processing of many time-varying signals. FT has certain limitations to characterize non-stationary signals. Though a time-varying, overlapping window-based FT known as Short-Time FT (STFT) is capable of processing the non-stationary signals, the selection of size and shape (rectangular, Gaussian and elliptic) of the window function is highly difficult, as it significantly affects the spectral resolutions. For example, a narrow window would give better time resolution, whereas a wider one yield could result in a better frequency resolution. In addition, the window size and shape could not be changed during analysis — once selected, the resolution is also set. Thus, the analysis of time series through STFT is completely dictated by the selection of the window function. Hence, its application on the analysis of the non-stationary and signals with sharp changes in its spectral characteristics along the time scale is highly questionable. The aforementioned drawbacks could well be overcome by the use of a window function of varying length or width. Such an analysis function would certainly be useful for analysis of signals with slowly varying characteristics with occasional sudden bursts. Wavelets are analysis mathematical functions that solve the above-said problems, the details of which are described in the following sections.

Wavelets were introduced in 1980 to analyze seismic data. Later, oceanographic applications such as turbulence, surface gravity waves, etc. were explored in a greater detail by wavelets. While FT provides averaged values of amplitude and phase, wavelet provides localized instantaneous estimate for amplitude and phase. Hence, wavelets are useful to analyze non-stationary time series.

4.6.1.   *Definition of wavelet*

A "wavelet" is a small wave which has its energy concentrated in time. It has an oscillating wave characteristic but also has the ability to allow simultaneous time and frequency analysis, which is a suitable tool for the analysis of the transient, non-stationary or time-varying phenomena. In order for a function to be called a wavelet, it must satisfy the following conditions.

(1) The wavelet must have zero mean. This condition, known as the admissibility condition, ensures the invertibility of the wavelet transform. Thus, the original signal can be obtained from the wavelet coefficients through the inverse transform.
(2) The integral of the wavelet function, usually denoted by ψ, must be zero.

$$\int_{-\infty}^{\infty} \psi(t)dt = 0 \qquad (4.44)$$

This assures that the wavelet function has a wave shape and is known as the admissibility condition.
(3) The wavelet function must have unitary energy, i.e.,

$$\int_{-\infty}^{\infty} |\psi(t)|^2 dt = 1 \qquad (4.45)$$

This assures that the wavelet function has compact support or has a fast amplitude decay enabling physical domain localization.

These wavelets split up a time-varying data into different frequency components, each of which are studied with a resolution that matches its scale. The wavelet transformation analysis is fairly advantageous in comparison to the traditional Fourier methods, where the signal contains discontinuities and sharp spikes. Waves are smooth, predictable and continuous, whereas wavelets are of limited duration, irregular and could be asymmetrical. The time-variant or stationary signals in a FT uses waves as its deterministic basis function. Wavelets can either be deterministic or non-deterministic in nature based on its application for the generation and analysis of the majority of natural signals to provide better time-frequency representation — the same cannot be achieved with waves using conventional FT.

4.6.2.   *Mother wavelet*

The wavelet analysis procedure is to adopt a wavelet prototype function, called an "analyzing wavelet" or "mother wavelet." A contracted, high-frequency version of the prototype wavelet is analyzed in time domain, while a dilated, low-frequency version of the prototype wavelet is analyzed in frequency domain. In simple terms, the kernel functions used in the transformations are obtained by scaling and translating the prototype function.

The kernels are obtained as,

$$\psi_{a,\tau}^*(t) = \frac{1}{\sqrt{a}}\psi\left(\frac{t-\tau}{a}\right) \tag{4.46}$$

where, a is the scaling factor and τ is the translating or shifting factor and $1/\sqrt{a}$ is the normalization factor to ensure that all wavelets have the same energy. ψ^* is complex conjugate of ψ. The scale variable, a, and the time variable, τ, are allowed to vary in $[-\infty, \infty]$. The wavelet analysis provides two-dimensional information (in the domain τ and a) from a one-dimensional time series.

Among the few mother wavelets, such as the orthogonal wavelets, Paul's wavelet or DOG wavelet (derivative of a Gaussian), the Morlet wavelet is widely adopted for oceanographic applications. The Morlet wavelet function is a Gaussian modulated complex-valued plane wave and hence widely adopted for the transformation of ocean wave signals. The Morlet wavelet function is defined as,

$$\psi(t) = \pi^{-(1/4)}e^{i\omega_o t}e^{-(t^2/2)} \tag{4.47}$$

where, ω_o is the non-dimensional frequency, which is taken as 6.0, in general for satisfying the admissibility conditions (Farge, 1992). Typical Morlet real and imaginary parts are shown in **Fig. 4.8**.

Using the above basis function, the time histories were multiplied by ψ^*, in which the parameters a and τ are continuously varying, thus leading

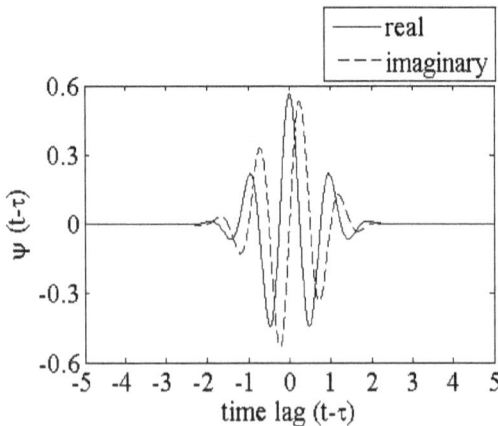

Fig. 4.8. The Morlet wavelet.

to a two-dimensional representation of the one-dimensional signal. For the Morlet wavelet with ω_o of 6.0, the Fourier period (or inverse of frequency) corresponding to the wavelet scale is 1.03a (Torrence and Compo, 1998).

4.6.3. *Continuous wavelet transformation*

The Continuous Wavelet Transformation $[\text{CWT}, W_x(a, \tau)]$ of a one-dimensional signal $x(t)$ is given as,

$$W_x(a, \tau) = \frac{1}{\sqrt{|a|}} \int_{-\infty}^{\infty} x(t) \cdot \psi^* \left(\frac{t - \tau}{a} \right) dt \qquad (4.48)$$

in which ψ^* is the scaled and translated version of basis function, ψ, given in Eq. (4.47). The variation of the size of the analysis function with respect to the time and frequency for the CWT is shown in **Fig. 4.9**. Equation (4.47) is wavelet transformation, which can be looked as mathematical microscope with the magnification of $1/a$ at position, τ. The optics is governed by the choice of wavelet, $\psi(t)$.

Upon applying CWT to the signals, one would obtain the real and imaginary parts of wavelet coefficients. Analogous to the Fourier energy density spectrum, a wavelet spectrum for a data $x(t)$ can be defined as,

$$S_x(a, \tau) = |W_x(a, \tau)|^2 \qquad (4.48)$$

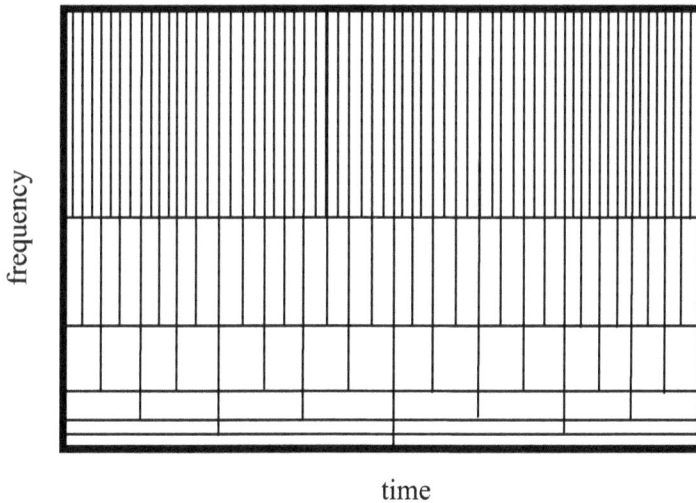

Fig. 4.9.   Size and shape of the analyzing function.

Wavelet analysis yields a measure of the localized amplitudes, a, as the wavelet moves through the time series with increasing τ.

4.6.4. Cross-wavelet transforms (XWT)

The analysis of the covariance of the two-time series has been carried out using the cross-wavelet transformation. The cross-wavelet transform of the two time series $X(t)$ and $Y(t)$ with wavelet transform W_x and W_y is defined as,

$$W_{xy}(\Upsilon, t) = W_x(\Upsilon, t) W_y^*(\Upsilon, t) \tag{4.49}$$

where the asterisk denotes complex conjugate. The phase angle of Wxy describes the phase relationship between X and Y, whereas, the power is obtained from abs (W_{xy}) in the time-frequency space. The main interest is the phase difference between the components of the two-time series, the mean and confidence interval of the phase difference were estimated. The circular mean of a set of angles within the 5% statistically significant region (or 95% confidence) is estimated following Zar (1999). As the confidence interval is difficult to interpret, it is determined using the circular standard deviation reported in Grinsted *et al.* (2004).

4.6.5. The wavelet transform coherence (WTC)

This is an estimation of the intensity of the covariance of the two given time series in time-frequency space, unlike the XWT power, which is the estimation of common powers. Coherence has been defined as (Torrence and Webster, 1999),

$$R^2(\Upsilon, t) = \frac{\left| S\left(s^{-1} W_{xy}(\Upsilon, t)\right)\right|^2}{S\left(s^{-1}|W_x(\Upsilon, t)|^2\right) \cdot S\left(s^{-1}|W_y(\Upsilon, t)|^2\right)} \tag{4.50}$$

where S is a smoothing operator, defined as, $S(W) = S_{\text{scale}}(S_{\text{time}}(W(\Upsilon, t)))$. The scales in time and frequency over which S is smoothing define the scales at which the coherence measures the covariance. Torrence and Webster (1999) reported the natural way to design the smoothing operator for the Morlet wavelet. The coherence significant level was estimated using Monte Carlo methods with red noise to determine the 5% statistical significance level (Grinsted *et al.* 2004). The circular mean as well as the confidence interval is calculated similar to XWT. The reason for the difference between

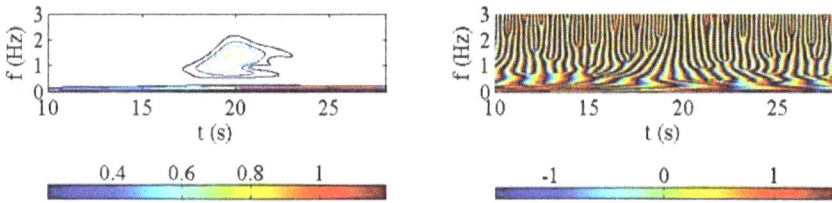

Fig. 4.10. Wavelet energy phase contours derived from the time history of a plunging wave.

the phase angles reported for the cross wavelet and coherence methods is because of the smoothing operator used in Coherence method.

4.6.6. *Prediction of breaking waves*

Let us consider the prediction of a breaking event from the measured surface elevation history. On applying CWT for the time history of a breaking wave, the wavelet spectral density contour and its corresponding phases can be obtained. The wavelet energy and phase contours are presented in **Fig. 4.10**. The locus of concentric contours of energy densities in the wave elevation time history indicates the breaking event, where the energy is very high. It is interesting to note that there exists a time, wherein the wavelet phases of all the time histories are nearly constant over a wider range of higher frequencies, indicating the energy concentration from all the frequencies occur at that particular instant.

4.7. Bi-Spectral Analysis

4.7.1. *Phase coupling*

The auto-spectral analysis yields energy distribution and is useful to understand energy distribution in the various frequency bands. However, it does not store the phase information. For a linear system, the phase variation between various frequency components does not play any role. However, nonlinear systems often contain dependent components.

$$\text{Let us take a simple nonlinear system, } y(t) = x^2(t) \qquad (4.51)$$

$$x(t) >> x^2 >> y(t)$$

Let,

$$x(t) = \text{sum of two frequency components}$$
$$= \cos(\omega_1 t + \theta_1) + \cos(\omega_2 t + \theta_2) \text{ where, } \theta_1 \text{ and } \theta_2 \text{ are random}$$
$$\text{and independent.}$$

Output, $y(t) = 1 + \cos[(\omega_1 + \omega_2)t + (\theta_1 + \theta_2)]$
$$+ \cos[(\omega_1 - \omega_2)t + (\theta_1 - \theta_2)]$$
$$+ 1/2 \cos(2\omega_1 t + 2\theta_1) + 1/2 \cos(2\omega_2 t + 2\theta_2)$$

Note that output $y(t)$ has components $\omega_1 + \omega_2$, $\omega_1 - \omega_2$, $2\omega_1$ and $2\omega_2$ but interestingly there are no fundamental frequency components, ω_1 and ω_2. And more importantly, the phase angles of $y(t)$ depend on the phase angles of input signal, i.e., the resultant components are phase-coupled. It is called quadratic (second-order) phase coupling. Phase coupling is typical of non-linear systems and it cannot be explored using "spectral density."

If we construct a new signal, $y'(t)$,

$$y'(t) = 1 + \cos(\omega_a t + \theta_a) + \cos(\omega_b t + \theta_b) + 1/2 \cos(\omega_c t + \theta_c)$$
$$+ 1/2 \cos(\omega_d t + \theta_d) \tag{4.52}$$

when, $\omega_a = \omega_1 + \omega_2$, $\omega_b = \omega_1 + \omega_2$, $\omega_c = 2\omega_1$ and $\omega_d = 2\omega_2$,

$$S_y(\omega) = S_{y'}(\omega)$$

However, the phase information is completely lost.

4.7.2. Bi-spectral analysis

Bi-spectral analysis is an advanced signal processing technique. It quantifies quadratic nonlinearities and deviations from normality. For a normally distributed (Gaussian distribution) signal, there is no nonlinearity. To parametrize the skewness and asymmetry of signal arising from wave-wave (triad or quadraplet) interactions, this technique is a useful tool.

Complex bi-spectral estimate of the signal, $x(t)$ is,

$$B(\omega_1, \omega_2) = E[X(\omega_1) \cdot X(\omega_2) \cdot X^*(\omega_3)] \tag{4.53}$$

where $E[\ ]$ is the expected value; $X(\omega)$ is the complex Fourier coefficient; $\omega_3 = \omega_1 + \omega_2$.

Note: $S_x(\omega) = E[x \cdot x^*]$ and the spectral density of wave surface elevation,

$$S_\eta(\omega) = E[x^2] \tag{4.54}$$

Equation (4.53) can be re-written as,

$$B(\omega_1, \omega_2) = |B(\omega_1, \omega_2)|e^{-i\beta}(\omega_1, \omega_2) \tag{4.55}$$

where $|B(\omega_1, \omega_2)|$ is the amplitude and $\beta(\omega1, \omega2)$ is the bi-phase.

$$\beta(\omega_1, \omega_2) = \tan^{-1}\left\{ \frac{\text{Im}(B(\omega_1, \omega_2))}{\text{Re}(B(\omega_1, \omega_2))} \right\} \tag{4.56}$$

The integral of real part of the bi-spectrum yields mean cube of a stationary random process (third moment) and the imaginary part of bi-spectrum is related to the vertical asymmetry of the waves. It is also of measure of the skewness of the temporal derivative of a time series.

$$\int \text{Re}[B(\omega_1, \omega_2)] = E[x^3] \tag{4.57}$$

Due to symmetry

$$B(\omega_1, \omega_2) = B(\omega_2, \omega_1) = B(\omega_1, -\omega_1, -\omega_2)$$

$$= B(-\omega_1, -\omega_2, \omega_1) = B(\omega_2, -\omega_1, -\omega_2) = B(-\omega_1, -\omega_2, \omega_2)$$

and, $\quad B(\omega_1, \omega_2) = B^*(-\omega_1, -\omega_2)$

Hence, due to symmetry, it is sufficient to evaluate in the domain defined by, $0 \le \omega_2 < \omega_1 \le |\omega_1 + \omega_2|$

$$\text{Skewness} = \frac{E\left[x^3(t)\right]}{E\left[x^2(t)\right]^{3/2}} = \frac{\int_{-\infty}^{\infty}\int_{\text{Re}[B(\omega_1,\omega_2)]}d\omega_1 d\omega_2}{\left[\int X(\omega)d\omega\right]^{3/2}} \tag{4.58}$$

$$\text{Wave asymmetry, } A = \frac{\int_{-\infty}^{\infty}\int_{\text{Im}[B(\omega_1,\omega_2)]}d\omega_1 d\omega_2}{E\left[x^2(t)\right]^{3/2}} \tag{4.59}$$

You may note that, in the absence of phase coupling, bi-spectrum tends to zero.

4.7.3. Bicoherence

Bicoherence (BIC) is a sort of normalized bispectrum. It represents the frequency dependency of correlation.

Bicoherence, $\mathrm{BIC}(\omega_i + \omega_j) = b^2(\omega_i, \omega_j)$

$$= \frac{|B(\omega_i, \omega_j)|^2}{\langle |X(\omega_i + \omega_j)|^2 \rangle \langle |X(\omega_i)X(\omega_j)|^2 \rangle} \qquad (4.60)$$

If $b^2(\omega_i, \omega_j) = 1$, complete quadratic coupling between frequency components at ω_i and ω_j.

If $b^2(\omega_i, \omega_j) = 0$, no coupling.

4.8. Extreme Value Analysis

4.8.1. General

The parametric frequency analysis approach is adopted to assess the risk of extreme events in extreme value analysis (EVA), i.e., the formulation is based on fitting a theoretical probability distribution to the observed extreme value series. The two extreme value models in EVA are the annual maximum series (AMS) method and the partial duration series (PDS) method, also known as the peak over threshold (POT) method.

In the AMS method, the maximum value in each year of the record is extracted for the EVA. The analysis year should preferably be defined for a period of the year where extreme events never or very seldom occur in order to ensure that a season with extreme events is not split in two. Alternatively, a specific season may be defined as the analysis year.

For estimation of T-year events, a probability distribution $F(x)$ is fitted from the extracted AMS data $\{x_{i=1,2,...,n}\}$ where n is the number of years of record. The T-year event estimate is given by,

$$x_T = F^{-1}\left(1 - \frac{1}{T}; \theta\right) \qquad (4.61)$$

where the values of θ are the estimated distribution parameters.

In the PDS method, all events above a threshold are extracted from the time series. PDS can be defined as threshold x_0 are considered $\{x > x_{0,i=1,2...n}\}\}$, implying that the number of exceedance n becomes a random variable. In threshold level becomes a random variable. If n equals the number of observation years, the PDS is referred to as the annual exceedance series.

In the present case, the AMS method is followed where the annual maximum of all years is picked up and used in the extreme value analysis.

- Chi-squared test statistic.
- Kolmogorov–Smirnov test statistic.
- Standardized least squares criterion.
- Long-likelihood measure.

It must be emphasized that the choice of probability distribution should not rely solely on the goodness-of-fit. The fact that many distributions have similar forms in their central parts but differ significantly in the tails emphasizes that the goodness-of-fit is not sufficient. The choice of probability distribution is generally a compromise between contradictory requirements. Selection of a distribution with few parameters provides robust parameter estimates but the goodness-of-fit may not be satisfactory. On the other hand, when selecting a distribution with more parameters, the goodness-of-fit will generally improve, but at the expense of a large sampling uncertainly of the parameter estimates. Besides an evaluation of the goodness-of-fit statistics, a graphical comparison of the different distribution with the observed extreme value series should be carried out. In this respect, the histogram/Frequency plot and the probability plot are useful.

4.8.2. *Distribution*

For AMS, the following eight different probability distributions are available.

- Generalized Extreme Value (GEV)
- Gumbel
- Weibull
- Frechet
- Generalized Pareto
- Gamma/Pearson Type3
- Log-Normal
- Square root exponential (SQE)

Selection of a particular function is based on past experience on similar data sets and personal preference. For present analysis, the data is fitted in to a particular frequency distribution (histogram). The distribution with the best fit is considered as the probability distribution for the data set considered.

4.8.3.   *Estimation methods*

The Method of Moments, Maximum Likelihood method, and Method of L-movements are the methods used. A suitable estimation method is chosen for the different probability distributions.

4.8.3.1.   *Independence and homogeneity tests*

For testing independence and homogeneity of the observed extreme value series, three different tests are available. These are (i) General non-parametric test for testing independence and homogeneity of a time series, (ii) Mann–Kendall test for testing monotonic trend of a time series and (iii) Mann–Whitney test for testing shift in the mean between two sub-samples defined from the time series. The current analysis applies Run Test for testing the independence and homogeneity.

4.8.3.2.   *Goodness-of-fit statistics*

For comparing the goodness-of-fit of the selected distribution and estimation method combinations, five goodness-of-fit statistics are provided by EVA.

- Chi-squared: Chi-square test statistic.
- Kolmogorov–Smirnov test statistic.
- SLSC: Standardized Least Squares Criterion.
- PPCCII: Probability Plot Correlation Coefficient based on the correlation between the ordered data and the corresponding order statistic medians.
- Long likelihood: Long-likelihood measure.

For the present analysis the Chi-squared and Kolmogorov–Smirnov test statistics has been applied.

4.8.3.3.   *Uncertainty calculations*

Two different methods are available for evaluating the uncertainty of quintile estimates, viz., Monte Carlo simulation technique and Jack-knife: Jack-Knife resampling. In this calculation, Monte Carlo simulations are carried out.

Table 4.3.   AMS series used for EVA.

Years of Occurrence	Wave Height (m)	Years of Occurrence	Wave Height (m)
1976	2	1991	3
1977	4	1992	2
1978	4	1993	3
1979	3	1994	1.5
1980	3	1995	3.5
1981	2	1996	3
1982	2.5	1997	1.5
1983	2.5	1998	1.5
1984	1	1999	2.5
1985	2.5	2000	1
1986	1.5	2001	1
1987	2	2002	2
1988	3	2003	3
1989	3	2004	4
1990	4	2005	2

Source: IMD: Data Set 1976–2005.

Table 4.4.   Wave heights under different return periods (based on Table 4.3).

	Wave Height for Various Return Period (m)		
Probability Distribution	10 Years	50 Years	100 Years
Log-normal	3.8	4.4	4.6

4.8.4.   *EVA — Wave data*

The wave data for the AMS series for the 30 years is given in **Table 4.3**. This data set on wave height forms the primary data for the EVA.

A similar exercise was carried out for the wave data collected from IMD using Weibull, truncated Gumble, Log-Normal, Log-Pearson and fitted with the data points. The Log-Normal was found to fit the best with the data. Hence the probability analysis was carried out using Log-Normal distribution and the results are given in **Table 4.4**.

References

Farge, M. (1992). Wavelet transforms and their applications to turbulence. *Annual Review of Fluid Mechanics*, 24, 395–457.

Griffin, O.M., Peltzer, R.D., Wang, H.T. and Schultz, W.W. (1996). Kinematic and dynamic evolution of deep water breaking waves. *Journal of Geophysical Research*, 101(C7), 16,515–16,531.

Grinsted, A., Moore, J.C. and Jevrejeva, S. (2004). Application of the cross wavelet transform and wavelet coherence to geophysical time series. *Nonlinear Processes in Geophysics*, 11, 561–566.

Huang, N.E., Long, S.R., Tung, C.C., Donelan, M.A. and Yuan, Y. (1992). The local properties of ocean surface waves by the phase-time method. *Geophysical Research Letters*, 19, 685–688.

Torrence, C. and Compo, G.P. (1998). A practical guide to wavelet analysis. *Bulletin of the American Meteorological Society*, 79(1), 61–78.

Torrence, C. and Webster, P.J. (1999). Interdecadal changes in the ENSO-Monsoon System. *Journal of Climate*, 12, 2679–2690.

Zar, J.H. (1999). *Biostatiscal Analysis*. Pearson Prentice Hall, Pearson Education. Inc.

Zimmermann, C.A. and R. Seymour. (2002). Detection of breaking in a deep water wave record. *Journal of Waterway, Port, Coastal, and Ocean Engineering*, 128(2), 72–78.

Chapter 5

Offshore and Nearshore Wave Measurement Techniques

Abstract

It is essential to have real-field measurements of the ocean waves both for understanding the environment as well as for validation/calibration of numerical modeling exercises. The pressure measurement of free surface elevation provides the temporal variation of wave height. Further, to resolve the direction of the wave, the vector information on propagation components is essential, which can be derived from the measurement of either velocity or wave slope in either direction. An in-situ measurement is spatially discrete. However, the remote sensing radar measurements complement for spatially rich data.

5.1. Introduction

The most important factors that dictate coastal morphology are the wave climate, orientation and alignment of the coast, sediment characteristics and the local bathymetry. The morphological changes resulting from the wave climate can be classified under two categories: (i) long-term changes, which may usually spread over years or decades, and (ii) short-term morphological changes due to the impact of the rough sea along with a rise in sea level resulting from, for instance, a cyclonic storm. Hence, knowledge is crucial to predict the morphology. The measurement of time-varying dynamical ocean wave climate is a challenging task. Continuous manual measurement is impossible in the middle of the ocean, and hence one has to rely on systems that automatically collect wave data. The different ways of wave measurement include wave staff or an array of gauges, underwater pressure or acoustic sensors, static current meters, moored buoys and satellite altimeters or high-frequency radars.

The wave climate can be classified based on seasonal and cyclonic activities, as both the classification has its own vital role to play in the coastal morphology. Sundar (1986) has carried out a detailed analysis of visually

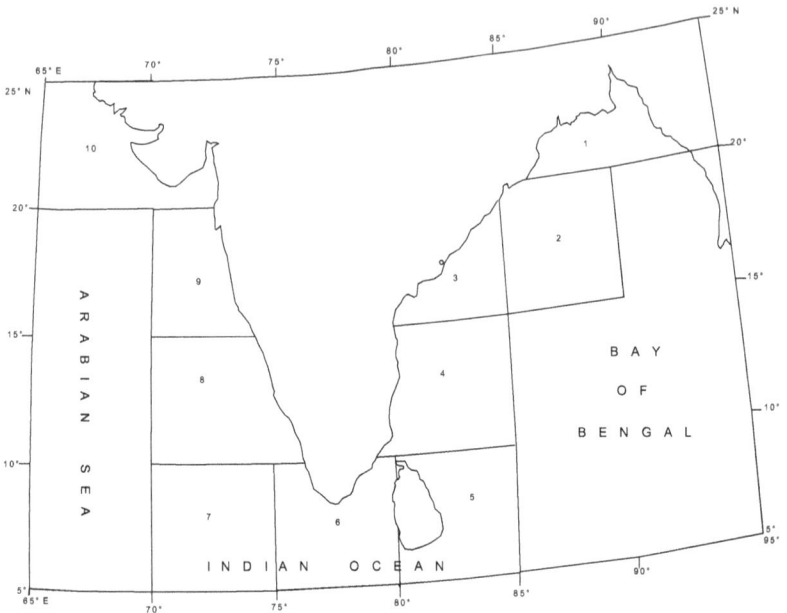

Fig. 5.1. Grids considered in the wave atlas of Chandramohan *et al.* (1991).

observed wave data off the Chennai coast, in southeast India, reported for a period of 10 years with five different standard spectral distributions to fit, and concluded that the wave characteristics off the Chennai coast did not follow any of the considered standard spectral models. The annual wave characteristics for our Indian coast, based on 19 years of ship-observed wave data sets developed as a wave atlas by Chandramohan *et al.* (1991) by dividing the Indian coast into 10 grids with the size of 5° Latitude by 5° Longitude, is shown in **Fig. 5.1**. The wave atlas data can be used at locations with no available data set. The average wave height along our Indian coast was reported by Narasimha Rao and Sundar (1982) to range between 1 and 2.8 m, whereas the wave period was found to vary between 5 and 8 s.

5.2.   Wave Generation

Waves are generated in an area by forces that disturb a body of water open to the atmosphere. Once the waves are generated, the forces of gravity and surface tension are activated, which allows the waves to propagate like a vibrating string. The majority of ocean waves are generated by wind, and depending on the magnitude of the force acting on the body of water, waves occur in different sizes and forms. The ultimate state of wave growth is primarily dependent on the wind factors, i.e., wind speed and wind duration or the length of time for which the wind blows, also known as fetch. The waves formed under the strong influence of winds, i.e., within fetch region, are called wind waves. These wind waves are characterized by a relatively sharp crest and flat trough. Once the waves leave the generating area of fetch, principally, they are smoother and lose their rough appearance, which are termed as swell waves, which are long waves with lesser height. In general, wind waves are short-crested random waves. On the other hand, swells are more uniform regular waves compared to the wind waves. For a given steady wind speed, the development of waves may be limited by the fetch or the duration of the wind. However, if wind blows for a sufficient length of time, in a more or less steady state of condition, any additional transfer of energy from the wind does not result the growth of waves; in other words, the steady state has been reached. This condition is called a fully developed sea (FDS).

5.3. Characteristics of Waves

5.3.1. Monochromatic waves

The characteristics of the waves can be described by their length (distance between two consecutive wave troughs or crests) L and its height (highest point of the surface elevation to its deepest point), H, and propagating over a depth, d. If the aforesaid parameters are known, the wave-induced water particle kinematics, like orbital velocity, acceleration and pressures, can be derived theoretically. Celerity, C, is the ratio of wave length over its period (i.e., $C = L/T$). Only water disturbances and the associated energy travel with the wave. In deep waters ($d/L \geq 0.5$), where the depth is greater than one-half of the wavelength, the celerity of the wave depends only on the wavelength and can be determined from the wave duration or wavelength. In shallow waters ($d/L < 0.05$), celerity, on the other hand, depends only on the water depth. Similarly, wave height depends on both depth and wavelength. When a wave propagates in shallow waters, they are subjected to shoaling (waves starts to feel the sea bottom), which leads to an infinite increase in their height and a reduction in their length. Eventually, the wave becomes unstable and breaks near the shore. Another breaking criterion in deep water is that the steepness is limited. In deep waters, the maximum height must be a function of the wavelength, above which the wave becomes unstable and breaks.

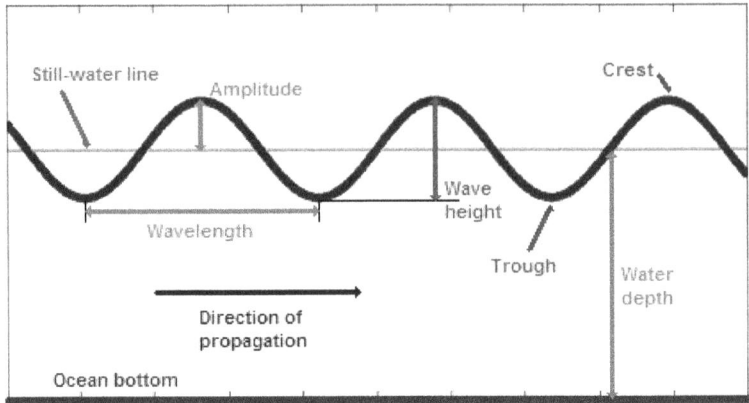

Fig. 5.2. Wave characteristics.

Table 5.1. Wave anatomy.

Still-Water Line	The reference axis describes the wave motion, i.e., the water surface level at calm conditions, excluding local variation caused by waves but including the effect of tidal variation and storm surge.
Wave Height	The vertical distance between crest and trough.
Wave Frequency	The number of waves passes a fixed point in a given amount of time.
Wave Amplitude	The vertical distance from either crest or the trough to the still-water line.
Depth	The distance from the ocean bottom to the still-water line.
Direction	The direction in which the wave propagates.

To understand simple wave motion, a two-dimensional scheme representation of a sinusoidal or monochromatic wave propagating in the x-direction is shown in **Fig. 5.2**. Its surface profile can be derived from sine or cosine function. Some important wave anatomies are briefly described in **Table 5.1**.

5.3.2. *Irregular wave*

The waves rarely appear monochromatic or sinusoidal in nature and do not travel in the same direction. Instead, the sea surface is composed of a number of sinusoids propagating in the different directions with different frequencies, phases and amplitudes. Through Fourier transformation, wave surface elevation can be considered as a linear superposition of a large number of sinusoids moving in different directions as shown **Fig. 5.3**. In fact, most regular waves generated in laboratories, which rarely appear in the

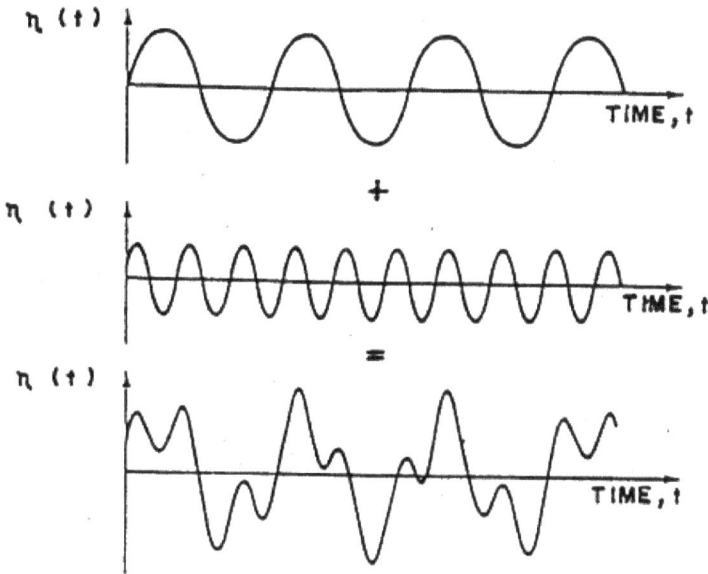

Fig. 5.3. Complex waveform resulting from the sum of two sinusoids.

real ocean environment, are found to be in coherence with the wave elevations dictated by the small amplitude or linear wave theory. As the wind
blows and the wave grows within the fetch distance, a continuous transfer of
energy from the wind to the water surface takes place, wherein a partially
developed sea state is attained. A fully developed sea state is achieved when
no further transfer of energy from winds can contribute to wave growth.
Although the swell is more or less regular in shape, it is fundamentally
irregular in nature, with some variation in its height and duration. Due
to the fundamental randomness of the sea surface, statistical and spectral
method of analysis are adopted, the former for the average characteristics, and the latter for the frequency domain description. The statistical
wave characteristics are adopted in the design of engineering structures in
a marine environment. Historically, the most important parameter that is
statistically derived from the recorded time series is the significant wave
height $H_{1/3}$, defined as the average of highest one-third of wave elevation.
Since the real sea condition is not constant over a long period, the wave
characteristics evaluation from short-term records is not reliable. The standard practice of real sea state measurement is once in every three hours for
a period of 20 minutes such that at least about 100 waves are covered.

The wave characteristics such as significant wave height ($H_{1/3}$), mean period (T_m) and mean direction (θ_m) of wave approach are derived from the field measurements and the phase-wise wave statistics. A comprehensive description of the analysis of ocean waves is discussed in detail by Sundar (2021).

5.4. Gauging Waves

5.4.1. *Wave measurement*

There are two main types of sensors that are being widely used to physically measure sea surface elevation: pressure sensors and wave buoys. The pressure sensors are usually mounted either at a fixed point over or close to the sea floor or on a solid structure like a pile. The pressure-based wave recorders are designed for use in water depths of less than 20 m. It is not feasible to adopt these gauge-based measurement systems in larger depths unless they are fixed on permanent structures and sited within 20 m from the surface or on a spar buoy. In the case of bottom-mounted pressure gauges in larger water depths, the propagation of the surface wave would not effectively be sensed by these sensors as the dynamic wave pressure decreases hyperbolically towards the seabed. In the shallower water depth regions, where wave breaking is dominant and the linear theory is no longer valid, it is impractical to use these gauges. In addition, the exact position of these gauges is very important for the determination of wave direction, which would be difficult to achieve from an array of bottom-mounted gauges. Measuring wave activity close to solid structures such as a vertical wall is difficult — it may not accurately represent the actual wave elevation time history. The analysis of surface elevations with pressure data sensed from bottom-mounted pressure gauges critically depends on measuring frequencies (Wolf, 1997). The effects due to wave currents in turbulence is difficult to be separated in the analysis of wave data.

Ever since Longuet-Higgins *et al.* (1963) proposed the methodology for estimating the directional wave characteristics from the responses of a floating buoy, measuring the orbital velocities as well as wave elevation. This method is called single point measurement. In addition to the above method an array of wave gauges are deployed, through which the directional spread can be ascertained. Due to its easy installation, precision in measurement and applicability over a wider range of water depths, the use of floating buoys, often termed "data buoys," for the ocean wave measurements has significantly increased world-wide. The radar-based wave data collection is

relatively a recent technology and is not cost-effective. The different types
of data buoys and their significance are detailed in the following section.

5.4.2. Gauging extreme events (storm surge, cyclone, tsunami)

Storm surge is an abnormal rise in seawater level with high damage potential, which is greatly amplified where the water depth is shallow or where the
coastal geo-morphology is funnel-like. It is caused mainly by atmospheric
weather systems such as tropical cyclones. The cyclone tracks along the
northern Tamil Nadu coast from 1891 to 2007 are shown in **Fig. 5.4**. The
prediction of sea state and its severity during a cyclone that usually occurs
in the deep ocean and its effect as it propagates towards the coast is of
paramount importance (Dean, 1976). It forms the design basis for all types
of marine structures, particularly along the coastal zone, since the effect of
both storm surges and storm waves could significantly influence the design
for their sustainability. For gauging such an extreme event, the sampling
interval and processing procedure differ from the average wind-generated
waves. However, an ordinary pressure sensor has also been used.

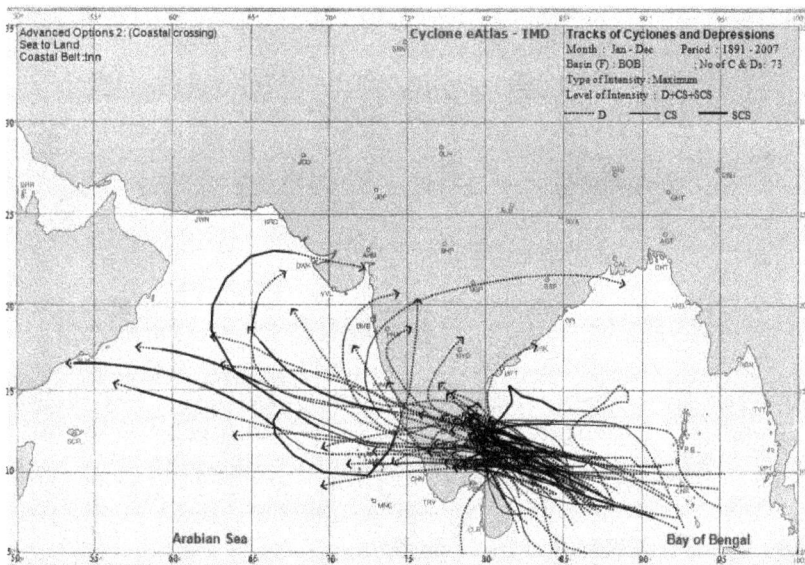

Fig. 5.4. Tracks of cyclones along the north Tamil Nadu coast (1891–2007).
Source: India Meteorological Department.

A tsunami is a long wave — caused by volcanic eruptions, earthquakes and landslides unlike the regular wind waves — that develops in the deep ocean and travels towards the land. A tsunami with a period of several minutes to several hours and wavelength exceeding 500 km behaves like a shallow water wave in very deep waters, such that its effect is felt over the seafloor. Since tsunami wave dynamics is highly in contrast to and significantly different from ordinary wind-generated waves, many existing measurement devices are not equipped enough to measure its complete characteristics; for instance, ordinary wave rider buoys cannot follow the extreme event due to mooring interference. Mostly, underwater pressure sensors fixed to the seafloor are capable of sensing periodic water surface displacements of very low frequency. They can be used for tsunami wave motions tracing over a long period, and an understanding of tsunami wave characteristics can be determined from this type of underwater pressure data. A real-time network of seismic stations, Bottom Pressure Recorders (BPR), tide gauges and 24 × 7 operational tsunami warning centres constitute the Indian Tsunami Early Warning Centre (ITEWS) to detect tsunamigenic earthquakes and monitor tsunamis. Severe storms could drive the sediment sands offshore, reaching deep waters from where it is impossible to recover them under normal wave conditions, or it could be moved alongshore. Both the aforesaid processes could irreversibly result in loss of sediments from the nearshore littoral zone. These storm events could significantly leave behind a huge impact on beaches with average low-energy waves. The said information is needed not only for the preparedness for evacuating the coastal community but also for the design of coastal structures. The behaviour of shoreline between a groin field due to the ingress of the 2004 Indian Ocean Tsunami has been discussed in detail by Sundar (2005).

5.4.3.   *Wave-measuring instruments*

5.4.3.1.   *Single-point pressure sensors*

A single-point pressure sensor measures the displacement of the water column passing across them. As the wave crest travels, the height of the water column increases, and as the trough approaches, the height of the water column decreases. By subtracting the depth of the sensor from the water column height, a record of sea surface elevations with tidal variation can be generated. By taking out the tidal variation in the time series using the linear regression of the tidal slope, wave pressure time series can be generated. A single-point pressure is generally used near the shore, where they are cabled with a land-based data acquisition system. Installation of such sensors are easy and can also be used for short- or long-term monitoring of wave elevation, particularly in the coastal waters.

5.4.3.2. *Arrays of pressure sensors*

Although the single-point pressure measurement is efficient enough to generate the energy spectrum, it cannot give information about the wave direction. Directional information is needed to determine the wave direction characteristics or to compute the directional spectrum. One way of generating a directional spectrum is to estimate it from the simultaneous time-series measurements of wave elevation from a number of pressure gauges placed either in a linear or a polygonal array. The direction can be determined by a number of methods.

5.4.3.3. *Acoustic Doppler Current Profiler*

An Acoustic Doppler Current Profiler (ADCP) has been initially developed for nearshore current measurement, but some ADCPs can be configured to measure the wave elevation and direction. An ADCP derives surface wave information from orbital wave velocity and pressure measurements and has an excellent ability to resolve multiple directional waves, even with a similar frequency.

5.4.3.4. *Directional wave recorder*

A directional wave recorder (DWR) is a sophisticated pressure-based wave recorder fitted with a pressure sensor (resonant quartz or strain gauge) to measure the pressure variations. It is also equipped with an electromagnetic current sensor to measure the current oscillations, with the direction referenced to an internal fluxgate compass.

5.4.3.5. *Wave Rider Buoy*

When measuring the waves far offshore or in remote locations at mid-sea, wave rider buoys are a viable choice. This type of deployment does not need a cabled connection from shore, and it can transmit the data with the help of a radio link or any other type of wireless communication.

In general, buoy systems are divided into three categories: surface, sub-surface and free drifting buoys. The surface and sub-surface buoy systems are moored to the seabed, and their functional difference is the former floats over the water surface, whereas the later floats at an elevation below the mean sea level. Free drifting buoys, simply called as drifters, are smaller in size and are allowed to wander over the sea surface, which obviously are continuously being displaced by the waves and currents. The drifters are mainly used for tracking ocean current direction and magnitude. The illustrations of the above-said buoy systems are provided in **Fig. 5.5**. Worldwide, there are more than 200 moored buoys installed and more than 1,300 drifters deployed in the ocean for the real-time collection of various oceanographic and meteorological parameters. The locations of wave buoys deployed globally shown in **Fig. 5.6**, clearly indicates the tremendous development in wave data measurement systems. Among the three types of buoy systems, the surface-floating buoy system is widely being adopted for the wave data collection because they can also be used for the measurement of meteorological parameters, such as characteristics of air (temperature and pressure), wind (velocity and direction), current (direction and magnitude) and water (temperature, conductivity and salinity). The generic forms of surface buoy hulls are shown in **Fig. 5.7**. The disk-shaped buoys have large waterplane areas but small displacements that naturally make these buoys follow the waves both in heave and slope. On the other hand, spar buoys have relatively small water planes and large displacements. These properties of spar buoys leaves behind a comparatively less wave excitation forces on it and hence tend to be surface decoupled. The intermediate buoy hull poses large water planes and moderate displacements leading to good following of the wave amplitude and poor tracking of the slope. Boat-shaped buoy hulls were introduced by the United States Navy in the late 1940s as an offshore autonomous meteorological platform and are named as Navy Oceanographic Meteorological Automatic Device (NOMAD). Though these boat-shaped buoys have good potential to be used as a dry stable platform, they can only be used for one-dimensional wave measurement and the directional wave information cannot be derived due to the non-symmetry of the buoy hull. The hull body contours and ballasting have to be done with care for proper functional aspects. Hereinafter, the data buoy denoted in the text refers to a surface-floating buoy system until otherwise specified.

The two primary classes of data buoys used in the ocean wave measurements are slope-following and orbital-following buoys. The schematic definition is shown in **Fig. 5.8**. In the former case, the heaving motion of the buoy is used to derive the non-directional spectrum, and from the

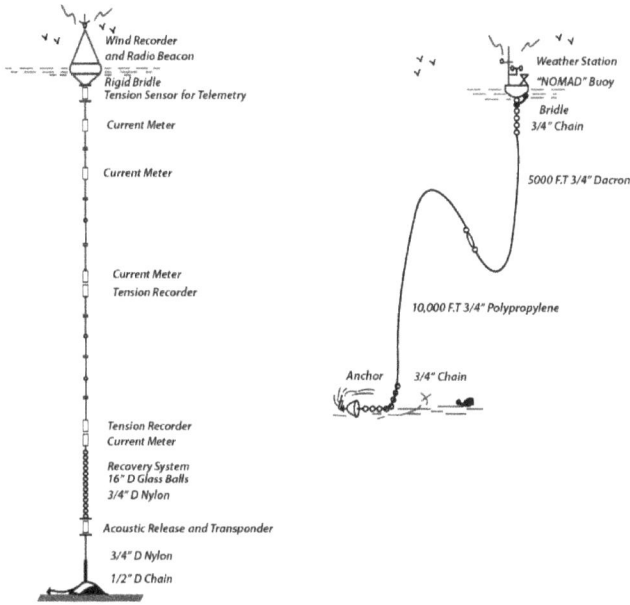

(a) Surface buoy systems

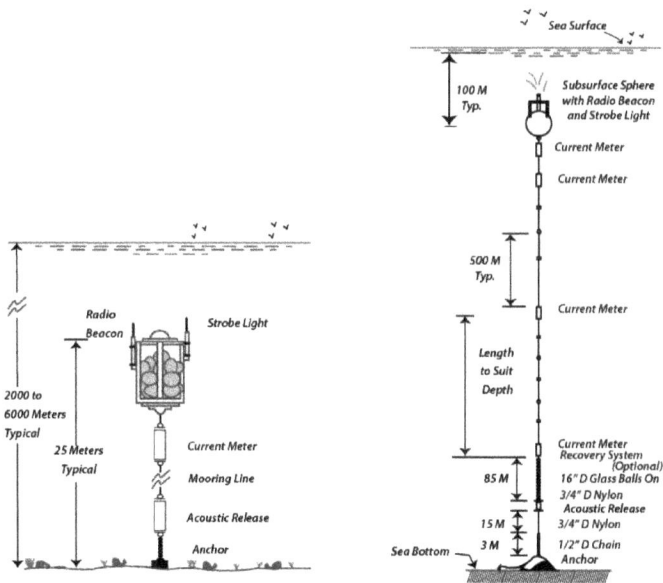

(b) Sub-surface buoy systems

Fig. 5.5. (*Continued*)

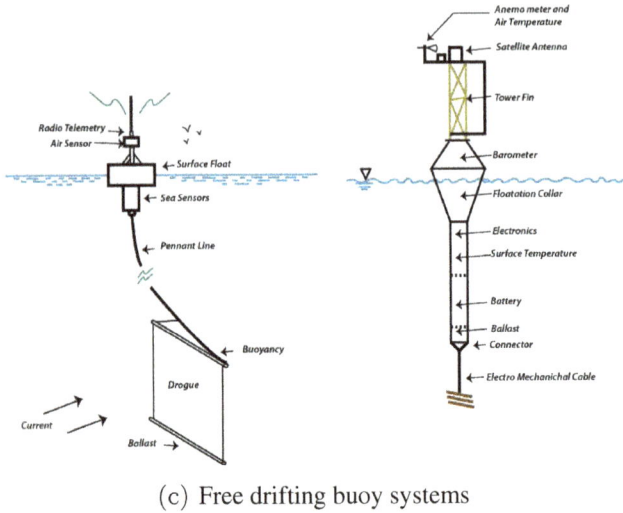

(c) Free drifting buoy systems

Fig. 5.5. Different types of oceanographic systems.
Source: Berteaux (1976, 1991).

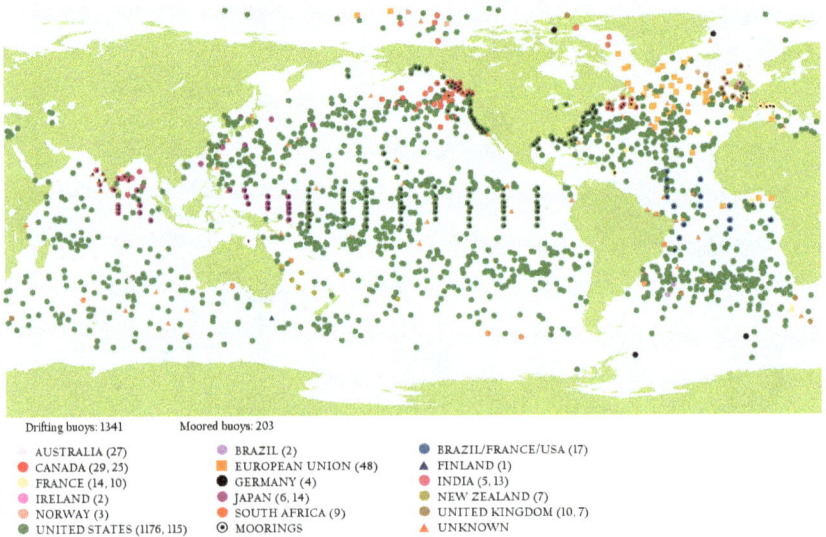

Fig. 5.6. Number of drifting and moored buoys as on the year 2006, country wise, as per the Data Buoy Cooperation Panel (DBCP) of the World Meteorological Organization.
Source: https://www.noaa.gov/.

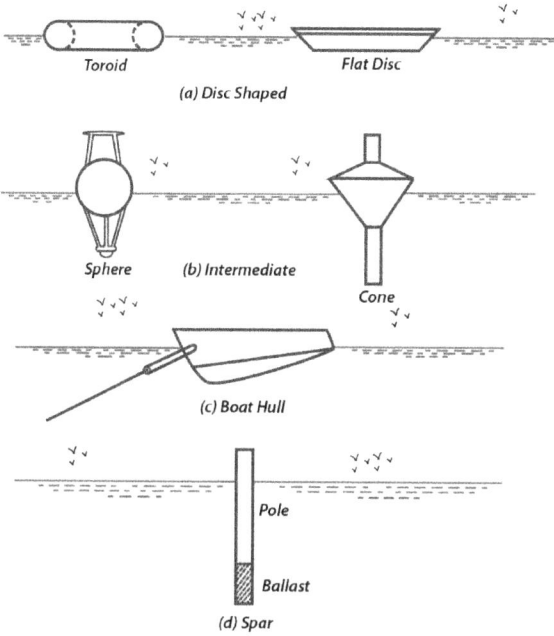

Fig. 5.7. Different generic forms of surface buoy hulls.
Source: Berteaux (1991).

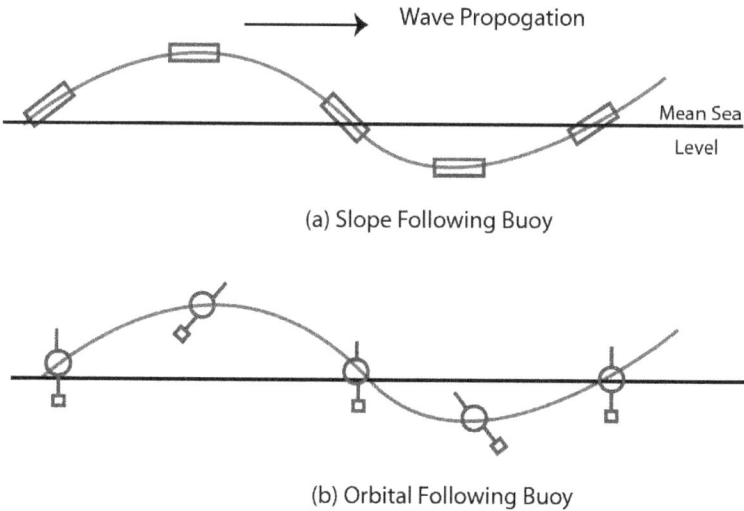

Fig. 5.8. Concept of slope- and orbital-following buoys.

tilting motion of the buoy (in magnitude as well as direction), the directional information can be deduced, assuming that these buoys closely follow the slope of the wave surface. The latter type of buoys measures the orbital motions of the water particle at the surface rather than the surface slope and accurately indicates the water particle movements. Particle-following buoy hulls perform well to measure the wave characteristics, however, the sturdier surface-slope–following principle is widely used in the world ocean. The surface-particle orbital displacements induced by waves can be measured using some of the recently advance buoys using satellite-based, position-fixing or global positioning systems.

5.5.   Indian Data Buoy System

Generally, the slope-following buoys are made of discus-shaped hulls as an aid to follow the water surface more easily. The diameter of these discus buoys varies depending upon the requirement; e.g., a 12-m-diameter discus data buoy (often called monster buoys) are sturdy even in the extreme wave conditions, whereas the installation of a 3-m-diameter buoy is relatively easy and cost-effective. The largest user of data buoys is the National Data Buoy Center (NDBC) of the US National Oceanic and Atmospheric Administration (NOAA). The schematic views of the different moored NDBC data buoys are shown in **Fig. 5.9**. The percentage of the different sizes of discus buoys used in the NDBC program projected in **Fig. 5.10** shows that about 59% of them are of 3 m diameter.

In order to collect the met-ocean parameters and monitor the marine environment over the Indian seas, the Ministry of Earth Sciences, Government of India, has established a National Data Buoy Program (NDBP).

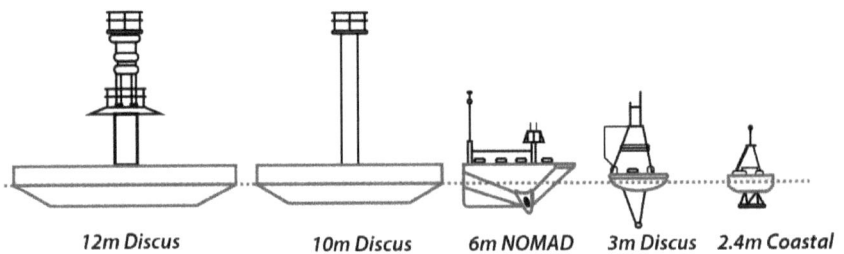

12m Discus   *10m Discus*   *6m NOMAD*   *3m Discus* *2.4m Coastal*

Fig. 5.9.   A selected fleet of NDBC's moored buoy programme.
Source: www.ndbc.noaa.gov.

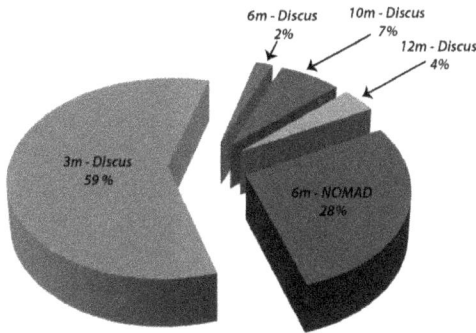

Fig. 5.10. Percentage occupation of individual buoys in NDBC programme.

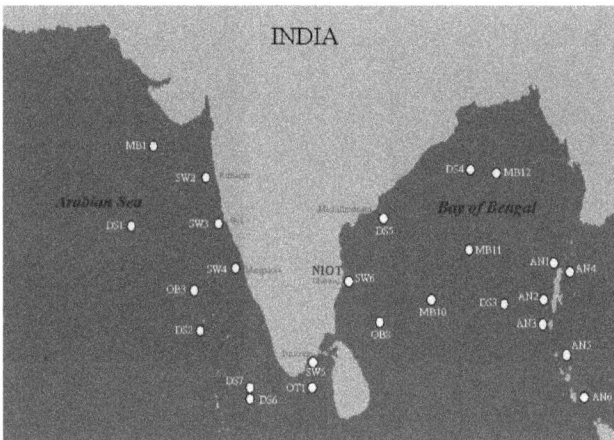

Fig. 5.11. Buoy locations around the Indian waters.
Source: www.niot.res.in.

Under this program, discus data buoys are deployed in the Bay of Bengal and the Arabian Sea to measure and supply data to improve the weather and ocean state prediction in Indian waters. All the data buoys deployed under NDBP are of 2.2-m-diameter, discus-shaped buoy hulls, which is a slightly modified version of the 3-m-diameter discus buoys that are widely adopted worldwide. These data buoys are deployed in shallow and deep water depths, the locations of which are shown in **Fig. 5.11**. These data buoys of different types (**Fig. 5.12**) collect various oceanographic and mete-orological parameters that are essential for the continuous monitoring of

12m Discus buoy

10m Discus buoy

6m NOMAD buoy

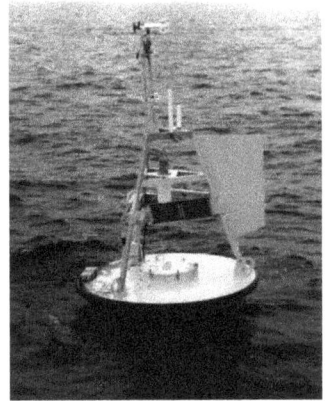

3m Discus buoy

Fig. 5.12. General views of the data buoys under operational conditions. *Source*: www.ndbc.noaa.gov.

the coastal and marine environment along the Indian coast. In the event of natural disasters such as cyclones or tsunamis, the meteorological data obtained from these buoys are vital for effective mitigation measures. In addition, the oceanographic data collected are useful for the understanding of ocean circulation. Apart from these data, various other parameters, directly or indirectly, are used to identify the potential fishing zones, develop predicting models, design marine structures and improve navigation aiding. The 2.2-m-diameter discus-shaped buoy of NDBP is shown in **Fig. 5.13**.

Fig. 5.13. View of NDBP data buoy under operation.
Source: www.niot.res.in

5.6. Remote Sensing Techniques

5.6.1. *General*

All traditional in-situ waves gauging techniques generally collect wave data at a defined location. Various remote sensing techniques to determine wave information may provide a synoptic view of wave fields at reasonable costs. The temporal and spatial evolution of the radar backscattered signals provide useful information of the sea surface parameters received within the accessible range of the coast. Direct values of wave heights cannot be read using remote sensing techniques due to the observation geometry where the ocean surface waves are captured from extreme grazing angles. It is difficult to avoid the effects due to diffraction and shadowing, and no rigorous radar backscatter model (i.e., a model derived from the Maxwell equations) exists to describe the relationship between surface waves and radar echo.

However, an empirical relation between the height of a wave and the signal-to-noise ratio of the radar image exists. The general possibility to use this relation for determining the significant wave height from nautical radar images has been shown in a number of cases (Vogelzang *et al.*, 2000). Every measuring system must be calibrated once using independent wave height data, e.g., from buoy data. The calibration using field data typically takes about a few days, however, the exact duration of time depends on the range of meteorological conditions. Post calibration of a radar mounted at a fixed position, the system can function as a stand-alone wave measuring device, nevertheless, calibration ought to be repeated if the radar is moved.

5.6.2.   *High-frequency radar*

The random sea waves scatter the transmitted signal in all directions. According to the Bragg scattering principle, the radar signal returns directly to its source only when it spreads from a wave which has exactly half the transmitted signal wavelength, and it progressively adds to form a strong return of energy at a precise wavelength. A Doppler shift is exhibited by the backscattered signal from the transmitted frequency. According to the deep-water linear wave theory, the returned signals are concentrated at a known position (first-order peak) in the absence of an underlying current. In the presence of ocean currents, however, an additional shift in the first-order peak is observed. This frequency shift is computed to arrive at the current velocity in a radial path (towards or away from the radar). The predicted Doppler shift from deep water linear wave theory in the absence of a current field is given a

$$\text{Doppler freq. shift } f_d = \frac{2V}{\lambda_r} \quad \text{and,}$$

$$\lambda_r = 2\lambda_w \sin\theta_i$$

where V is the relative velocity, λ_r is the radar wavelength and λ_w is the ocean wavelength.

The high-frequency radars (HFRs) network off the Indian coast has 10 CODAR-Mac systems. Of these, six are located on the east coast, two in the Gulf of Khambhat and two in the Andaman Islands, as shown in **Fig. 5.14**. These HFR networks continuously record the ocean surface parameters (Jena *et al.*, 2019). An HFR system consists of a transmitting and receiving antenna, signal generating and receiving circuit and data processing software. The HFR systems are commonly installed in the vicinity of the coast,

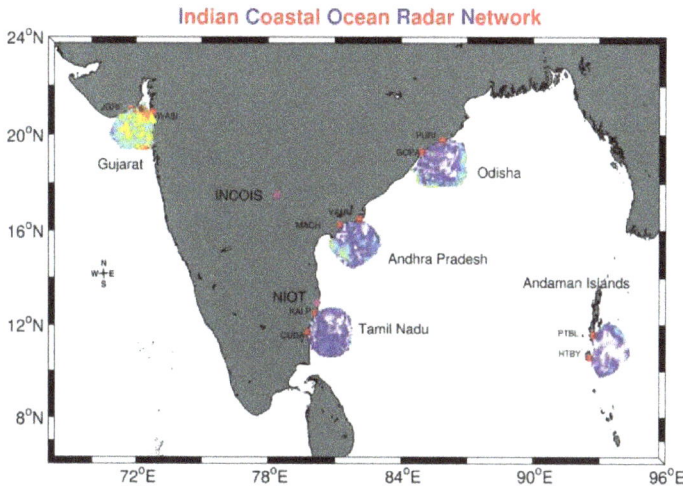

Fig. 5.14. Indian Coastal Ocean Radar Network: The radar stations are depicted by the solid red squares, and the 2D surface current maps are plotted at each paired site. The central data servers at NIOT and INCOIS are marked by magenta colour diamond markers.

Source: Jena *et al.* (2019).

to measure the surface oceanographic parameters. The extent of this measurement is dependent on the transmitting frequency of the device — here, measurements up to 200 km were recorded with a 4.4 MHz frequency device. Every single HFR installation is capable of measuring the wave parameters, while it requires a pair of radars to measure the surface current vector (John *et al.*, 2015). In the study done by John *et al.* (2015) on Cyclone Phailin that made landfall along the Gopalpur coast in Odisha, HFRs installed at the Gopalpur port jetty that measured the wave height at different range cells, viz. 6, 12 and 18 km at 27, 48 and 49 m water depths, respectively. The significant wave height (H_s) was determined by averaging the highest one-third of waves in the measurement period. The maximum wave value in a wave record is called maximum wave height (H_{max}). The observed H_s and $H_{max}(1.86 \times H_s)$ of the buoy deployed in 17 m water depth off Gopalpur has been compared with those of the first three cells of the radar (**Fig. 5.15**). The wave height reported by the HFR was found to closely match with the maximum wave height (H_{max}) reported by the buoy. The significant wave height measured by the buoy was about 7 m. The maximum wave height measured by the buoy (13 m) closely matched that of HFR (14 m).

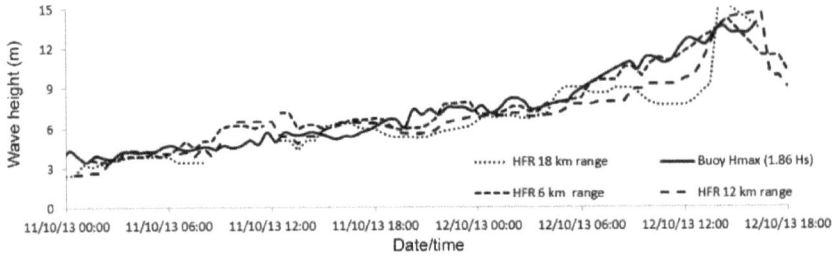

Fig. 5.15. Time series plot between HFR wave heights at different range cells and buoy wave heights.

Source: John *et al.* (2015).

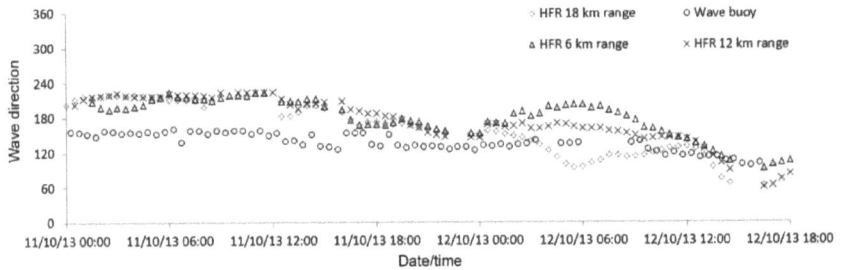

Fig. 5.16. Time series plot between HFR wave directions at different range cells and buoy wave directions.

Source: John *et al.* (2015).

The wave direction from the HFR during the cyclone period matched well with the measured buoy data (**Fig. 5.16**).

5.6.3. *Synthetic aperture radar*

The synthetic aperture radar (SAR) images have made a potential contribution in measuring wavelengths and directions and estimating wave heights of ocean waves under varied weather conditions. An ocean SAR image contains fine resolution images with approximate signatures of oceanographic phenomena such as waves, currents and inward waves (Chaturvedi *et al.*, 2013). Linear system approaches to the SAR imaging mechanism (Alpers, 1983) provide the basis for a process for estimating the wavelength spectrum from SAR intensity images. The effectiveness of the linear system approach to SAR wave imaging can be explored through numerical simulation analysis. The backscattered radar energy from ocean surface depends on wavelength, polarization, geometry, attenuation by the atmosphere and ocean surface roughness. Chaturvedi *et al.*'s (2013) study describes the estimation of spherical wave parameters in SAR images acquired over the coast of Chukk, Micronesia. The main causes for the interaction of SAR signals with ocean waves can be retrieved through the Bragg scattering mechanism.

5.6.4. *Satellite radar altimeters*

For more than three decades, a simple remote sensing technique has been providing a global coverage of wind speeds and significant wave height. This technique is adopted to arrive at two basic geometric measurements: (i) distance between satellite and the sea surface (determined by measuring the round-trip travel time of microwave pulses emitted by the satellite's radar) and (ii) independent tracking system to compute the satellite's three-dimensional position relative to the fixed Earth's coordinate system. These two measurements can be used to arrive at the profiles of sea-surface topography or sea levels, with respect to the reference ellipsoid (a smooth geometric surface that approximates the shape of the Earth) as pointed out by Cheney (2001).

In order to calibrate the altimeters in a consistent fashion, a long-term, high-quality database of the buoy in situ measurements of wind speed and wave height is required. These data should span a range of different meteorological environments and geographic regions. In addition, such data should be relatively far from land, so as to avoid contamination of altimeter measurements due to land/islands within the altimeter footprint. A total of 33 years of altimeter data (wave height and wind speed) have been calibrated and validated against National Oceanographic Data Center (NODC) buoy data by Ribal and Young (2019).

5.7. Summary

The operational and design perspectives of all the maritime projects depend on the availability of a large amount of wave data at the site location. With the present computational power, we have the capability of a hindcasting large number of datasets through numerical wind-wave modeling. However, the field measurements of wave climate are of prime importance in view of calibration and assimilation of numerical model coefficients and state variables. Further, the projects of priority always prefer to have the real-time wave measurements to supplement the predicted data. Hence, the physical and remote sensing capturing of wave climate both in the offshore and nearshore locations play a vital role. In this chapter, the salient techniques and instrumentation for wave measurements are discussed in detail.

References

Alpers, W. (1983). Monte Carlo simulation for studying the relationship between ocean waves and SAR image spectra. *Journal of Geophysical Research*, 88, 1745–1759.

Berteaux, H.O. (1976). *Buoy Engineering*. New York: Wiley.

Berteaux, H.O. (1991). *Coastal and Oceanic Buoy Engineering*. Woods Hole, MA.

Chandramohan, P., Sanil Kumar, V. and Nayak, B.U. (1991). Wave statistics around the Indian coast based on ship observed data. *Indian Journal Of Marine Science*, 20, 87–92.

Chaturvedi, S., Shanmugam, P., Yang, C. and Guven, U. (2013). Detection of ocean wave parameters using Synthetic Aperture Radar (SAR) data. *Journal of Navigation*, 66(2), 283–293.

Cheney, R.E. and Steele, J.H. (2001). Satellite Altimetry. In *Encyclopedia of Ocean Sciences*, Academic Press, pp. 2504–2510.

Dean, R.G. (1976). Beach erosion: Causes, processes, and remedial measures. *CRC Review of Environmental Control*, 6, 259–296.

ITEWC (2011). Indian Tsunami Early Warning Centre User Guide Version-2. Indian National Centre for Ocean Information Services Ocean Valley.

Jena, B.K., Arunraj, K.S., Suseentharan, V., Tushar, K. and Karthikeyan, T. (2019). Indian Coastal Ocean Radar Network. *Current Science*, 116, 372–378.

John, M., Jena, B.K. and Sivakholundu, K. (2015). Surface current and wave measurement during cyclone Phailin by high-frequency radars along the Indian coast. *Current Science*, 108, 405–409.

Longuet-Higgins, M.S., Cartwright, D.E. and Smith N. D. (1963). Observation of directional spectrum of sea waves using the motion of a floating buoy, *Proceedings of Conference in Ocean Spectra*, Easton, MD, May 1–4, 1961, Prentice-Hall, NJ, USA, pp. 111–132

Narasimha Rao, T.V.S. and Sundar, V. (1982). Estimation of wave power potential along the Indian coastline. *Energy*, 7(10), 839–845.

Ribal, A. and Young, I. (2019). 33 years of globally calibrated wave height and wind speed data based on altimeter observations. *Scientific Data*. 6(77). DOI: 10.1038/s41597-019-0083-9.

Sundar, V. (1986). Wave characteristics off the South East coast of India. *Ocean Engineering*, 13(4): 327–338.

Sundar, V. (2005). Behaviour of shoreline between groin field and its effect on the tsunami propagation. *Proceedings of Solutions to Coastal Disasters Conference of ASCE*, 8–11 May, Charleston, SC, USA.

Sundar, V. (2021). *Ocean Wave Dynamics for Coastal and Marine Structures. Advanced Series on Ocean Engineering* Vol. 52, World Scientific, Singapore. https://doi.org/10.1142/12268.

Vogelzang, J., Boogaard, K., Reichert, K., and Hessner, K. (2000). Wave height measurements with navigation radar. *International Archives of Photogrammetry and Remote Sensing*, XXXIII(Part B7).

Wolf, J. (1997). The analysis of bottom pressure and current data for waves. *Proceedings of the 7th International Conference on Electrical Engineering in Oceanography*. Southampton, June 1997. Conference Publication 439. IEE, pp. 165–169.

Chapter 6

Morphodynamic Observations in Coastal Areas

Abstract

Sandy beaches and alluvial estuaries are some of the most dynamic environments on the earth, and inspecting coastal morphodynamic behaviour through multidisciplinary approaches and techniques will throw significant light on its physical process. In addition, understanding the physical geomorphology of the coast in response to the hydrodynamic driving forces, it is also an important benchmark in shaping both the visible coasts and the invisible underwater/seabed features. The changes in the coastal geomorphology can be identified by taking appropriate, continuous and repeated measurements. The rate of changes can be reckoned by comparing measurements of the first with the last/latest. Controlled measurement and monitoring of changes in water depths and quantum of water and sediments in a water body could go a long way in understanding the morphodynamics behaviour and respective physical processes.

6.1. Background

Coastal areas around the world are mostly composed of fine sandy beaches, which are one of the most dynamic environments in the earth, and inspecting coastal morphodynamic behaviours relying upon multidisciplinary approaches and techniques presents a real challenge to the coastal scientists and engineers. It is believed that the longshore sediment dynamics affect most of the coastlines. Naturally, the beach attains and maintains its equilibrium, however, artificial interventions through various approaches (such as implementation of seawall, groynes, jetty, breakwaters, beach nourishment, etc.) leads to the sediment deficit or deposition along other regions. The host of parameters responsible for the morphodynamic changes in the nearshore often overlap and over power each other alongside the coastal dynamics. This makes it difficult to understand the role of natural processes on the coastal morphodynamics.

Monitoring the beaches will provide important information about the state of the coastal system, such as the required input into the statistical descriptors and numerical models adopted for understanding the behaviour of waves, tides and the underlying hydrodynamics which dictate the characteristics of the marine environment. The coastal zone is where wave and tidal energy are expended to set the sediments in motion. The coastal zone is subject to constant spatial and temporal changes by unstoppable tidal action, waves breaking, seasonal changes in wave approach, tide with daily high and low, monthly spring tide, neap tide, yearly tidal cycles, long-term sea level changes, increased frequency of cyclones and storm surges in a year and climate changes with unpredictable occurrence of catastrophic events such as tsunami, underwater earthquakes and volcanic eruptions. An in-depth knowledge on the above-stated driving forces controlling the characteristics of the coastal morphodynamics is required for the sustainable development in the coastal zone.

In addition, understanding the physical geomorphology of the coast in response to the above-stated hydrodynamic driving forces is an important benchmark in shaping the visible coasts and the invisible underwater/seabed features. The changes in the coastal geomorphology can be identified by taking appropriate continuous and repeated measurements.

In this chapter, the underwater coastal morphodynamics is explained through examples of field measurements and analysis to understand the morphodynamics. The changes in depth contours as an effect caused by the accumulated sediment load due to natural sediment motion through control volumes are analyzed. This will provide a clear idea about the natural and anthropogenic impacts and also its involvements for the understanding of underwater seabed morphodynamics.

6.2. Monitoring of Underwater Morphology

Seabed monitoring can be achieved through standard conventional methods such as ship-borne bathymetry survey. For good-quality data, the bathymetry surveys are carried out using narrow single-beam echo sounder (SBES) and multi-beam echo sounder along with additional mandatory equipment such as Global Positioning System (GPS), Motion Reference Unit (MRU), Sound Velocity Profiler (SVP) and Tide Gauge (TG). The equipment-based seabed monitoring method adopted was accepted by the global coastal and shipping sectors and mariners worldwide. Depending on the specific project requirements, the bathymetry surveys for the seabed monitoring can be chosen either through multi-beam echo sounder or single-beam echo sounder systems. A transducer having dual-frequency

SBES and/or single-frequency SBES may be used depending on requirements. During processing of bathymetric data, the following checks need to be performed:

- Draft measurements to correct single-beam echo sounder system.
- Correct vessel offsets applied to the correct sensors.

The processing of survey data consists of two main functions: rejection of outliers and statistical combination of the remaining soundings into binned survey data. Seabed mapping based on single-beam transects relies substantially on the spatial interpolation methods, often guided by manual expert interference. The Hypack processing software is designed to provide the tools required by a surveyor to perform efficient processing of the large amount of data collected while maintaining good quality control of the final output. During processing, the very high-resolution ping-by-ping data is converted into a binned bathymetric chart — typically, 1-m binning is used in shallow water surveys. The bathymetry records from individual lines can be re-examined to observe specific features at a higher resolution, if required. Immediate feedback on survey results is important so that the surveyor can perform quality check on the data as it comes in and ensure the required accuracy standards is maintained. The data quality control tools are an important part of the software provided with bin-by-bin measure of the data density, distribution and confidence, both online and in post-processing. The processed data from each line is added to the survey window sheet to give the confidence that the coverage is complete and that overlapping data from different lines matches. The online display of the bathymetry also allows the surveyor to immediately identify the features that may be of interest for more details.

Reliable data and understanding are produced by accurate horizontal positions. The coordinates of the collected soundings are marked using GPS. The horizontal position of the survey vessel can be captured continuously and followed during the survey from GPS receiver system. The GPS can achieve Differential Global Navigation Satellite System (DGNSS) positioning with sub-metre precision using Radio Technical Commission for Maritime Services (RTCM). DGNSS corrections are either broadcast free by International Association of Marine Aids to Navigation and Lighthouse Authorities (IALA) MSK Beacon stations, from a Networked Transport of RTCM via Internet Protocol (NTRIP) source, from SBAS (satellite based augmentation systems) such as Wide Area Augmentation System (WAAS),

European Geostationary Navigation Overlay Service (EGNOS) and Multifunctional Transport Satellites (MTSAT) Multi-functional Satellite Augmentation System (MSAS) or via an external radio from a local reference station. The RTCM correction stream from an MSK source can be passed to other DGNSS receivers using the Repeat RTCM function. The positioning data received have high reliability and integrity. The GPS/DGPS system shall always be calibrated prior to mobilizing the system in the field. The available technical specification of the GPS is given in **Table 6.1**. The quality of the observations depends strongly on the effective compensation of the ship motion and the high quality of the GPS signals.

The survey vessels/boats will be built to withstand all loads and stresses induced by waves and wind from all directions, turning forces, structural stresses, etc. Ships, boats, survey vessels or any other craft in the open ocean experiences six degrees of freedom (DOF), i.e., a rigid object can move through three dimensions of space, or six DOF defines the number of directions a body can move in the open ocean. If these motions of the ship

Table 6.1.   GPS technical specifications for hydrographic survey.

Description	Specifications
Satellite Tracking	IRNSS, GPS L1, GLONASS L1, SBAS, BeiDou, Galileo, QZSS, etc.
No of Channels	More than or equal to 220 channels
Accuracy:	
Horizontal accuracy	0.30m + 1 ppm (post processing mode)
Vertical accuracy	0.50m + 1 ppm (post processing mode)
Ports	One serial port for hand held device, One serial port for external radio/ modem, external power port, network and Bluetooth.
Operating temperature for all major components	−40°C to +65°C
Storage temperature for all major components	−40°C to +80°C
Position update rate	10 Hz
Operation time on internal battery	6–8 h
Water and dust proof	IP66 & IP67
Keys	Integrated bluetooth, One serial port or connecting receiver
Power source for operations	Internal & external (both automatic and manual)

and/or forces on the ship are not monitored precisely, it will end up in an unrealistic ship navigation and results. In general, the motions of ships are classified as: (i) rotational motion and (ii) translation motion.

During the bathymetry survey, the vessel motions will be computed by fixing the MRU equipment at the centre of the vessel, which is one of the best for the applications of altitude determination (bathymetry survey) and motion compensation. The MRU is connected to the echo sounder system using the data collection software for compensating the ship motion encountered during data collection, and the values will be recorded and stored for further ready reference.

6.3.   Bathymetry Survey Control

6.3.1.   *Geodesy*

The Differential Global Positioning System (DGPS), referred to as the World Geodetic System — 1984 (WGS-84) uses a spheroid as the position-fixing system for the bathymetry survey vessel. It has an accuracy of ± 2 m. The raw horizontal position data was logged in the standard Universal Transverse Mercator (UTM) projection in the local UTM Zone to match the geodetic system in which the area coordinates has been fended manually. **Table 6.2** provides the grid and spheroid parameters adopted during the survey.

6.3.2.   *Navigation software*

HYPACK is a Windows-based software for hydrographic surveyors which provides a user-friendly, turnkey solution for all types of marine navigation, positioning and surveying activities. HYPACK hydrographic survey software also provide the surveyor with all the tools needed to design the survey and collect, process and reduce data to generate the final product, with the processing-enabled real-time imaging, terrain modelling and statistical reporting. Initially, in the planning stage with the Hardware Setup program, a Template Database containing all survey configuration parameters pertinent to the project can be created and verified by the survey engineer. The survey area and the cross-sections can be surveyed using HYPACK max software showing the interface of GPS and echo sounder. The survey plan area will be continuously displayed on computer monitor while undertaking the bathymetry survey.

Table 6.2. Geodetic parameters and conversion.

Geodetic Parameters	
Datum	WGS 84
Ellipsoid	WGS 84
Grid	UTM North
Semi-Major Axis (a)	6378137.000 meters
Zone	Zone 44 (78E-84E)
Inverse Flattening (1/f)	298.2572236
Projection Parameters	
Grid Projection	Universal Transverse Mercator
Central Meridian (CM)	81.0° East
Origin Latitude (False Lat.)	00.0°
Delta Scale	0.0
False Easting (FE)	50,0000.0 m
False Northing (FN)	0.0 m

Prior to the commencement of the seabed morphology mapping through survey, all computer hardware systems (Laptop, Monitor, Desktop, etc.), equipment (transducers, GPS, etc.) and software (Windows, Hypack, etc.) need to be tested for the operability of the system and the receipt of the survey data from the various positioning and survey systems. A consistency check needs be carried out by the DGPS by repeated observations at a same reference survey point. Offset measurements in terms of latitude, longitude, East, North (x, y and z) needs to be checked and the position of the know points with respect to the survey vessel need to be defined.

6.3.3. *Survey trials*

Testing of online computer systems need to be done to ensure that all navigation systems have been logged and computed, with positions being displayed for the helmsman. Verification of the navigation system offsets is performed by observing the positions derived for all systems on different headings.

The Single-Beam Echo Sounder (SBES) system is used for mapping of the seabed morphology of the Cochin Port regions. The Cochin Port is located on the southwestern Indian coastal state of Kerala at latitude 09° 58.4′ N and longitude 76° 15.2′ E, as shown in **Fig. 6.1**. The Cochin Port is located next to the East–West trade route, 11 nautical miles from the

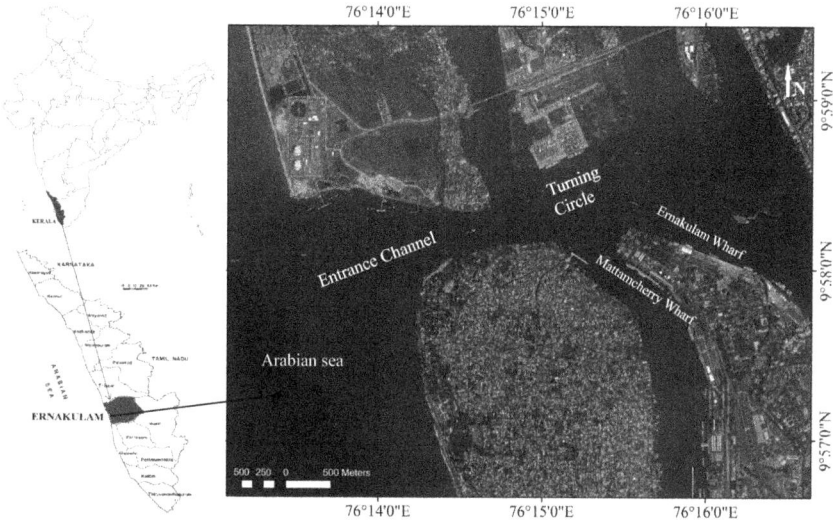

Fig. 6.1. Location map of Cochin, Ernakulum, Kerala, India.

direct Middle East–Far East sea-route. There is no other Indian port with such strategic geographic proximity to the major maritime sea routes.

The entrance to the port is through Cochin between the headland of Vypin peninsula and Fort Kochi, where after the channel splits into the Mattancherry Channel and the Ernakulum Channel (as shown in **Fig. 6.2**). The CoPT entrance to the harbour is by a 16,500 m long and 260 m wide 15.95 m deep outer approach channel marked with eight sets of buoys. The inner harbour is divided into two navigational channels — the Ernakulum Channel that is 2800 m long and 300–500 m wide with depths from 9.75–13.5 m and Mattancherry Channel that is 2200 m long and 180–250 m wide with a depth of 9.75 m. The Mattancherry Channel, which extends from the west end of the Gut to Q1 berth has a length of 2.6 km with draft ranging from 9.14–10 m. The length of the Ernakulam Channel from west end of the Gut to Q9 is about 5 km, with an available draft ranging from 9.14–12.5 m.

The seabed monitoring at Cochin Port and its surroundings were carried out through bathymetry survey as follows:

- Complete coverage of bathymetry survey in the Inner Harbour Area and Outer Channel Area during the pre-monsoon and post-monsoon periods.

Fig. 6.2. CoPT Inner Harbour, Turning Circle, Inner Channels.

- Sounding lines are the trajectory of the survey lines along which the sounding measurement were taken.
- Grid lines are lines perpendicular to the sounding lines.
- Grid points are the points of intersection of sounding lines and grid lines, where soundings are taken for preparing the sounding charts.
- The location of the sounding lines shall be fixed such that they are at an interval not more than 50 m, measured along the longitudinal directions of the channels. Grid lines shall be not more than 25 m apart. A grid line shall be established along the design toe of each side slope and also near the wharf/jetty frontage. Thus, the grid points on the cross section shall be at, close to but not exceeding 25 m interval.

The horizontal position of soundings shall be obtained at intervals not exceeding 25 m along a sounding line, and these locations shall be marked on the echo trace chart as fix marks. Fix marks shall be obtained where a sounding line crosses the toe and the top of dredged side slope, where feasible. Intermediate soundings representing the shallowest depth in a length of not more than 10 m shall be obtained by interpolation between fix marks.

The horizontal accuracy of each position fix shall be:

- +3.0 m along the sounding line and
- +3.0 m perpendicular to the sounding line.

A detailed bathymetry survey was carried out during the year 2020 for the pre-monsoon and post-monsoon periods in and around the Cochin region in Kerala. The bathymetry survey was carried out along the Cochin Port Basin Area, Inner Channels and Outer Harbour Area at 50 m spacing to identify the changes due to scoring during the pre-monsoon and shoaling during the post-monsoon (two seasons), which were presented in **Fig. 6.3**.

6.4.   Quantum of Sediment Motion

6.4.1.   *Acoustic Doppler Current profiler*

The volume of sediment motion could be related to the total rate of movement of water mass and sediment concentration in the water column. The discharge is the volume of water moving down a stream or river per unit of time, commonly expressed in cubic feet per second or gallons per day. The measurement of discharge rate and direction of the flow has been carried out for the Cochin Port area. The discharge across the channel with tidal fluctuations obtained from these measurements was used for estimating the overall discharge into the Cochin Port region. The Acoustic Doppler Current Profiler (ADCP) is also used to measure stream flow. An ADCP measures water velocity by transmitting sound waves, which are reflected off sediment and other materials in the water. The ADCP works by transmitting "pings" of sound at a constant frequency into the water, the data collected from which can then be used for bathymetric mapping. The pings are so high-pitched that humans and even dolphins cannot hear. As the sound waves travel, they ricochet off particles suspended in the moving water and are reflected back to the instrument. Due to the Doppler effect, the sound waves reflected from a particle moving away from the profiler have a slightly lower frequency when they return. Particles moving towards the instrument send back higher frequency waves. The difference in frequency between the waves the profiler sends out and the waves it receives is called the Doppler shift. The instrument uses this shift to calculate how fast the particle and the water around it are moving. Views of an acoustic

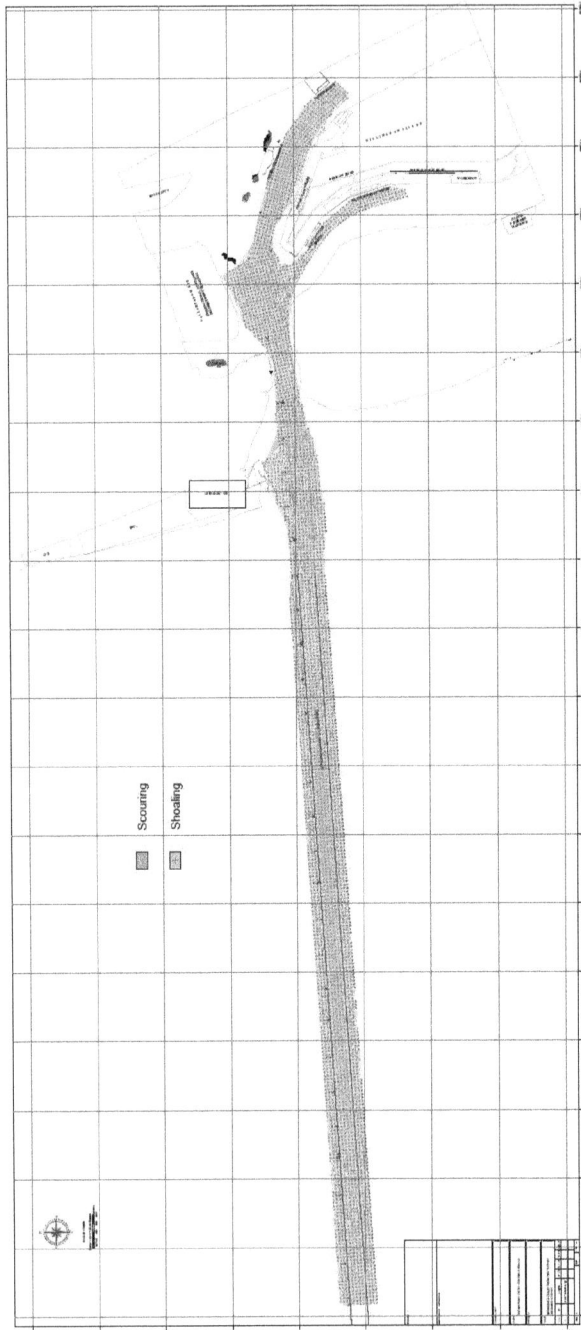

Fig. 6.3. Cochin port area depth differential.

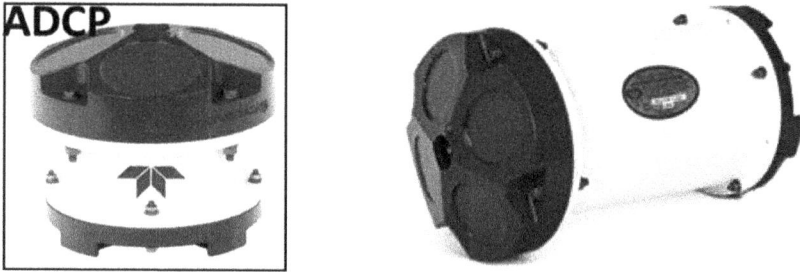

Fig. 6.4. Acoustic Doppler current profiler.

Doppler current profiler (ADCP) capable of measuring vertical cross section of current speed and direction are shown in **Fig. 6.4**.

6.4.1.1. *ADCP calibration*

The ADCP compass calibration verification is an automated built-in test which measures and shows the compass calibrations. This procedure involves the measurement of compass parameters at every 5 degree of rotation for a full 360 degree rotation. The data collected across all the calibrated directions are projected in the final results. This is a crucial exercise to be carried out with utmost care because the presence of any magnetic/electric field in the vicinity of the device could potentially tamper the output. The 5 degree value is chosen to minimize the overall error involved in the measurement.

6.4.2. *ADCP discharge measurements*

Shallow water discharge measurements have been taken at cross-sections C16 and C17 locations (**Fig. 6.5**) with RDI River Ray ADCP during full tidal cycles for 16 days and (alternating between C16 on even days and C17 on odd days) during each of the 3 seasons (covering a full spring–neap–spring tide cycle). The measurements provide the flow speed in the vertical direction and along with the bathymetry. The discharge across the channel with tides has been obtained from these measurements. These measurements, in conjunction with current measurements, are used in estimating the overall discharge into the port region. The discharge measurements have been carried out at the provided location using a Rio Grande 600 KHz

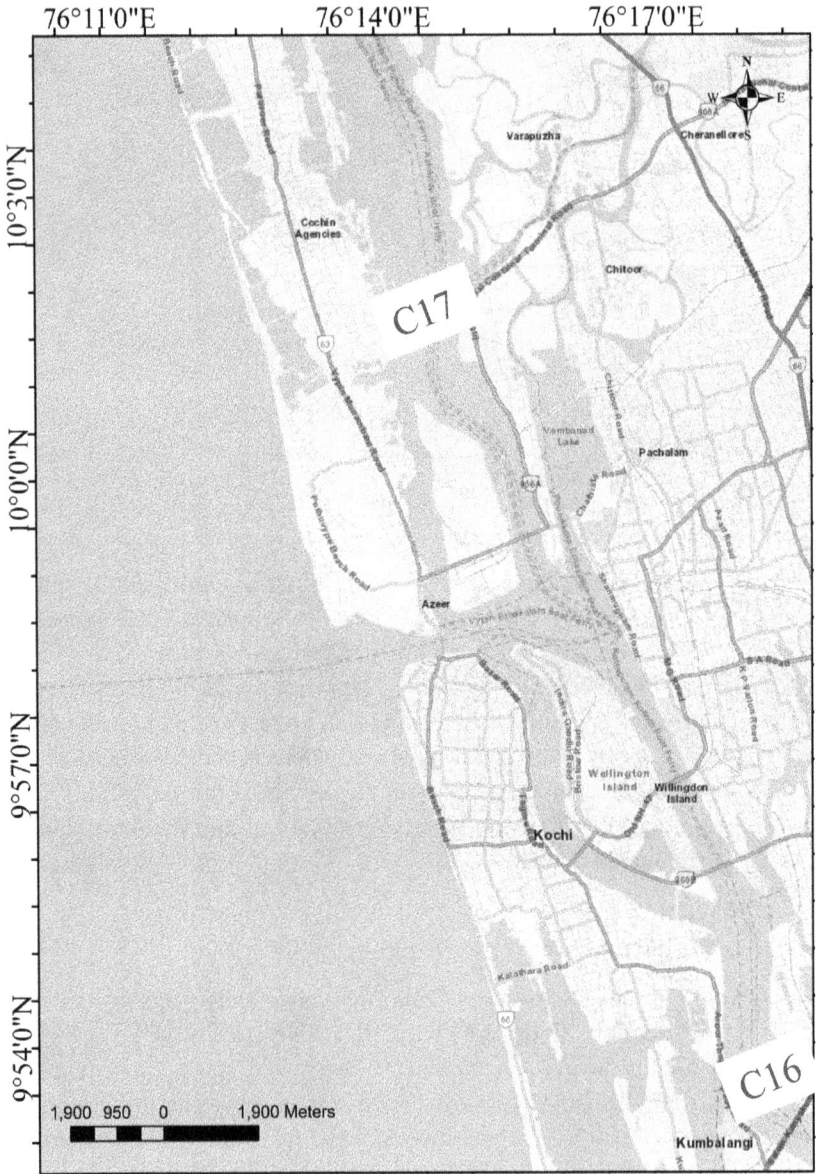

Fig. 6.5. ADCP discharge measurements locations.

ADCP. The ADCP is interfaced with DGPS, and transects have been carried out at the location for 14 days during all seasons. The measurements have provided a clear understanding of the sediment movement processes (**Tables 6.3** and **6.4**) and are useful in understanding the process of siltation in the basin and management of dredging. In addition, it provides details of the sediment transport rate which is important for the dredging philosophy (the quantity of sediments to be dredged, the selection of dredgers as well as dredged spoil disposal).

6.5.   Case Study — Application at Hooghly Estuary

6.5.1.   *Study area*

The Hooghly estuary is deemed to be one of the largest estuaries in the world. The funnel-shaped estuary has a breadth and cross-sectional area at the mouth of 25 km and 156,250 sqm, respectively. The mixing zones of the estuary extends up to Diamond Harbour, about 80 km upstream, with a dynamic system in terms of hydrodynamics and morphodynamics. The depth of the estuary varies from 4 to 20 m. The estuary experiences semidiurnal tides (M2), with spring tide ranges of the order of 4.27–4.57 m and neap tidal ranges of about 1.83–2.83 m. In the major portion of the estuary, the flood currents varied from 0.5 to 2.0 m/s, with an average of 1.2 m/s. As the estuary has two ports in the upper reaches and some 3,000 vessels call at the ports every year, monitoring this estuary morphodynamically is of significance. The study area is shown in **Fig. 6.6**.

6.5.2.   *Discharge and current measurements*

Discharge measurements have been carried out at two sections, L1 (Section AB) and L2 (Section CD) between 17 March 2021 and 30 March 2021 using Teledyne RiverRay ADCP at Haldia, West Bengal. The section length of L1 is of 1381 m, located in the upstream side of Haldia Dock entrance. The section length of L2 is of 1800 m, located in the downstream side of Haldia Dock entrance. The basic parameters to achieve good signal/data were configured on the software panel. The rub-test was performed to ensure the ping/beam transmission. Subsequently, the ADCP was installed on the starboard side of boat and connected to an external battery. The DGPS antenna was installed above ADCP, and connections were established to the survey laptop. The trial run was performed and data was verified. Then, the data acquisition was carried out on pre-planned sections L1 and L2 (from

Table 6.3. Measured sediment transport in metric tons/day at location C16 and C17 during the non-monsoon period.

Date		12.05.16	14.05.16	16.05.16	18.05.16	20.05.16	22.05.16	24.05.16	
C16	Tidal range (m)	0.36–0.84	0.40–0.71	0.41–0.70	0.34–0.81	0.28–0.91	0.25–0.96	0.24–0.94	
	Discharge (m³/sec) (flood flow)	−641	−2,802	−4,085.88	−3,721.3	−3,504.8	−8,056	−2,520	Most of the sediments are directed towards the river.
	Average TSS (mg/lit)	145	145	145	145	145	145	145	
	Sediment transport (metric tons/day)	−8,030	−35,103	−51,187	−46,620	−43,908	−1,00,925	−31,570	
	Discharge (m³/sec) (Ebb flow)	7,851	3,401	1,643	2,137	5,941	5,263	3,564.99	
	Average TSS (mg/lit)	145	145	145	145	145	145	145	
	Sediment transport (metric tons/day)	98,357	42,607	20,583	26,772	74,428	65,934	44,662	
	Net sediment balance (metric tons/day)	90,327	7,504	−30,604	−19,848	30,520	−34,991	13,092	

Date		13.05.16	15.05.16	17.05.16	19.05.16	21.05.16	23.05.16	25.05.16	
C17	Tidal range (m)	0.40–0.78	0.67–0.43	0.38–0.76	0.31–0.87	0.27–0.95	0.24–0.96	0.25–0.92	
	Discharge (m³/sec) (flood flow)	−16,424	−3,988	−8,560.8	−1,359	−3,486	−4,084.29	−10,051	Most of the sediments are directed towards the sea.
	Average TSS (mg/lit)	148	148	148	148	148	148	148	
	Sediment transport (metric tons/day)	−2,10,016	−50,995	−1,09,468	−17,377.8	−44,576.	−52,226	−1,28,524	
	Discharge (m³/sec) (Ebb flow)	19,234	8,606	12,895.3	10,579	19,387.5	21,146	22,161	
	Average TSS (mg/lit)	148	148	148	148	148	148	148	
	Sediment transport (metric tons/day)	2,45,949	1,10,046	1,64,894	1,35,275	2,47,911	2,70,398	2,83,377	
	Net sediment balance (metric tons/day)	35933	59051	55426	117897.2	203335	218172	154853	

Table 6.4. Measured sediment transport in metric tons/day at location C16 and C17 during monsoon.

Date	21.07.2016	23.07.2016	25.07.2016	27.07.2016	29.07.2016	31.07.2016	02.08.2016	
C16								
Tidal range (m)	0.17–0.89	0.15–0.85	0.19–0.76	0.23–0.71	0.29–0.70	0.29–0.78	0.21–0.85	
Discharge (m³/sec) (flood flow)	−3355	−3861	−2276	−1873	−2727	−3184	−3143	**Most of the sediments are directed towards the sea.**
Average TSS (mg/lit)	158	158	158	158	158	158	158	
Sediment transport (metric tons/day)	−45,799	−52,707	−31,070	−25,568	−37,226	−43,465	−42,905	
Discharge (m³/sec) (Ebb flow)	6600	2380	3889	2694	4028	3501	3657	
Average TSS (mg/lit)	158	158	158	158	158	158	158	
Sediment transport (metric tons/day)	90,097	32,489	53,089	36,776	54,987	47,792	49,922	
Net sediment balance (metric tons/day)	44298.92	−20217.4	22019.39	11207.63	17760.21	4327.43	7016.72	

Date	22.07.2016	24.07.2016	26.07.2016	28.07.2016	30.07.2016	01.08.2016	03.08.2016	
C17								
Tidal range (m)	0.16–0.87	0.16–0.8	0.26–0.73	0.25–0.69	0.3–0.73	0.26–0.82	0.18–0.86	
Discharge (m³/sec) (flood flow)	−14636	−12430	−11640	−6990	−5471	−7593	−18988	**Most of the sediment are directed towards the sea.**
Average TSS (mg/lit)	188	188	188	188	188	188	188	
Sediment transport (metric tons/day)	−2,37,735	−2,01,903	−1,89,070	−1,13,540	−88,866	−1,23,334	−3,08,425	
Discharge (m³/sec) (Ebb flow)	16871	18973	15929	3810	13196	13342	8342	
Average TSS (mg/lit)	188	188	188	188	188	188	188	
Sediment transport (metric tons/day)	2,74,039	3,08,182	2,58,737	61,886	2,14,345	2,16,716	1,35,500	
Net sediment balance (metric tons/day)	36304	106279	69667	−51654	125479	93382	−172925	

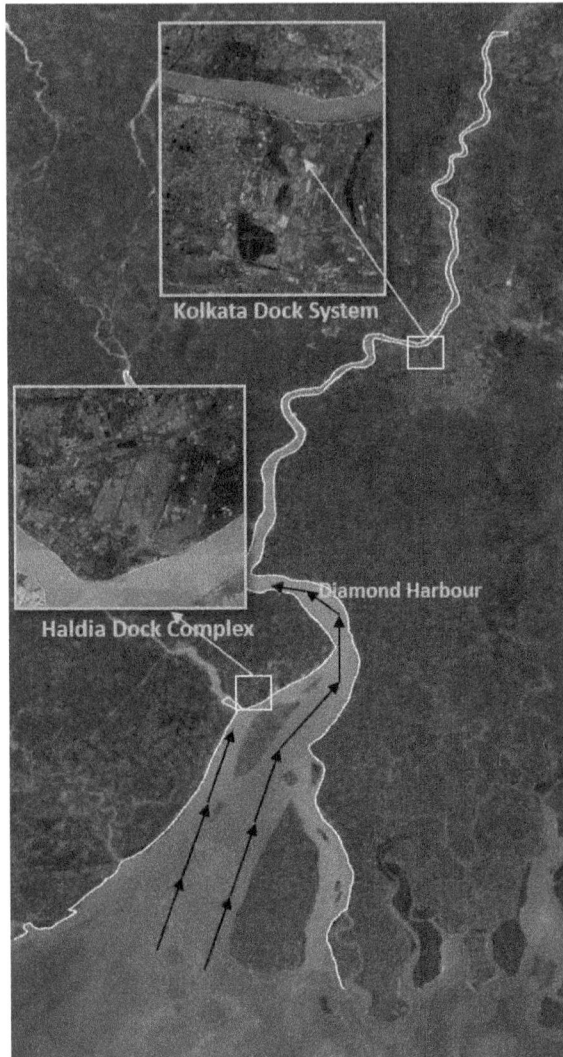

Fig. 6.6.   Layout of the study area, Hooghly.

right to left and vice versa). The details of transects carried out between bank to bank for two sections are presented in **Tables 6.5** and **6.6**.

The current measurements were carried out at four locations: C1 and C2 (from 16 March 2021 to 23 March 2021) and C3 and C4 (from 24 March 2021 to 30 March 2021) at Haldia, and the details are given in **Table 6.7**.

Table 6.5. The geographical details of L1 section.

Location (Section AB)	Geographical		UTM (Zone 45)	
	Latitude	Longitude	X (m)	Y (m)
Right-Bank	22° 02′ 03.5299″ N	88° 06′ 06.7501″ E	613714.16	2437035.28
Left-Bank	22°01′ 25.5100″ N	88° 06′ 32.4099″ E	614458.30	2435871.48

Table 6.6. The geographical details of L2 section.

Location (Section CD)	Geographical		UTM (Zone 45)	
	Latitude	Longitude	X (m)	Y (m)
Right-Bank	22° 01′ 03.0701″N	88°04′ 14.1899″ E	610500.17	2435153.18
Left-Bank	22° 00′ 13.8000″ N	88° 04′ 48.0901″ E	611482.88	2433644.96

Table 6.7. The geographical details of four current measurement locations.

Location	Geographical		UTM (Zone 45)	
	Latitude	Longitude	X (m)	Y (m)
C1	22° 00′ 43.1555″ N	88° 04′ 32.5430″E	611030.71	2434544.50
C2	22°01′ 00.4760″ N	88° 05′ 13.1337″E	611814.23	2435082.67
C3	22° 01′ 34.0900″ N	88° 05′ 33.2800″E	612761.07	2436123.10
C4	22° 01′ 48.3500″ N	88° 06′ 32.8200″E	614464.96	2436573.90

The discharge and current measurement locations near the Haldia Dock Complex (HDC), Haldia, are given in **Fig. 6.7.**

6.5.3. *Details of discharge measurements*

6.5.3.1. *Section L1*

The discharge measurements were carried out at section L1, covering 688 transects during the period of 14 days between 17 March 2021 and 30 March 2021. The maximum discharge of (+) 12,886 cu.m has been observed during the ebb tide. During flood tide, the maximum discharge was around (−)17,931 cu.m. It has been observed that the flow speed was a maximum of 2.1 m/s. The flow direction mainly remained at 246° during ebb tide and 65° during flood tide. The graphical representation of maximum flow velocity contour at section L1 during flooding and ebbing tide, and day-wise discharge values during ebbing and flooding between 17 March 2021 and 30 March 2021 are presented in **Figs. 6.8** to **6.10.**

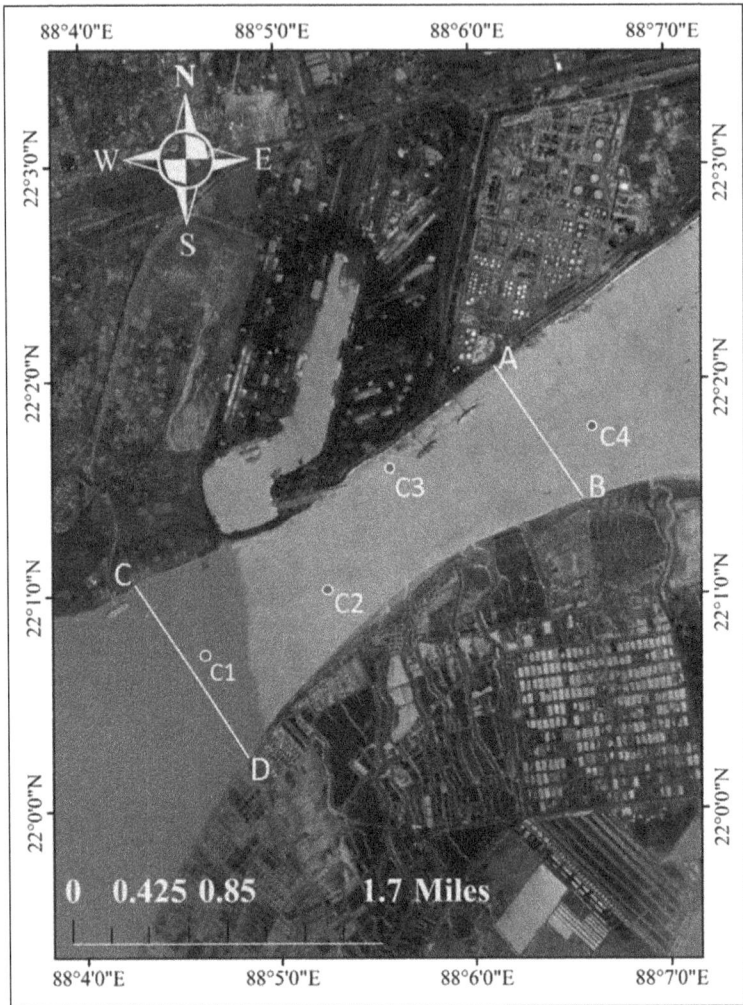

Fig. 6.7.   Discharge and current measurements locations near Haldia.

6.5.3.2.   *Section L2*

The discharge measurements were carried out at section L2, covering 838 transects during the period of 14 days between 17 March 2021 and 30 March 2021. The maximum discharge of (+) 14,713 cu.m has been observed during the ebb tide. During flood tide, the maximum discharge was around (−) 22,316 cu.m. It has been observed that the flow speed was maximum

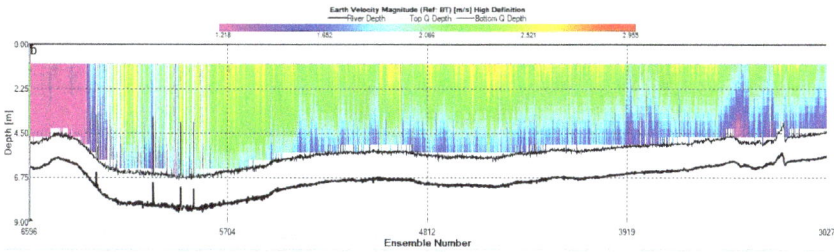

Fig. 6.8. Flow magnitude contours during Flood Tide at "L1" section.

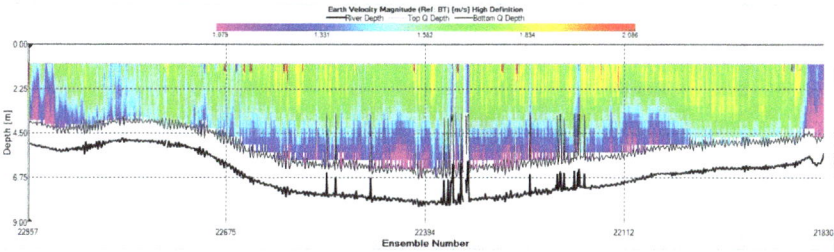

Fig. 6.9. Flow magnitude contours during Ebb Tide at "L1" section.

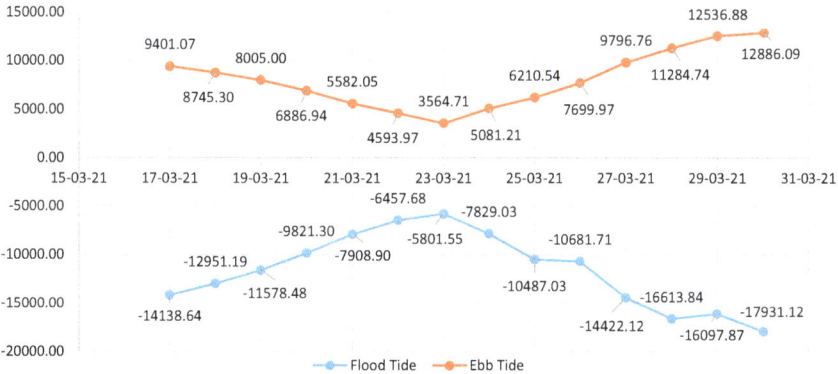

Fig. 6.10. Day-wise maximum and minimum discharge values at L1 section.

of $1.9\,\text{m/s}$. The flow direction mainly remained at $238°$ during ebb tide and $60°$ during flood tide. The graphical representation of maximum flow velocity contour at section L2 during flooding and ebbing tide, and day-wise discharge values during ebbing and flooding between 17 March 2021 and 30 March 2021 are presented in **Figs. 6.11** to **6.13**.

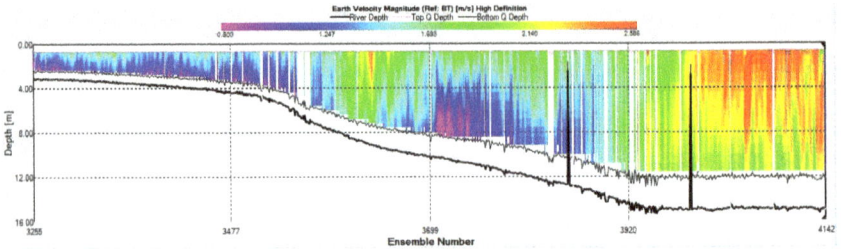

Fig. 6.11. Flow magnitude contours during flood tide at "L2" section.

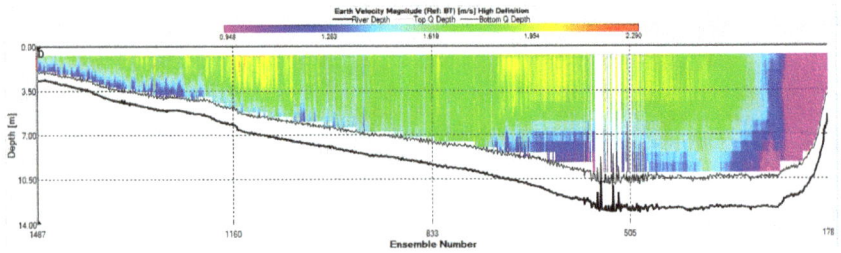

Fig. 6.12. Flow magnitude contours during Ebb Tide at "L2" section.

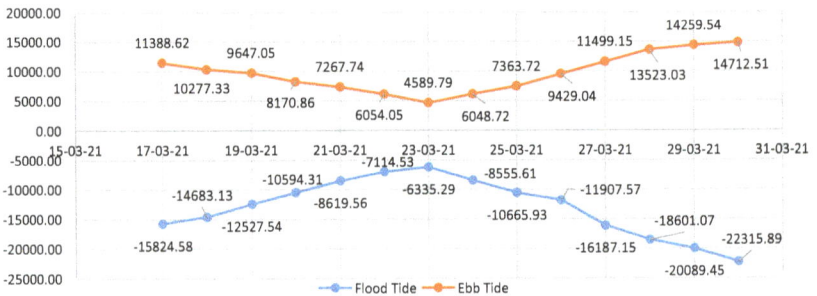

Fig. 6.13. Day-wise maximum and minimum discharge values at L2 location.

6.5.4. *Details of current measurements*

6.5.4.1. *Station-C1*

The current speed and directions measured in Station-C1 are shown in
Figs. 6.14 and **6.15**. The current speed varied between 0 and 1.7 m/s during the measurement period. Unidirectional currents were observed, which
mainly remained between 230° during flood-tide and 50° during ebb-tide.

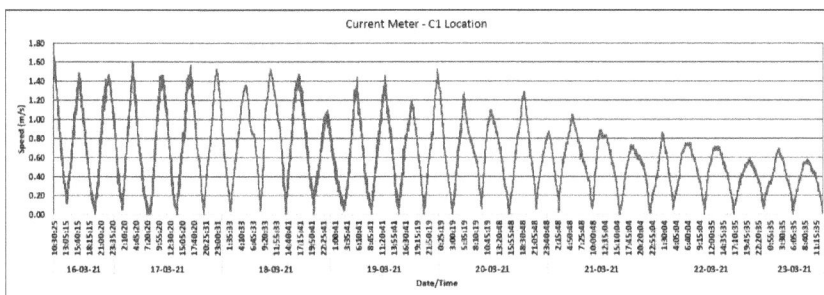

Fig. 6.14. Variation for current speed at Station-C1 from 16 March 2021 to 23 March 2021.

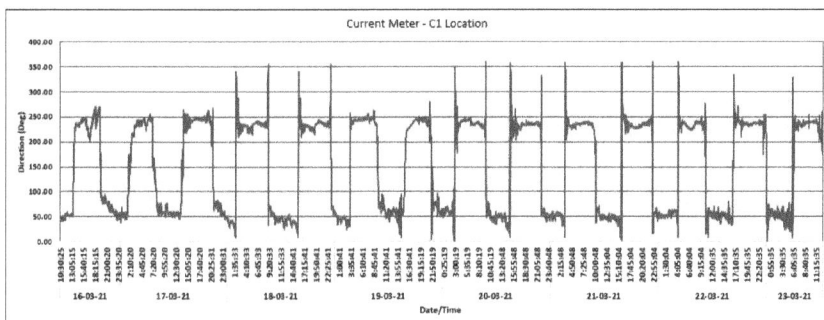

Fig. 6.15. Variation for current direction at Station-C1 from 16 March 2021 to 23 March 2021.

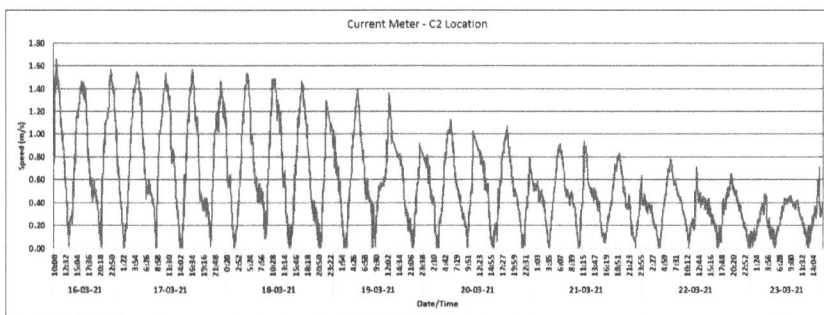

Fig. 6.16. Variation for current speed at Station-C2 from 16 March 2021 to 23 March 2021.

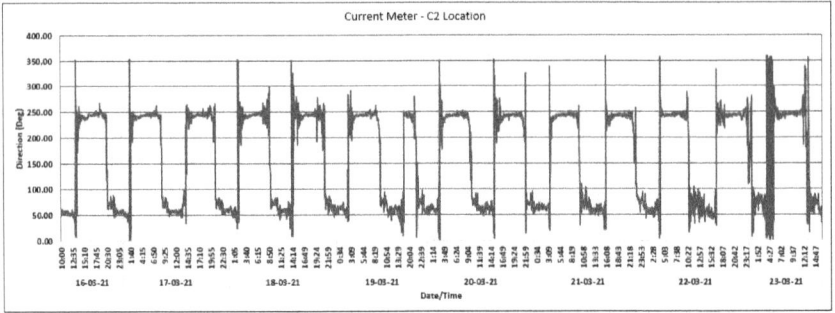

Fig. 6.17. Variation for current direction at Station-C2 from 16 March 2021 to 23 March 2021.

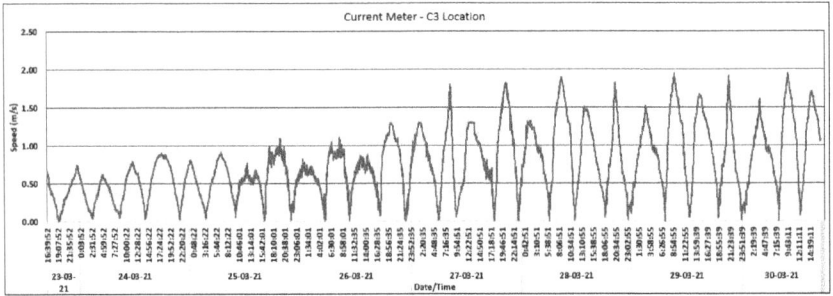

Fig. 6.18. Variation for current speed at Station-C3 from 23 March 2021 to 30 March 2021.

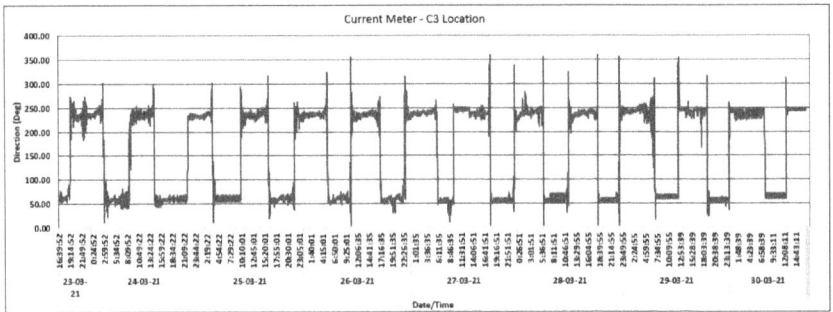

Fig. 6.19. Variation for current direction at Station-C3 from 23 March 2021 to 30 March 2021.

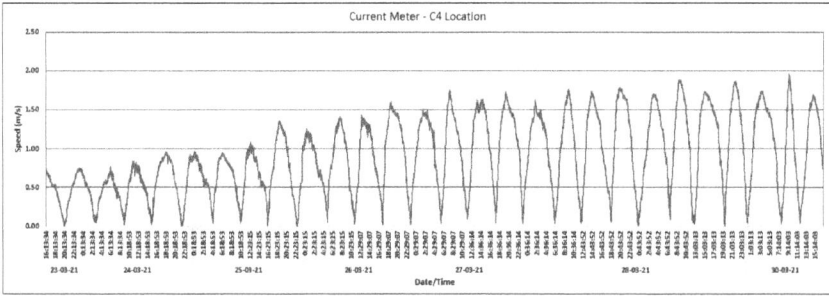

Fig. 6.20. Variation for current speed at Station-C4 from 23 March 2021 to 30 March 2021.

6.5.4.2. *Station-C2*

The current speed and directions measured in station C1 is shown in **Figs. 6.16** and **6.17**. The current speed varied between 0 and 1.7 m/s during the measurement period. Unidirectional currents were observed which mainly remained between 240° during flood-tide and 60° during ebb-tide.

6.5.4.3. *Station-C3*

The current speed and directions measured in station C1 is shown in **Figs. 6.18** and **6.19**. The current speed varied between 0 and 1.95 m/s during the measurement period. Unidirectional currents were observed which mainly remained between 240° during flood-tide and 60° during ebb-tide.

6.5.4.4. *Station-C4*

The current speed and directions measured in station C1 is shown in **Figs. 6.20** and **6.21**. The current speed varied between 0 and 1.96 m/s during the measurement period. Unidirectional currents were observed which mainly remained between 230° during flood-tide and 60° during ebb-tide.

6.5.5. *Measurement of bathymetry and application of differential depths*

It is interesting to see that the following useful analysis could be derived on morphodynamics of this case study. The overall changes within certain

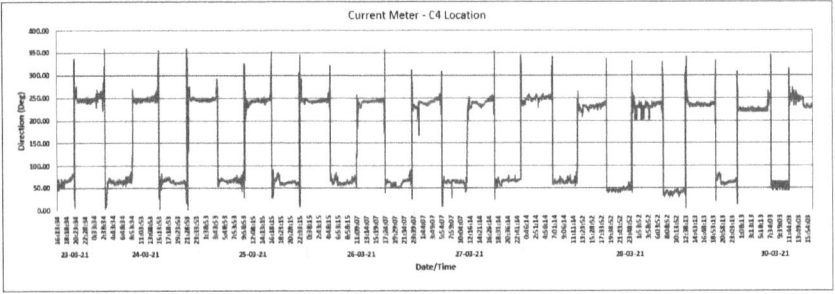

Fig. 6.21. Variation for current direction at Station-C4 from 23 March 2021 to 30 March 2021.

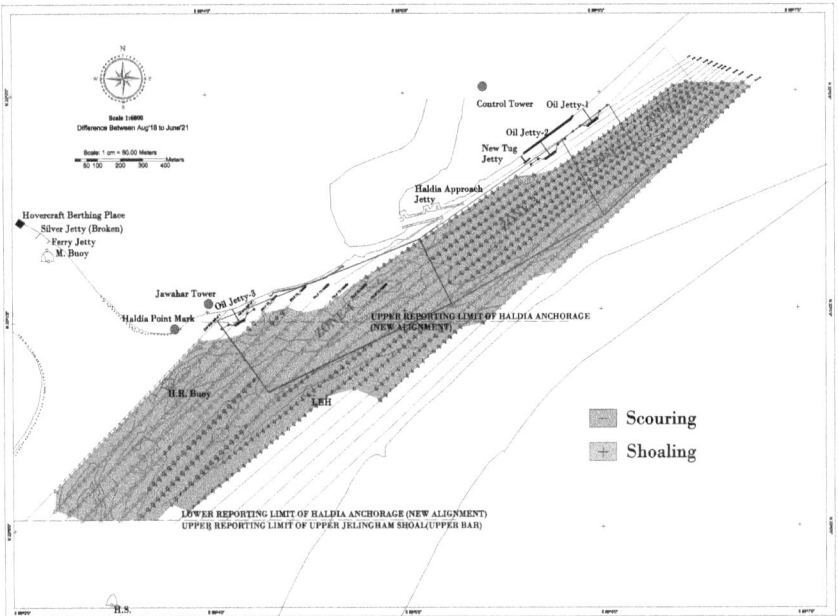

Fig. 6.22. Differential depths observed in the study area.

duration could be depicted in terms of differential depths (**Fig. 6.22**). Such details also could be extracted numerically and will be useful to look at it over time between tide to tide (**Fig. 6.23**) or cumulative data (**Fig. 6.24**). These details would go a long way in understanding the physical behaviour of the water body/coastal area.

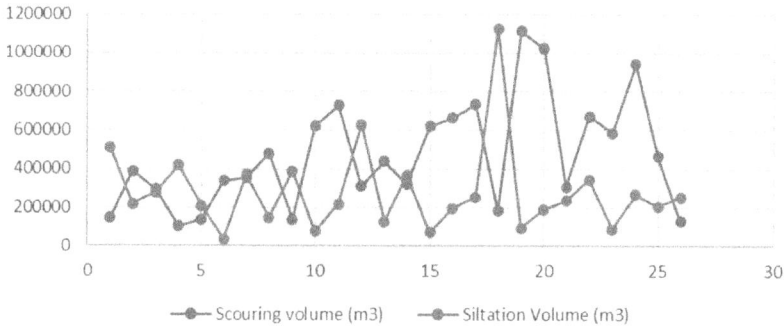

Fig. 6.23.   Volume of sediments accreted or eroded over time (no. of observations) from each spring–neap tidal cycle.

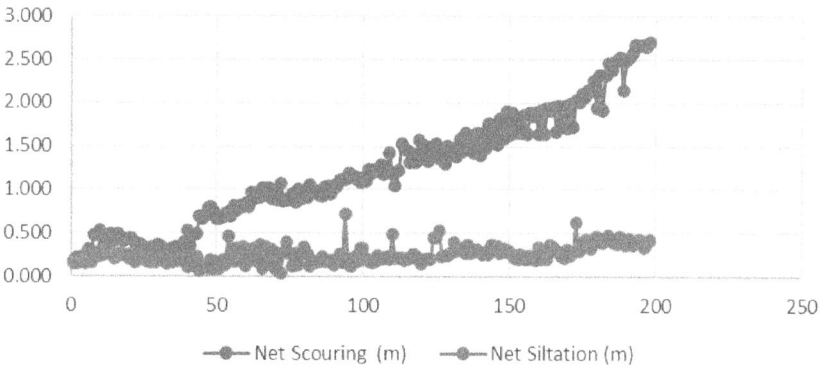

Fig. 6.24.   Average siltation or scouring rate between each observation.

6.6.   Summary

The discharge measurements were carried out at two sections on the upstream and downstream side of the HDC. From the discharge measurements, it is inferred that the flux entering the section L2 is reduced while it is leaving section L1 during flooding. The flux leaving the section L1 is increased while it is entering L2 during ebbing. This incurs addition of volume of water in between these two sections. Such a difference in volumes is the main reasons for siltation within the region of interest. Hence, measurement of discharges helps in the studies related to understanding the water levels, sediment discharge, bed evolution, etc. The measurement of currents facilitates the understanding of the hydrodynamics, and these data shall be applied for calibration and validation of numerical models.

Understanding the morphodynamics of a domain, especially in estuaries and rivers, helps to manage the migration of navigation channels, protection of river banks, maintenance dredging, etc. The measurement of bathymetry, estimation of siltation and accumulation volume and patterns, history of minimum depth in the channels and correlation with the hydrodynamics helps in sediment management in the study domain. Based on the results, it is observed that eroding volumes are higher than deposition volumes along Haldia. The navigation depths appear to be increasing from 2016 to 2020, as per the bathymetry charts.

Chapter 7

Shoreline Change Monitoring Techniques: Past to Present

Abstract

Shorelines are of great importance to coastal engineers and planners. They are the line of action for planning coastal protection measures and act as the first line of defence during extreme weather events. The intensity of a storm on the coast can be estimated by calculating the changes in the shoreline. The prominence of shoreline changes is further stressed in the planning of ports and harbours and other coastal installations. There are various sources of data available for shoreline extraction from field surveys to satellite image extraction. The analysis of collected data is performed using statistical methods of Linear Regression Rate, End Point Rate and Weighted Linear Regression Rate. In this chapter, the various sources of data collection and analysis and the role of remote sensing and Geographical Information System (GIS) in shoreline change studies are discussed using case studies. In addition, the advantages of various shoreline mapping techniques and the errors and uncertainties involved in performing a remote sensing study are discussed.

7.1. Introduction

Coastal scientists, engineers and managers are interested in the status of the shoreline and its change over time, the knowledge of which is essential for effective coastal management and engineering design as any human intervention on the littoral drift may often lead to adverse effects on the shoreline changes. Additionally, a study of shoreline information is required in the design of coastal protection in order to calibrate and verify numerical models, predict sea-level rise, establish hazard zones, formulate policies to regulate coastal development and assist with legal property border and coastal research and monitoring. Shoreline position can provide information on shoreline reorientation near structures, as well as beach width and volume, and it is used to quantify historical rates of change (Sundar *et al.*, 2022). A functional definition of shoreline is necessary to examine shoreline variability and trends. The satellite image definition chosen must consider the coastline in both temporal and spatial scales, as well as the variability's dependence on the time scale under investigation, due to the dynamic nature of the land–water boundary. For practical reasons, the particular definition used is usually less important than the capacity to quantify how a shoreline indication corresponds to the physical land–water boundary in a vertical/horizontal sense. The objective is to create a technique that is both robust and consistent enough to enable such detection of the desired coastline feature within the available data set. The detection methods differ on the data source and the shoreline definition selected (Boak and Turner, 2005).

7.2. Shoreline Definition

7.2.1. *General*

Shoreline is defined as the physical boundary between land and water in an idealized sense (Dolan *et al.*, 1980). In actuality, despite its seeming simplicity, this term is difficult to employ as the coastline location fluctuates over time due to the cross-shore and alongshore sediment movement in the littoral zone, as well as the dynamic character of water levels at the coastal boundary (e.g., waves, tides, groundwater, storm surge, setup, run-up, etc.) (Boak and Turner, 2005). As a result, the coastline must be examined in a temporal context, with the time scale selected based on the context of the investigation. For instance, while examining the swash zone of a beach front, it is required to collect 10 samples per second, although sampling every 10–20 years may be enough for a long-term study. The instantaneous coastline is the location of the land–water contact at any given time. Many researchers suggest that the most common and incorrect assumption in the coastal studies is considering the immediate shoreline to reflect "normal" or "average" shoreline positions. Depending on the beach slope, tidal range and prevailing wave/weather conditions, the horizontal/vertical position of the shoreline can change from millimeters to meters (or more) over time (Komar, 1998). The coastline is a time-dependent phenomenon with significant short-term volatility, which must be considered while choosing a single shoreline point (Morton, 1991). When determining the coastline, the variance along the beach must be considered. The majority of shoreline change studies monitor individual transects or places through time to see how they change. However, this simple approach may add to the level of uncertainty.

7.2.2. Shoreline indicators

The dynamic nature of coastal interface due to the action of tides, and the long-term effects of sea-level and sediment transport, makes it difficult to consistently map the true shoreline position (Su and Gibeaut, 2017). The "true" shoreline is often represented by a shoreline indicator (van der Werff, 2019). The definition of a shoreline depends on the indicator chosen for the analysis. Shorelines derived from tidal data of the location are termed as mathematical indicators whereas the berm crests, vegetation and dunes are classified under morphological indicators, as illustrated in **Fig. 7.1.**

The interpretation and definition of a border through historical maps and topo sheets are common, but more important in the coastal geomorphology is defining an indicator for shoreline analysis. The indicators are important for various kinds of datasets, including aerial photographs, satellite imagery, light detection and ranging technology (LiDAR) data and as well as for field surveys. The field surveys carried out regularly collect the indicators of low tide level and high tide level of that particular day of the survey. The shoreline dedicated surveys follow this data collection but must be accounted for repeatability and representation of the specific feature. Modern data collection using LiDAR and aerial photographs should follow consistent morphological indicators as mathematical indicators change with respect to time. Topo sheets and historical maps give an outline for the position of the shoreline rather than specify a particular indicator. Historical analysis of shoreline changes using satellite images in a long-run in terms of a common indicator is improbable. In the long term, the indicators witness various changes due to either sea-level rise or the impact of the extreme weather events.

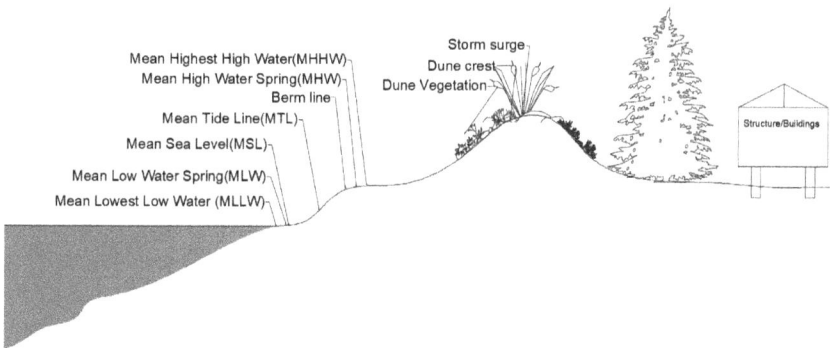

Fig. 7.1. Different indices used for the analysis of shorelines.

The shorelines derived from satellite imagery choose the high water line or the wet/dry boundary (Crowell *et al.*, 1991; Boak and Turner, 2005), based on which the shorelines are delineated on a historical basis. To verify the accuracy of the delineated shoreline, a corresponding field survey needs to be carried out by noting down the date and time of acquisition of the satellite image, through which errors in the satellite image delineation can be reduced. For example, in a 30-year shoreline analysis study by Dhananjayan *et al.* (2022) for the Ennore coast, the delineated shoreline was correlated with the shoreline data collected through field surveys. To understand the wet/dry boundary properly, a screen digitization technique was performed by Tyagi and Rai (2020) for delineating the shoreline.

7.3. Data Sources

Shoreline analysis depends on the amount of data available, and sources range from maps or atlas to modern-day LiDAR and video imaging techniques. Coastal data from maps and charts are limited and restricted to very few places of availability. The Topo sheets were then created by the national mapping agency of the individual country and made it available to public.

Boak and Turner (2005) classified the data sources and listed according to chronological order. A similar kind of classification has been followed here and are listed as follows.

7.3.1. *Coastal land-based photographs*

Although there are several land-based photographs available for a long time, the data on the coastal features and the tidal lines could not be extracted due to the coarser image or improper imaging or the lack of coastal features in the region. Therefore, the land-based photographs find limited application in the analysis of the shoreline changes.

7.3.2. *Maps and charts*

There have been many kinds of charts and maps developed over time which have been constantly updated with changing boundaries. The cartographers earlier used to give prominence to both decoration as well as data by making them useful for historical shoreline analysis. These charts contain data on the coastal features as well as the water line. The charts can be digitized by the coordinates and shoreline delineated accordingly for the analysis.

In India, the charts date back as early as 18th century, with the Survey of India preparing maps and charts. These charts have many limitations as they have been marked for a cluster of villages rather than an entire area. Also, in the earliest charts, the shoreline and the coastal features have been given little importance, and a few charts find no mention of the sea level. The modern topo charts with the updated boundary details can be found from the 20th century. These charts have all the required data for a coastal mechanism.

Potential errors associated with historical coastal maps and charts include errors in scale; datum changes; distortions from uneven shrinkage, stretching, creases, tears, and folds; different surveying standards; different publication standards; projection errors and partial revision (Kankara *et al.*, 2015). These errors are overshadowed by the fact that the maps and charts provide long-term data.

7.3.3.   *Aerial photographs*

Aerial photographs are a snapshot of locations. These photographs are used for delineating the shoreline from the photographs. Though the photographs have been around over a long time, the delineation of the shorelines began late in the 20th century. The photographs have to be corrected for orthogonality and delineated for shorelines. The aerial photographs are site-specific and are limited both spatially and temporally. A digitally scanned pair of aerial pictures can be transformed into a three-dimensional digital terrain model and a geo-rectified ortho-photo using modern softcopy photogrammetry. The addition of datum-referenced elevation data makes determining tidal-datum–based shorelines simple and accurate. Aerial photography is the most popular data source for identifying past shoreline positions.

7.3.4.   *Beach surveys*

The beach surveys form a prominent data set for shoreline analysis. This kind of data are the most accurate reference of a shoreline. The collection of field data is very site-specific and can be attributed to the site. This data is useful in correlating digitized shorelines from aerial or satellite imagery. Though the data is of the highest quality, it also has its own drawbacks, such as the labour-intensive work required and the limited length of shoreline being collected. The shoreline data collected through field surveys accurately define the high and low water lines and the accurate representations of the coastal geomorphology.

7.3.5. *Satellite imagery*

With the advancements in remote sensing, satellite images play a major role as a data source for extracting the shoreline details. The satellite images from LANDSAT cover a greater swathe on the earth, covering greater distances. Remote sensing techniques for classification of the images provide with easier ways of delineating the shorelines. Analysis can be carried out for longer periods of time, and a particular pattern can be established, which is helpful for the coastal engineers to plan further developments accordingly.

7.3.6. *Airborne light detection and ranging technology*

In a short duration of time, airborne LiDAR can cover hundreds of kilometres of a coastal stretch. The time taken for the emitted laser beam to return its reflection after exiting the device is the principle of LiDAR technology. Knowing the speed of light, one can efficiently determine the distance and using differential GPS, specific location can be pinpointed. Tidal-datum–based shorelines, such as high tide line or low tide line, can be detected by fitting a function to cross-shore profiles of LiDAR data. Because of its high cost, this data source has limited temporal and spatial availability. The fundamental benefit of LiDAR data is that it can quickly cover wide areas. A sample of data collection being used in LiDAR is presented in **Fig. 7.2.** With the altitude help, wide range of data can be collected. A sample data collected using LiDAR is shown in **Fig. 7.3.**

7.4. Shoreline From Remote Sensing

The vast data available through satellite images and the increasing resolution makes them the best fit for shoreline changes that capture annual- to decadal-scale dynamics. The major factors concerning this type of analysis are the pixel resolution, ortho-rectification of the satellite images, and the magnitude of change detectable in succeeding imagery. LANDSAT dates back to as early as 1970s with a pixel resolution of 60 m. Initially, Multi Spectral Scanners (MSSs) were used, with a resolution of 60 m and enhanced to 30 m by the introduction of thematic mapper sensors. The enhanced thematic mapper sensors launched subsequently with a panchromatic band further increased the resolution to 15 m. The data was made publicly available by the USGS in 2008 along with archive data. The use of LANDSAT data sets in historical shoreline analysis has been vividly

Fig. 7.2.   Data collection using LiDAR (USGS).

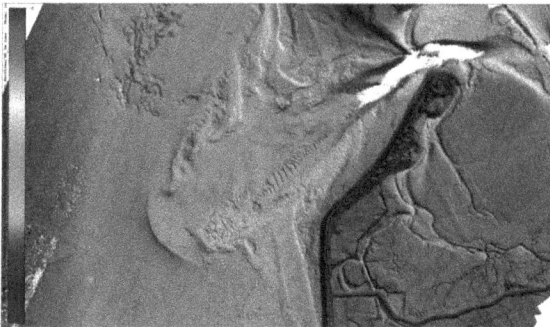

Fig. 7.3.   Sample coastal data collected using LiDAR.
Source: Fairley *et al.* (2019).

discussed by Sundar *et al.* (2021) and Gracy *et al.* (2020) by performing a shoreline change study for 20 years along the city of Chennai, situated in the southeast coast of the Indian peninsula.

Due to its high pixel resolution (10 m in visible and near-infrared wavelengths) and temporal frequency (10 days revisit time reduced to 5 days), the Sentinel-2 programme, which has been active since 2015, is being recognized as a key instrument for investigating sub-annual shoreline dynamics. The rest of the wavelengths are collected at a resolution of 20 m, and this data can be used to delineate more discrete shoreline features that represent changes in water volume and vegetation cover. Similar to LANDSAT, unrestricted access to these data allows for extensive shoreline change analysis at different scales around the world. Several satellite systems use sensors that collect ultra-high-resolution images for ultra-high-resolution imaging (Jackson and Short, 2020).

7.5. Digital Shoreline Analysis System Tool for Shoreline Changes

7.5.1. *General*

The digital shoreline analysis system (DSAS) tool has been developed to estimate the rate of change calculation of shorelines. This tool is used in the ArcGIS software to perform the analysis. Flow chart of the methodology is given in **Fig. 7.4**. First, we have to create a database of the shorelines and baselines. Then the transects are cast perpendicular to the shoreline from baseline and then the rate of change of shoreline is calculated. For further methodology on DSAS it is recommended to read Himmelstoss *et al.* (2021).

The quantification of shoreline change is based on the derivation of five key measures, namely Net Shoreline Movement (NSM), Shoreline Change Envelope (SCE), End Point Rate (EPR), Linear Regression Rate (LRR) and Weighted Regression Rate (WLR) (Himmelstoss *et al.*, 2018).

7.5.2. *NSM*

NSM is the distance between the latest and the oldest shoreline (**Fig. 7.5**). The value is a horizontal distance, hence the units are in meters (m).

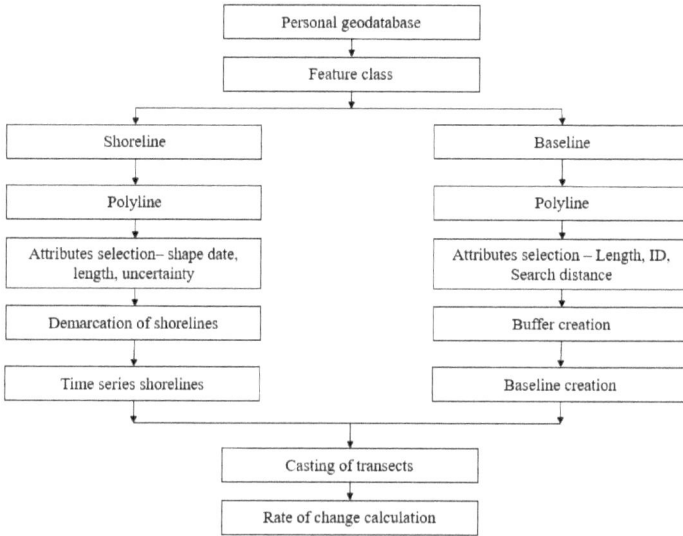

Fig. 7.4.   Flowchart for shoreline and baseline creation.

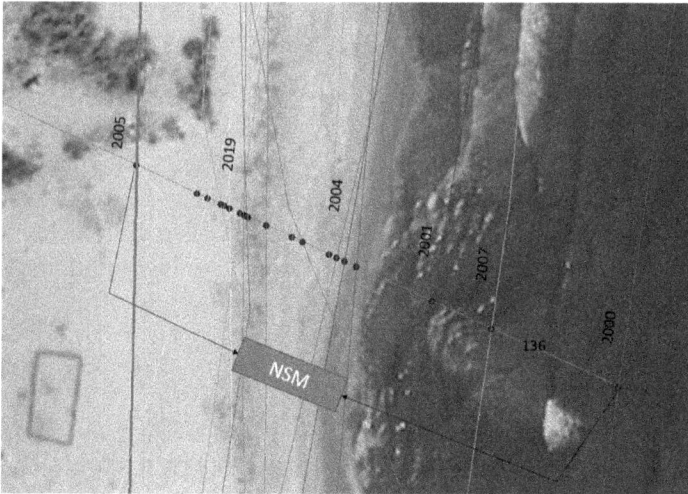

Fig. 7.5.   Net shoreline movement.
Source: Gracy *et al.* (2021).

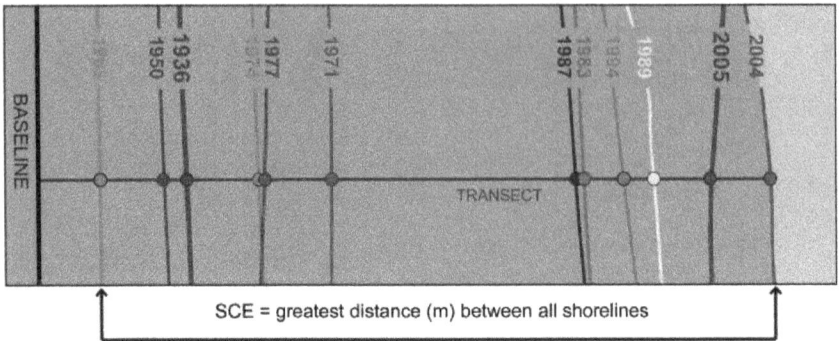

Fig. 7.6.   Shoreline change envelope.
Source: Himmelstoss *et al.* (2018).

7.5.3.   *SCE*

SCE reports distance (in meters), not rate, and the SCE value represents the greatest distance among all the shorelines that intersect a given transect **(Fig. 7.6)**. As the total distance between two shorelines has no sign, the value for SCE is always positive (Himmelstoss *et al.*, 2018)

7.5.4.   *EPR*

EPR is the ratio of the distance between latest and oldest shoreline to the time lapse in years between the two shorelines **(Fig. 7.7)**. Unit of EPR is m/yr. The main advantages of EPR are its simplicity of computation and the fact that it just requires two shoreline dates. The drawback is that in circumstances where there is more data available, the additional data is ignored.

7.5.5.   *LRR*

A LRR statistic can be determined by fitting a least-squares regression line to all shoreline points for transect **(Fig. 7.8)**. The LRR is the slope of the line. The method of linear regression includes the following features: (1) all the data are used, regardless of changes in trend or accuracy, (2) the method is purely computational, (3) the calculation is based on accepted statistical concepts, and (4) the method is easy to understand (Dhananjayan *et al.*, 2022). However, the LRR method underestimates the actual shoreline in few cases when compared to EPR (Himmelstoss *et al.*, 2018).

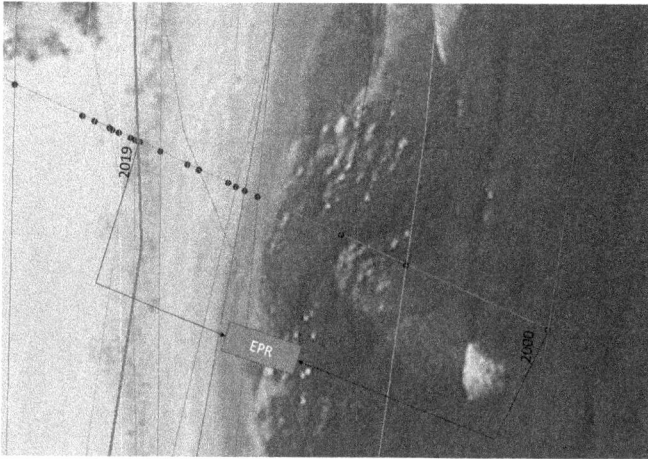

Fig. 7.7.   End point rate.
Source: Gracy *et al.* (2021).

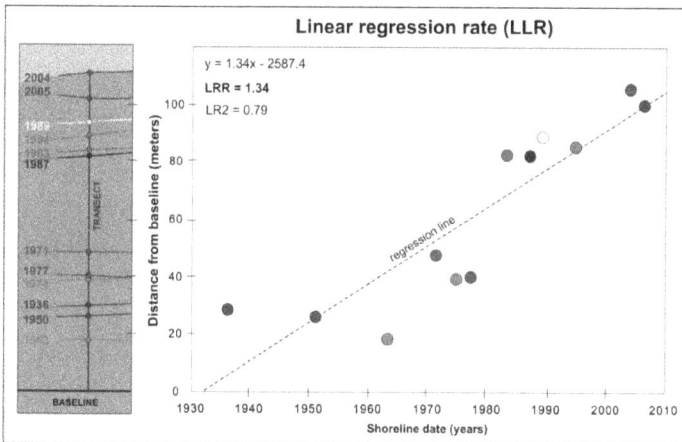

Fig. 7.8.   Linear regression rate.
Source: Himmelstoss *et al.* (2022).

7.5.6.   *WLR*

In WLR, the more reliable shoreline is given greater emphasis towards a best-fit line. For calculating the rate of change, shorelines with greater emphasis are placed in the data points for which the uncertainty is

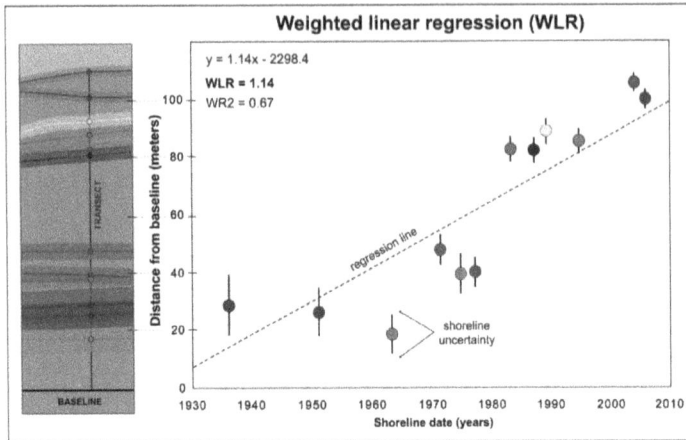

Fig. 7.9. Weighted linear regression.
Source: Himmelstoss *et al.* (2018).

smaller (**Fig. 7.9**). The weight (w) is defined as a function of the variance in the uncertainty of the measurement (e) (Himmelstoss *et al.*, 2018).

7.6. Analysis of Shoreline Data

With advancements in remote sensing and the extensive availability of satellite images, three shoreline extraction methodologies have been developed, namely manual, automatic and semi-automatic, each of which is explained with case studies in the subsequent sections.

7.7. Manual Shoreline Extraction

7.7.1. *General*

Manual shoreline extraction is easy and can be performed for any type of data sources, ranging from Topo sheets to LiDAR to satellite images. In this method, the user manually draws the shoreline through the land–water boundary in the satellite images, and in Topo sheets, the image is digitized and geo-referenced and the shoreline is drawn on the land–water boundary. This process is easy, costless and depends on the assessment skills of the user. The pixels used for mapping in the satellite images have to be followed throughout. The major advantages in this type of shoreline mapping is that shorelines can be demarcated in segments, different time scales, structures can be avoided, and the length of the shoreline can be determined. A study conducted for the coast of Chennai located along the southeast coast of India explains the use of manual shoreline mapping techniques. The study has been divided into three segments of Adyar (13° 0′45.67″N, 80° 16′39.23″E) to Cooum (13° 4′5.20″N, 80° 17′25.11″E) Royapuram groyne field (13° 10′56.91″N, 80° 18′53.36″E), and North of Chennai coast (13° 17′12.39″N, 80° 20′50.28″E). The Adyar to Cooum study has been analyzed by Gracy *et al.* (2020) and the Royapuram groyne field shoreline analysis has been performed by Sundar *et al.* (2021), while the North Chennai coast shoreline analysis is carried out by Dhananjayan *et al.* (2022). The methodology that has been followed by the researchers is the same and is shown in **Fig. 7.10.**

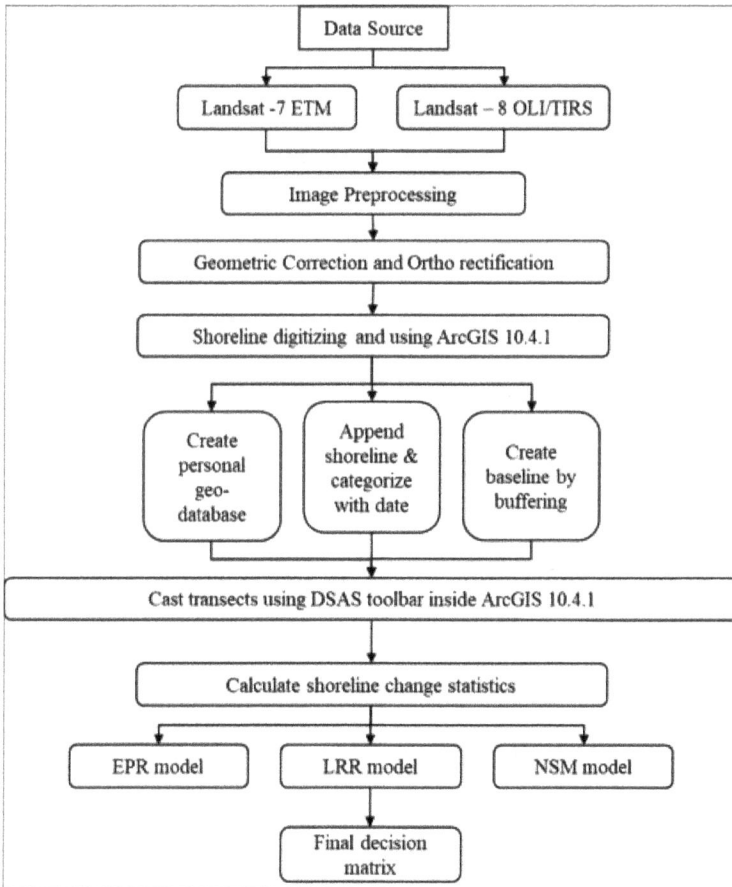

Fig. 7.10. Methodology of manual shoreline extraction.
Source: Gracy *et al.* (2020).

7.7.2.   Analysis of shoreline change between inlets along the coast of Chennai, India

This study has been performed by Gracy *et al.* (2020) for the inlets of Cooum and Adyar River in the Metropolitan of Chennai.

7.7.2.1.   Study area

This study focuses on the shoreline change detection along the east coast of Chennai from south of Cooum (13° 4'5.74"N, 80° 17'8.7000"E) to north of Adyar (13° 00'51.7300"N, 80° 16' 19.1500"E) (**Fig. 7.11**) using geographic

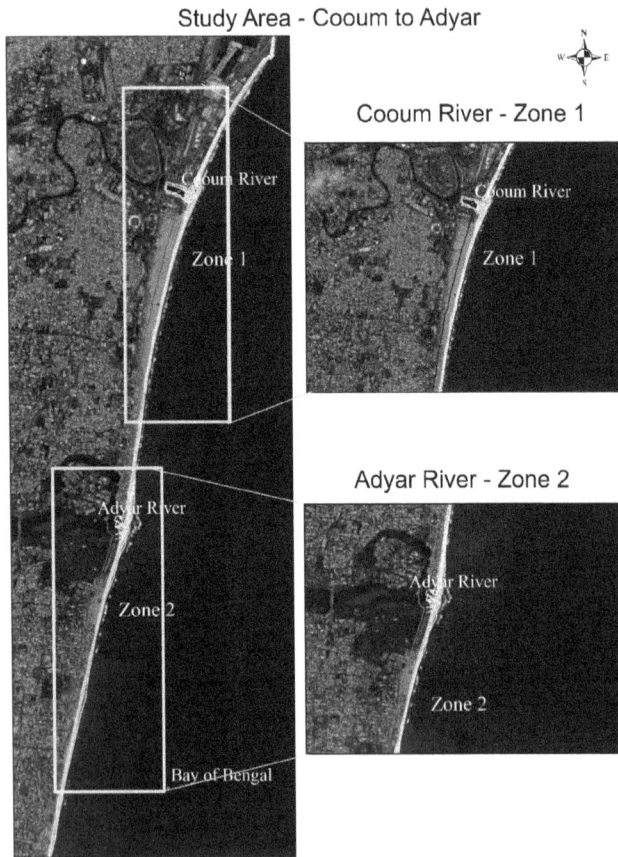

Fig. 7.11.   Study area highlighting Zones 1 and 2.
Source: Gracy *et al.* (2020).

information system and digital shoreline analysis system (DSAS) during the elapsed period from 2000 to 2019. This study has been divided into two zones, namely Cooum River (Zone 1) and Adyar River (Zone 2). The shoreline is divided depending on the coastal vulnerability into Zone 1 that covers the Cooum, which is the shortest classified river draining into the Bay of Bengal. Zone 2 is the region between the south of Cooum and the mouth of Adyar. The Adyar mouth is situated about 11 km from south of Chennai harbour breakwater and 6.25 km from the training walls of the Cooum River.

7.7.2.2. *Analysis of shorelines*

From 2000 to 2019, the historical shoreline position along the Chennai coast was evaluated using profile variation and is detailed here. The LRR, EPR and NSM from DSAS are analyzed along the Zones 1 and 2 from south of Cooum to north of Adyar. The DSAS results show that Zone 1 has the most accretion while Zone 2 has the most erosion (**Fig. 7.12**). The changes in the shoreline in Zone 1 from transect 175 to transect 274 over a distance of 5 km from Chennai Port to Marina Beach are displayed.

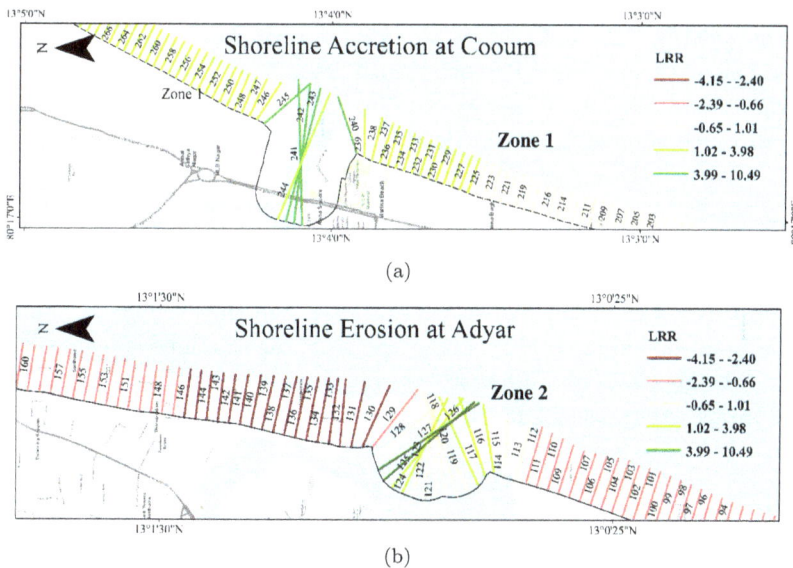

(a)

(b)

Fig. 7.12.   (a) Shoreline Zone 1 and (b) Shoreline in Zone 2.
Source: Gracy *et al.* (2020).

Due to the installation of a training wall along the Cooum River, Transect 245 has experienced maximum accretion to the extent of around 10.5 m. The transects from 1 to 174 depict the shoreline variations in Zone II over a distance of 9 km from Marina Beach to Adyar River. At transect ID 134, this zone experiences greatest erosion to the extent of 4.15 m. The Adyar River mouth is considered dynamic in nature because of the continual sediment movement near the Adyar river entrance, which traps sand and causes erosion to the north. In 2005, shortly after the tsunami of 2004, the area eroded at Srinivasapuram reached a maximum of roughly 50 m.

7.7.2.3. *Statistics adopted*

LRR

The maximum accretion observed at transect 245 shows an accretion of 10.46 m/yr, which is close to the statistical value from DSAS as shown in **Fig. 7.13**. This transect is located on the Cooum River, just south of the northern training wall, which was constructed in 2008. After the construction of training walls, there was extensive accretion in the updrift side of the mouth. As a result, the width of the Marina Beach has increased at a higher rate of growth. In Srinivasapuram, at transect 134, maximum erosion **(Table 7.1)** is observed as stated earlier. The least mean square value calculated for the LRR of −4.15 m, which is considered to be the maximum erosion rate at Adyar inlet in Zone II. This transect is located at a distance of 60 m to the north of the Adyar river mouth.

EPR

The baseline is roughly 275 m away from the beginning period of 2000. The shoreline receded during the tsunami of 2004, and the current shoreline is 183 m from the baseline. Throughout the year, the shoreline is seen to fluctuate. The greatest erosion rate at the transect 134 zone II Adyar estuary is −4.98 m/yr. At transect 245, the maximum rate of accretion is 8.83 m/yr.

NSM

For the past 19 years, the average net coastline movement has been −12.26 m. The shoreline degraded to a maximum length of about −95 m at the transect 134 in Zone II Adyar River after the tsunami of 2004. Transect 245 saw a maximum accretion of 140 m near the Cooum mouth

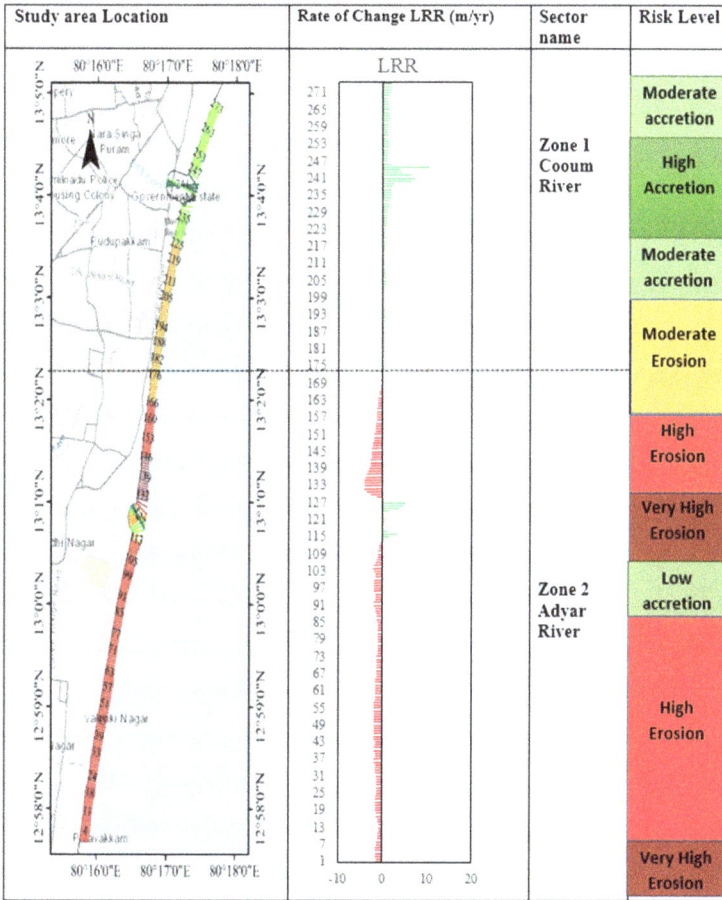

Fig. 7.13. LRR decision matrix.

Source: Gracy *et al.* (2020).

Table 7.1. Shoreline change based on EPR, LRR and NSM.

Category	Rate of shoreline change (m/yr)	Shoreline classification
1	<-2.5	Very high erosion
2	>-2.5 to <-0.5	High erosion
3	>-0.5 to 0	Moderate erosion
4	0	Stable
5	>0 to <1	Moderate accretion
6	>1 to <3	High accretion
7	>3	Very high accretion

Source: Gracy *et al.* (2020).

training walls during the same time period. **Table 7.1** shows the classification of shorelines based on the rate of sediment accretion/erosion. The danger level based on the designated shoreline classification is indicated by the overall regional mean of evolution from 2000 to 2019. The maximum reported coastal erosion/accretion kinematics for the Cooum and Adyar river beaches, respectively, are 0.33/+10.49 and 4.15/+5.28 m/yr.

7.7.3. *Shoreline changes due to construction of groyne field in north of Chennai Port, India*

Shoreline change analysis for the construction of groyne field has been carried out by Sundar *et al.* (2021).

7.7.3.1. *Study area*

This research is being carried out to the north of the Chennai Port and to the south of the Ennore stream, at Royapuram fishing harbour (**Fig. 7.14**). With the construction of Chennai Port, this coastal strip saw significant

Fig. 7.14. Study area for groyne field.

erosion for a period of time. The 2004 Tsunami added to the problem by exacerbating erosion. To fight the persistent coastal erosion problem, a plan was created to build a series of transitional groynes, which were built in three parts from 2004 to 2014, as illustrated in **Fig. 7.15**. The construction history of the transitional groyne field, which lasted 10 years, is the basis for this classification.

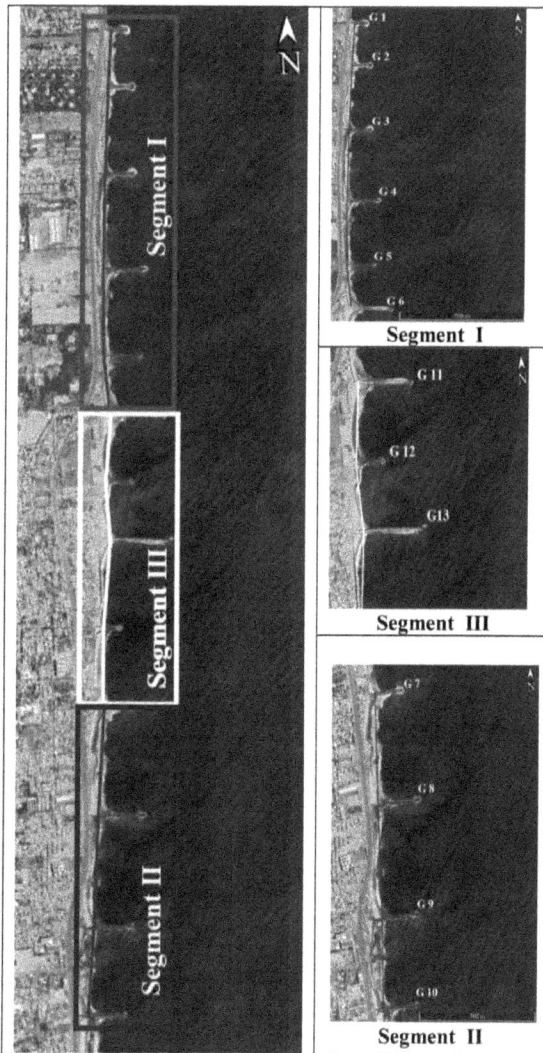

Fig. 7.15. Shoreline analysis based segments.

7.7.3.2. Analysis of shorelines

The analysis of shorelines have been classified into three segments, based on the construction timeline of the groynes. The study is further classified into monsoon and non-monsoon for individual segments and analysis is carried out. The details of which are discussed below.

Segment I

A few of the groynes in segment I were built before the 2004 Indian Ocean Tsunami that struck the Chennai coast. Sundar investigated the impact of shoreline alterations in the region of the groynes using field data (2005). In addition, in 2005, a total of six groynes with a total length of around 2 km were completed. Because a national highway runs next to the shore connecting two large ports, Chennai Port and Kamarajar Port, this stretch was accorded top priority. Before 2003, this portion was protected by sea walls to prevent erosion, but the results were substandard. The annual rate of change in plotted in **Fig. 7.16**. The measured rate of accretion

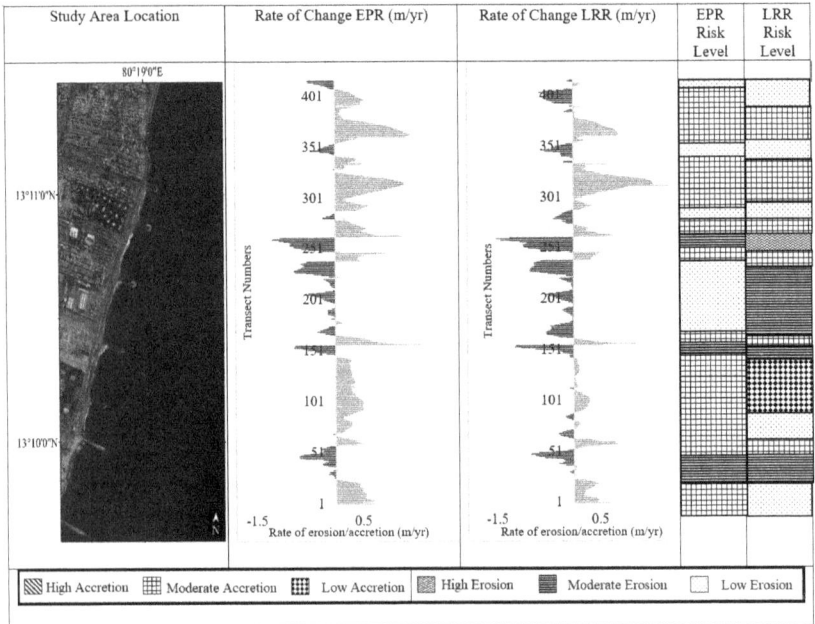

Fig. 7.16. Annual rates of shoreline change in Segment I.
Source: Sundar et al. (2021).

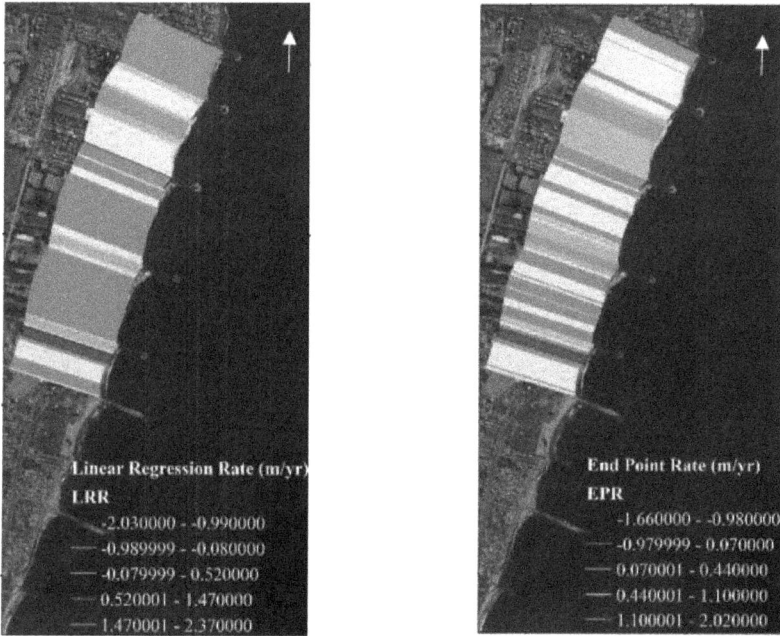

Fig. 7.17(a). Rate of shoreline changes in Segment I for non-monsoon season.

per year is slightly greater than that of erosion using EPR and LRR for coastal data gathered between 2008 and 2020 for the non-monsoon season **(Fig. 7.17(a))** (i.e., January to May). Similarly, the observed rate of accretion per year for shoreline data sampled between 2012 and 2020 during the southwest monsoon season **(Fig. 7.17(b))** (i.e., June to September) is nearly twice as great as that of erosion using EPR, whereas the rate of accretion per year for the same input data set is slightly greater than that of erosion using LRR.

Segment II

Segment II runs for roughly 2.3 km from the Royapuram fishing harbour to the north. In Segment II, construction of four groynes began in 2005 and was completed in 2006. Because of its proximity to the fishing harbour, this stretch of the north Chennai coast is one of the most heavily populated areas. It serves as a local crossroads for thriving fisheries-related trade. The current circulation in this location is further dominated by diffracted and refracted waves from the adjacent breakwater arm, exposing the coast to

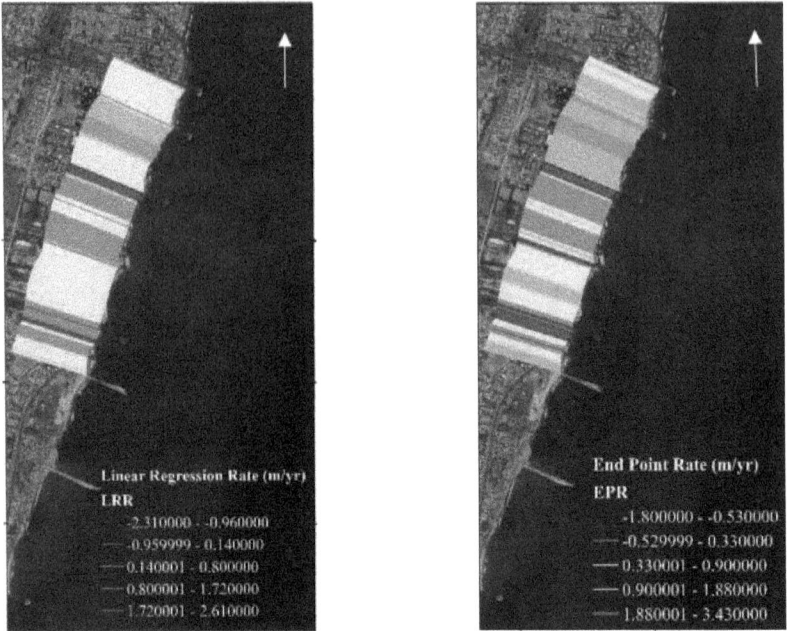

Fig. 7.17(b). Rate of shoreline changes in Segment I for southwest monsoon season.

coastal erosion. Shoreline change on a yearly scale is plotted in **Fig. 7.18.** The measured rate of accretion per year is exponentially greater than that of erosion using EPR and LRR for coastline data gathered between 2009 and 2020 throughout the non-monsoon season **(Fig. 7.19(a))** (i.e., January to May). Similarly, the observed rate of accretion every year is roughly twice as big as the rate of erosion using both EPR and LRR for shoreline data gathered between 2014 and 2020 for the northeast monsoon season **(Fig. 7.19(b))** (i.e., October to December).

Segment III

Segment III is a 1.5-km-long coastline stretch that runs between segments I and II. The coastal protection programme was delayed in this section until 2014, over a decade after the start of the Segment I groynes, because there was no large structure or habitation on the landward side and the national highway was slightly distant from the seashore. However, after the development of groyne fields in segments I and II, this 1.5-km length

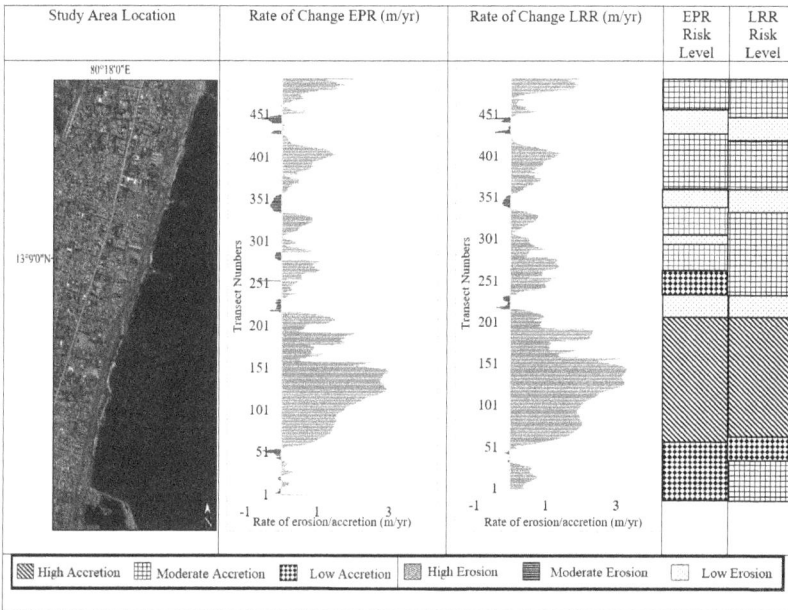

Fig. 7.18.   Annual rates of shoreline change in Segment II.
Source: Sundar *et al.* (2021).

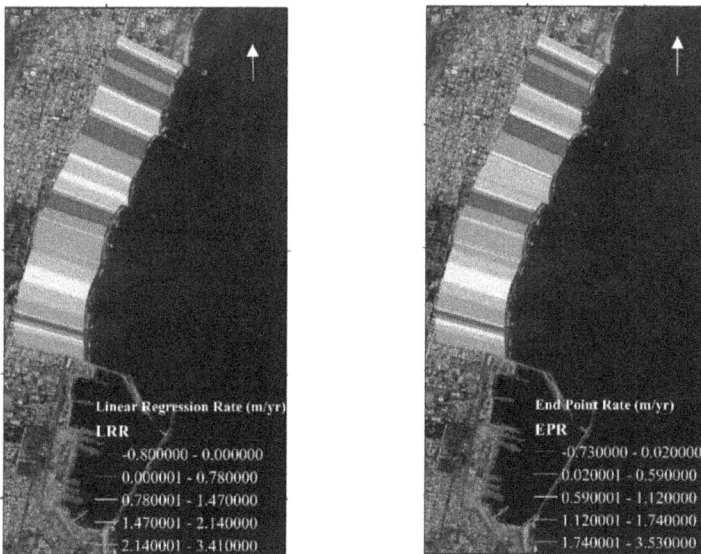

Fig. 7.19(a).   Rate of shoreline changes in Segment II for non-monsoon season.

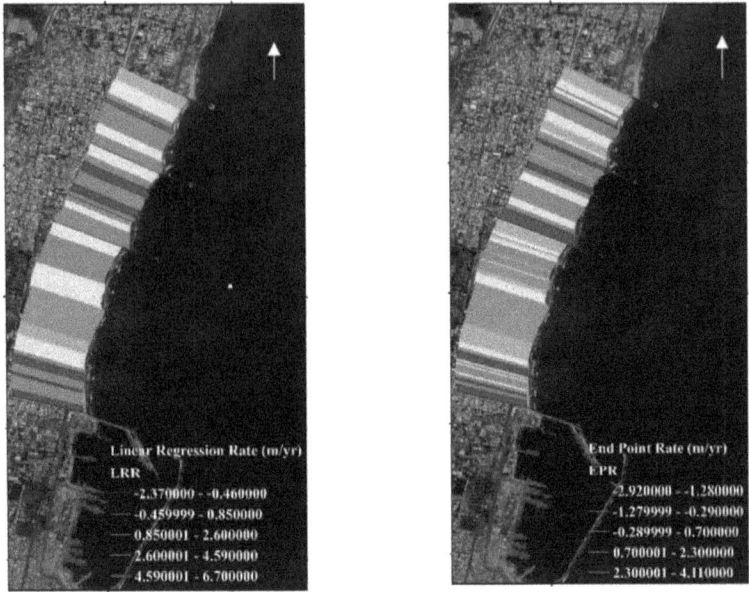

Fig. 7.19(b). Rate of shoreline changes in Segment II for northeast monsoon season.

had seen significant erosion **(Fig. 7.20)**. Three groynes were installed in 2014 to defend this segment III. For the shoreline data sampled between 2014 and 2020 for the non-monsoon season **(Fig. 7.21(a))** (i.e., January to May), the observed rate of accretion per year is significantly greater than that of erosion using EPR as well as LRR. The transects showing accretion and erosion trends are given in Fig. 7.10. Similarly, for the shoreline data sampled between 2014 and 2020 for the southwest monsoon season **(Fig. 7.21(b))** (i.e., June to September), the observed rate of accretion per year is over twice as greater than that of erosion using both EPR and LRR.

While there is a substantial amount of literature on the shoreline dynamics along the north Chennai coast that considers short-term or seasonal effects as well as coastal hazards such as cyclones, storm surges or tsunami, studies on the long-term effects of groyne fields as a coastal protection measure are scarce. As a result, the current research is focused on this critical topic, which is explained in full in this chapter. The study's seasonal variation demonstrates the dynamic nature of the coast, which varies throughout the year. Nonetheless, the development of transitional groynes has helped

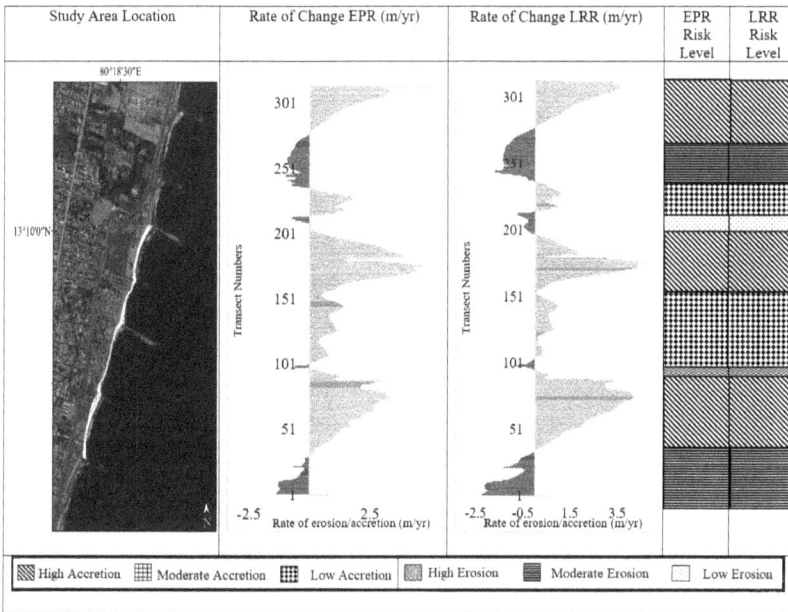

Fig. 7.20.   Annual rates of shoreline change in Segment III.
Source: Sundar *et al.* (2021).

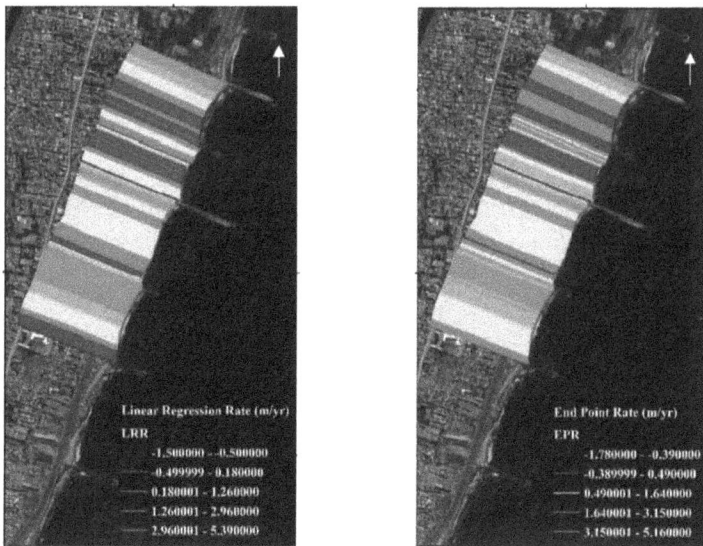

Fig. 7.21(a).   Rate of shoreline changes in Segment III for non-monsoon season.

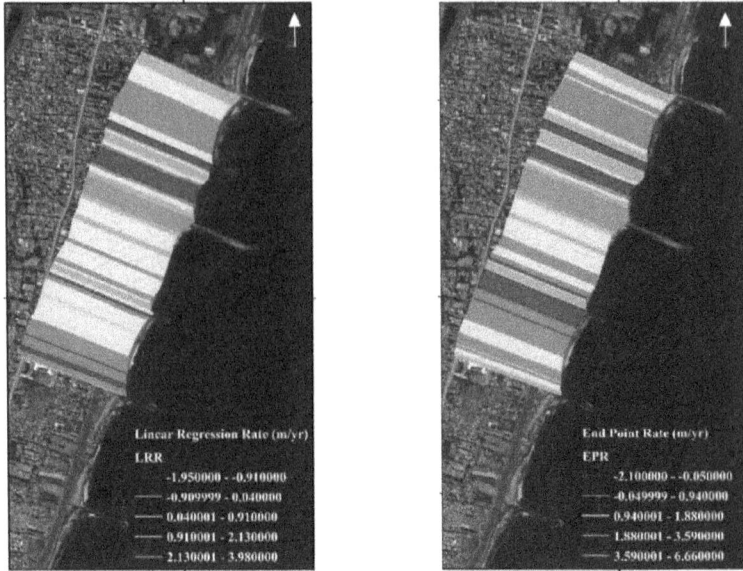

Fig. 7.21(b). Rate of shoreline changes in Segment III for southwest monsoon season.

to alleviate the persistent erosion problem by increasing beach width and reclaiming lost sediments.

7.7.4. *Historical shoreline analysis and field monitoring at Ennore coastal stretch along the southeast coast of India*

This study has been carried out by Dhananjayan *et al.* (2022) using DSAS for historical shoreline analysis and RTK-GPS for field monitoring.

7.7.4.1. *Study area*

The research region is on the east coast of India, in the northernmost portion of the Tamil Nadu coast. One of Chennai's key industrial hubs is the 12-km-long coastline stretch between Ennore creek (13° 14′01.0″N 80° 19′47.8″E) to Karungali village (13° 21′36.5″N 80° 20′23.1″E). It is 21 km north of Chennai's urban area. This region is a vital asset to the Chennai metropolitan area due to the presence of oil refineries, fertilizer industries, chemical factories, cement factories and some of the city's oldest thermal power plants. In 1999, a port was proposed in Ennore Creek to provide better water transport facilities to the neighbouring regions, in response to the region's industrial expansion. In order to assist the Ennore Port, another port at Kattupalli has been constructed. This study focuses on the shoreline changes due to anthropogenic activities and also aims to establish a database for future coastal engineers.

7.7.4.2. *Analysis of shorelines*

Different timelines were chosen for this analysis based on the introduction of coastal infrastructures. The first timeline spans the years 1991–1999, when the shore was mostly peaceful and devoid of any artificial activity. The second timeline spans the years 1999–2009, when the region underwent a transition that began with the construction of a satellite port at Ennore in 1999, which set the stage for the region's rapid industrial growth. The third timeline runs from 2009–2019, when the second timeline's efforts on the coast were extended. The shoreline change is studied separately during each phase, and a comprehensive analysis is conducted from 1991 to 2019 to identify a trend in the shoreline change.

7.7.4.3. *First timeline — 1991–1999*

Without any infrastructure construction along the coast from 1991 to 1999, the majority of the shoreline showed an accretion pattern. The coast near the Ennore Port location, where pre-construction work began in the 1990s, was one of the few areas that underwent erosion. The maximum erosion rate was around −11.1 m/yr, while the maximum accretion rate was around +9.9 m/yr in the Karungali settlement further north. **Figure 7.22** shows the LRR (m/yr) for this era. The coast to the north of the creek was accumulating slowly but steadily. With a total accretion distance of +87.3 m, the maximum NSM (in m) has been reported near Karungali village **(Table 7.2)**. With a landward coastline movement of −64.2 m, the maximum erosion (negative distance) was reported near Kattupalli village. During this time period, EPR's overall average reveals a +2.7 m/yr accretion.

7.7.4.4. *Second timeline — 1999–2009*

During this time period, the average rate of change of the shoreline was −4.1 m/yr. The average rate of erosion is −11.82 m/yr, with a high rate of

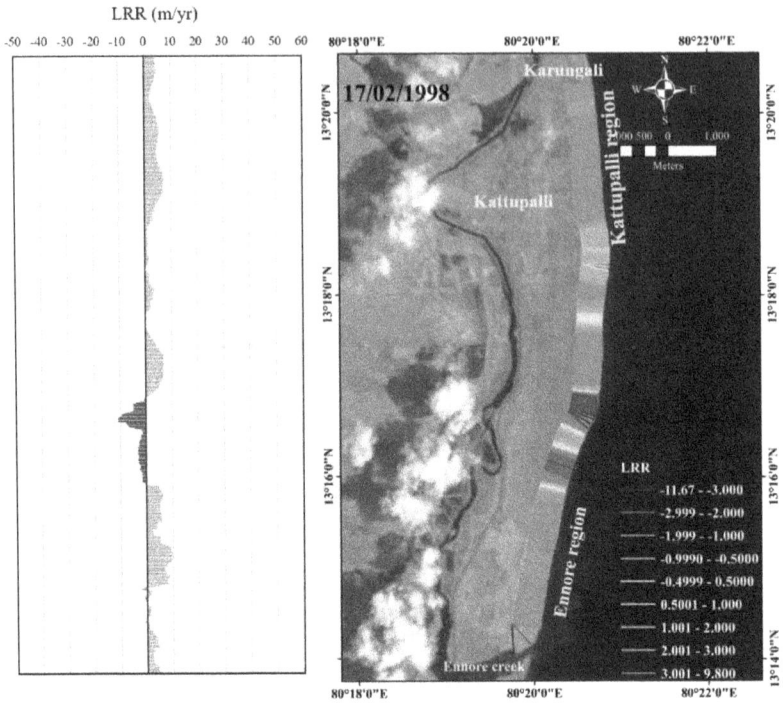

Fig. 7.22. LRR for the first timeline.

Source: Dhananjayan *et al.* (2022).

Table 7.2. Shoreline change rates comparison for the different timelines.

	Timeline		
Properties	1991–1999	1999–2009	2009–2019
Total transects	587	493	348
Mean shoreline change (m/yr)	+2.97	−4.1	−2.37
Mean erosion rate (m/yr)	−2.82	−11.02	−11.34
Mean accretion rate (m/yr)	+4.26	+5.82	+4.97
Maximum shoreline erosion (m/yr)	−11.10	−43.15	−32.79
Maximum shoreline accretion (m/yr)	+9.82	+53.92	+47.3
Erosional transects	143	286	128
Accretional transects	444	200	219
Overall trend	Accretion	Erosion	Erosion

−43.15 m/yr at Ennore. The average accretion rate is +5.82 m/yr, with a maximum rate of +53.92 m/yr seen to the south of the Ennore Port. The rate of change in the shoreline in the Ennore region is +7.77 meters per year, with a maximum erosion rate of −8.77 m/yr and a maximum accretion rate of +53.92 m/yr. The average rate of shoreline change in the Kattupalli region is −4.00 m/yr. The highest rate of erosion is −9.7 m/yr, whereas the maximum rate of accretion is +2.5 m/yr. The average NSM throughout this time period is −19.73 m, with the highest positive or accretion of +434.4 m to the south of the Ennore Port **(Fig. 7.23)** and the highest negative or erosion of −338.4 m to the north. The average rate of coastal movement in the Ennore and Kattupalli regions over this period is +81.38 m and −46.77 m, respectively.

7.7.4.5.   *Third timeline — 2009–2019*

Erosion was recorded across a longer stretch of the coast during this time. Erosion transects accounted for 57% of all transects, while accretional

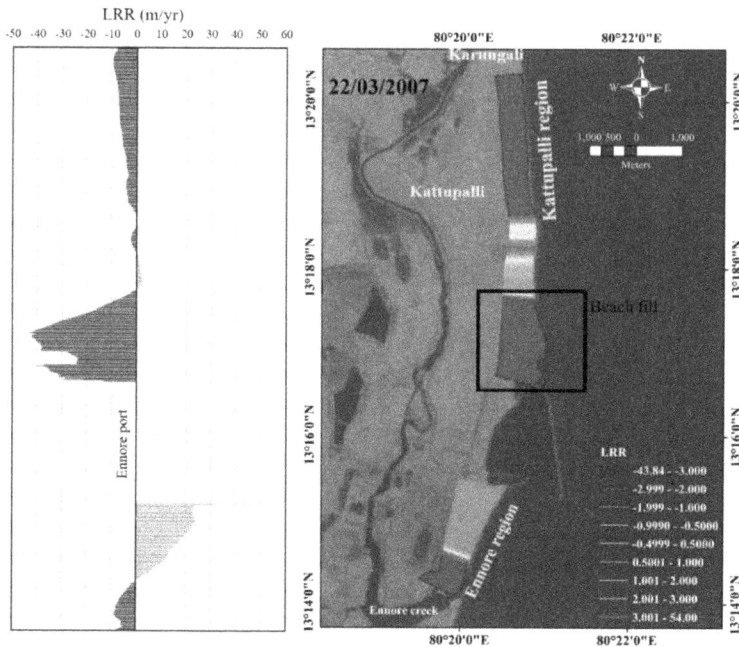

Fig. 7.23.   LRR for the second timeline.
Source: Dhananjayan *et al.* (2022).

transects accounted for 40% (**Table 7.2**). A 2,000-m-long coastline section between the two ports has been under accretion since 2011, compared to erosion throughout the prior era. Since 2011, the rate of accretion between the two ports has been +1.5 million per year. This section of coastline is located updrift from Kattupalli port. The shoreline in Ennore (updrift) began to accrete at a pace of +5.5 m/yr, whereas the coast in Kattupalli (downdrift) began to erode at a rate of −16.4 m/yr. The rate of change of shoreline along the coast is depicted in **Fig. 7.24**. During this time period, the NSM is −19.78 m. The largest accretion can be found in the Ennore region (updrift) at −316.67 m, while the maximum negative or erosion can be found in the Kattupalli region (downdrift) at −301.65 m.

7.7.4.6.   *Overall timeline — 1991 to 2019*

The Kattupalli region is undergoing serious erosion, according to trends observed during the investigation. The statistics in this timeline are shown

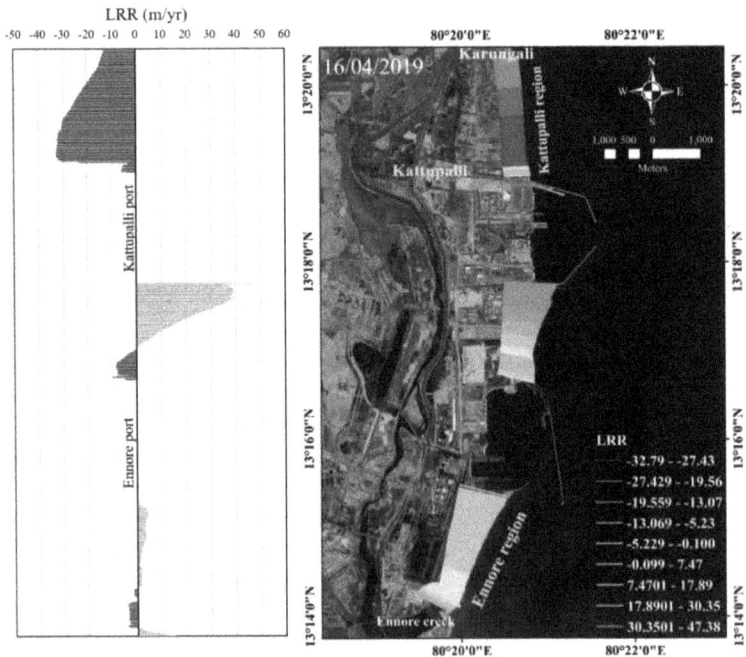

Fig. 7.24.   LRR for the third timeline.
Source: Dhananjayan *et al.* (2022).

Table 7.3.   Shoreline change rates properties from 1991 to 2019.

Properties	Timeline: 1991–2019
Total transects	587
Mean Shoreline Change (m/yr)	+2.18
Mean Erosion rate (m/yr)	−5.88
Mean accretion rate (m/yr)	+6.46
Maximum Shoreline erosion (m/yr)	−11.82
Maximum Shoreline accretion (m/yr)	+26.74
Erosional transects	206
Accretional transects	381
Overall trend	Accretion

in **Table 7.3**. The erosion rate in this area is −16.37 m/yr. Based on the constructions along the coast, the coastline stretch between the two ports has acted differently. For a portion of the shore, the overall trend has been accretion. This coast's average rate of change is +7.6 m/yr, with a maximum rate of change of +17.6 m/yr and a minimum rate of change of −0.56 m/yr. While the beach in the Ennore region has the highest accretion rate in the entire region, erosion is much greater than accretion. **Figure 7.25** shows the LRR plot for the shoreline changes. The NSM value for the entire coast at this moment is +55.71 m. All erosional transects have a mean of −142.88 m, while all accretional transects have a mean of +176.4 m. With an NSM of −308.98 m, the coast in the Kattupalli region experienced the most erosion. The Ennore region has the most positive or accretion, with an NSM of +496.72 m.

These studies go into greater detail about the usage of manual shoreline extraction and the analysis that goes along with it in the DSAS tool. The investigation was carried out in a variety of zones, segments and timelines. This type of analysis can aid in the creation of a database that coastal engineers can use to plan future actions, and the manual shoreline extraction method aids in the comprehension of the studied area's features. However, this type of shoreline analysis has a few drawbacks as well. Because this is manually mapped, a consistent pixel must be used throughout the study, and the land–water interface or wet/dry boundary may not be apparent in satellite images. Due to these constraints, uncertainties must be defined before to conducting the study and must be provided as input before statistics can be calculated.

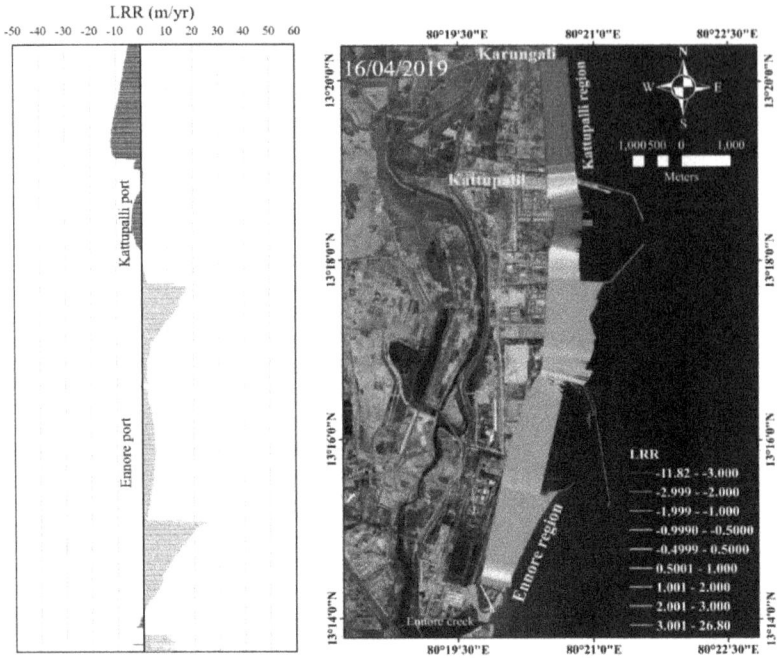

Fig. 7.25. LRR for the overall timeline.
Source: Dhananjayan *et al.* (2022).

7.8.   Automatic Shoreline Extraction

Automatic shoreline extraction in satellite images helps a great deal in identifying the shoreline with utmost accuracy. In order to extract shorelines automatically, various techniques are available, of which a study by Hossain *et al.* (2021) is discussed below.

7.8.1.   *Indices used in automatic shoreline extraction*

Various indices exist depending on the vegetation cover and water extent. Few of the generally used indices are discussed as follows.

7.8.1.1.   *Normalized Difference Water Index*

The Normalized Difference Water Index (NDWI) (Gao, 1996) is a satellite-derived indicator based on the Near-Infrared (NIR) and Short-Wave Infrared (SWIR) channels. The SWIR reflectance in vegetation cover reflects changes in both vegetation water content and porous xylem structure, whereas the NIR reflectance is influenced by leaf internal structure and leaf dry matter content but not by water content. The use of the NIR in conjunction with the SWIR eliminates changes caused by leaf internal structure and leaf dry matter content, enhancing the accuracy of vegetation water content retrieval.

$$\text{NDWI} = \frac{Green_{BOA} - NIR_{BOA}}{Green_{BOA} + NIR_{BOA}} \tag{7.1}$$

where $Green_{BOA}$ and NIR_{BOA} represent reflectance in the green and near infrared bands, respectively.

7.8.1.2. Normalized Difference Vegetation Index

In order to quantify vegetation, the Normalized Difference Vegetation Index (NDVI) evaluates the difference between NIR (which vegetation strongly reflects) and red light (which vegetation absorbs). The NDVI value is in the range between -1 and $+1$. Each and every type of land cover does not have its own specific limiting value. The negative readings on the NDVI certainly represent the presence of water, and if the NDVI value is near $+1$, it can be inferred as dense vegetation. However, when the NDVI is near 0, there are no green vegetation and the region may be urbanized.

$$\text{NDVI} = \frac{NIR - Red}{NIR + Red} \tag{7.2}$$

whereas, Red and NIR stand for the spectral reflectance measurements acquired in the red (visible) and NIR regions, respectively.

7.8.1.3. Modified Normalized Difference Water Index

This index is intended to optimize water reflectance by using green wavelengths, decrease low NIR reflectance by water features and take advantage of high NIR reflectance by vegetation and soil characteristics (Xu, 2006). As a result, water features typically have positive values and are thus amplified, whereas vegetation and soil typically have zero or negative values and are thus repressed. However, in water zones with a built-up land background, the NDWI does not achieve its aim as planned. In some areas, the retrieved water data was frequently mingled with land noise. As a result, many built-up land features in the NDWI image exhibit positive values. However, a close study of the figure's signatures indicates TM band 5, which represents SWIR radiation and has a substantially higher average digital number than TM band 2 (Green band). As a result, if NDWI uses a SWIR band instead of the NIR band, the built-up land should have negative values. The NDWI is changed based on this assumption by substituting the SWIR band for the NIR band. The MNDWI (modified NDWI) can be written as follows:

$$\text{MNDWI} = \frac{Green - SWIR}{Green + SWIR} \qquad (7.3)$$

7.8.2. *Automatic shoreline extraction and change detection: A study on the southeast coast of Bangladesh*

This study uses a 10-year interval to automatically extract the shoreline and changes in coastal location owing to accretion and erosion along Bangladesh's southeast coast from 1980 to 2020. For autonomous shoreline extraction, the generic edge detection algorithms Threshold, Sobel, Prewitt, Canny and Robert are utilized, including Canny's ability to accurately detect the coastline being particularly impressive. To statistically measure shoreline changes, the DSAS employs NSM, EPR, and LRR. The flowchart of the methodology is shown in **Fig. 7.26**.

7.8.2.1. *Methodology*

Remotely sensed indices (e.g., NDWI) are often employed for coastline extraction. The NDWI was calculated using multi-temporal LANDSAT images (MSS, TM/ETM+, and OLI sensors) in this work. The NDWI index is a useful tool for detecting water bodies on the surface of the planet. As indicated in the equation 7.1, the NDWI was determined using the green and near-infrared bands. NDWI is explained in Section 7.8.1.1. The value of the NDWI ranges from -1 to $+1$. In most cases, the NDWI image yields a positive result for the water feature and a negative result for the non-water feature (McFeeters, 1996). The non-water and water objects were separated by assigning them 0 and 1 (i.e., following the binary image classification) to demarcate the line of separation as a shoreline (Ji *et al.*, 2009).

7.8.2.2. *Edge detection*

In this study, canny edge detection is used to detect the shorelines. In coastal investigations, canny edge detection is regarded as an accurate and effective technique for extracting the border between water and non-water

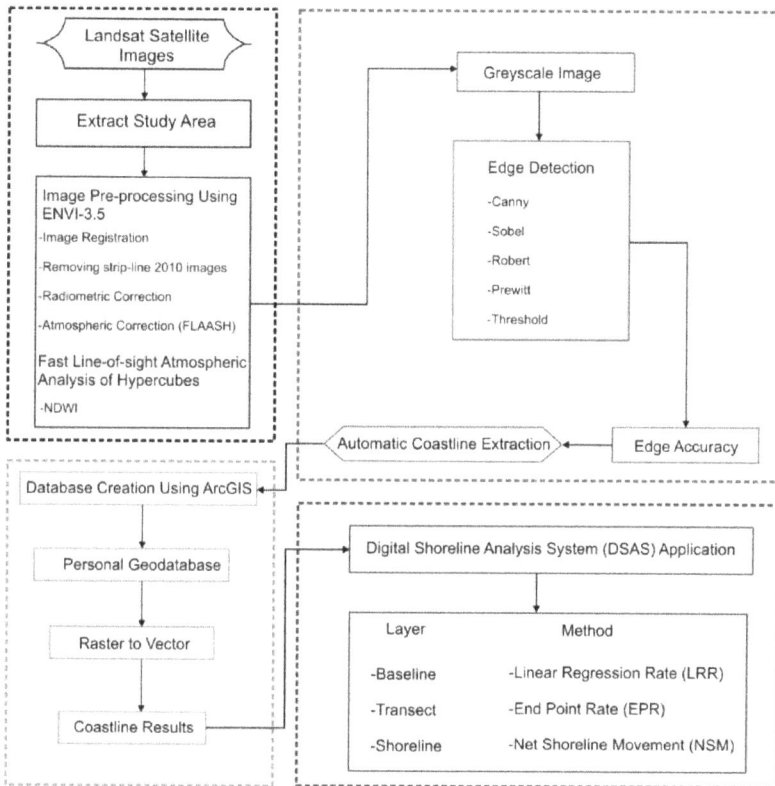

Fig. 7.26. Flowchart of methodology adopted by Hussain *et al.* (2021).

(Hossain *et al.*, 2021). The advantage of canny edge detection over other edge detection approaches (such as Robert, Prewitt edge detection, Sobel edge detection and threshold techniques) has been noted by a number of researchers. Furthermore, after a few repetitions, the canny edge detection process accurately demarcates the shoreline in a short amount of time (Hossain *et al.*, 2021).

7.8.2.3. *Analysis of Shorelines*

The analysis has been carried out in short-term and long-term basis from 1980 to 2020. Long-term basis for 40 years from 1980 to 2020 and short-term on a decadal scale ranging from 1980–1990; 1990–2000; 2000–2010; 2010–2020.

Analysis on long-term basis (1980–2020)

Between 1980 and 2020, the coastal movements were assessed using the LRR approach (40 years). By matching the least square regression at each of the transects to all of the coastline locations from the oldest to the newest, this programme provides the rate of change statistics. Positive numbers indicate coastal accretion, whereas negative values indicate coastal erosion (**Fig. 7.27**). The LRR cumulative average rate suggests an accretional tendency (4.39 m/yr) for the coastline over a 40-year period, indicating that Bangladesh's southeast shoreline is predominantly vulnerable to accretion. The greatest accretion distance is 162.91 m/yr, while the maximum erosion distance is 45.91 m/yr.

Shoreline changes (1980–1990)

The rates of coastal position fluctuations derived by the NSM and EPR methods for this time period (1980–1990) suggest that the coastline is largely exposed to accretion and minor erosion, respectively (**Fig. 7.28**). The total average EPR rates show a 2.33 m/yr erosion trend, whereas the NSM distance data show the reverse pattern with a 23.48 m accretion rate. The EPR maximum accretion rates are 469.58 m/yr, while the NSM maximum accretion distances are 4689.4 m. The maximum erosion distance for NSM is 2255.1 m, and the maximum erosion rate for EPR is 225.85 m/yr. This result demonstrates that the southeast coast of Bangladesh is affected

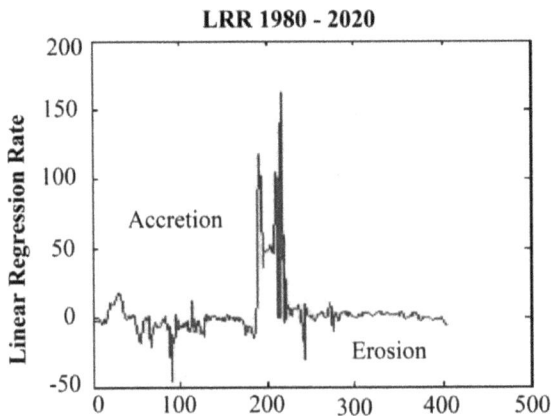

Fig. 7.27.   Long-term shoreline change.

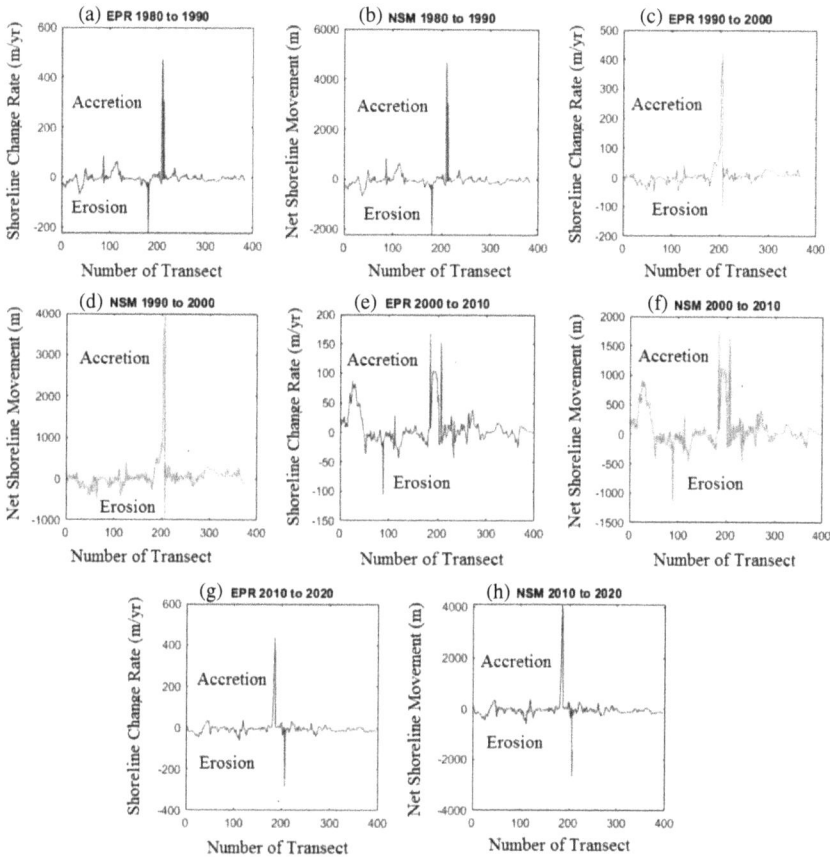

Fig. 7.28.   Short-term shoreline analysis with EPR and NSM.

by both erosion and accretion processes, though the pace of erosion is far slower.

Shoreline changes (1990–2000)

The rates of change in the placement of the coastline over this time, as estimated by the EPR and NSM approaches, indicate that the coastline is largely prone to accretion **(Fig. 7.28)**. The total average EPR rates are increasing at 7.06 m/yr, and the NSM distance values are increasing at 65.5 m. The greatest accretion distance (NSM) is 3950.8 m, with a maximum accretion EPR of 425.9 m/yr, whereas the highest erosion distance (NSM) is 996.95 m, with a maximum erosion EPR of 107.6 m/yr. During this second phase of detection, the shoreline appears to be more accretive and less erodible. In comparison to the prior observation period, this period's average accretion rate is higher than that of 1980–1990.

Shoreline changes (2000–2010)

The main recurring trend is still the same in the third coastal change observation period, which runs from 2000 to 2010. In the field of research, the incidence of accretion still holds sway. The NSM distance values, which are 79.03 m, follow the same patterns, and the overall average EPR rates, which are 7.33 m/yr, show an accretional trend. The greatest accretion distance (NSM) is 1807.68 m, with a maximum accretion EPR of 167.77 m/yr, while the maximum erosion distance (NSM) is 1144.38 m with an EPR of 106.27 m/yr.

Shoreline changes (2010–2020)

The total average EPR rates in the research region reveal a 2.92 m/yr erosion trend, while the NSM distance data suggest a 27.48 m accretion trend. The greatest accretion distance (NSM) is 4126.72 m, with a maximum accretion rate of 439.12 m/yr. The maximum erosion distance (NSM) is 2673.64 m, while the annual erosion rate (EPR) is 284.49 m/yr. This research indicates that when erosion is less, accretion has a greater impact on the coastline. Furthermore, compared to past years, the accretion and erosion that occurred at this period are considerably higher. Numerous changes in land use that occur in the study region cause this phenomena.

7.9.   Semi-automatic Shoreline Extraction

The semi-automatic method of shoreline extraction involves band ratio techniques, where the images are colour-specified and the shorelines are demarcated manually. A study conducted by Sutikno *et al.* (2016) in the Rokan Estuary describes the semi-automatic shoreline extraction method for shoreline analysis.

7.9.1.   *Integrated remote sensing and GIS for calculating shoreline change in Rokan Estuary*

7.9.1.1.   *Methodology*

Cropping the image, image enhancement, geometric correction, digitization, and overlapping are all steps in the LANDSAT data analysis and interpretation process. Cropping the photograph was done to reduce storage space in the computer memory by removing the research focal area. Image enhancement was a set of bands used to make the line between land and sea more distinct, making the process of digitizing the coastline easier. The use of band 2, 4, and 5 to directly separate water and land is proposed for LANDSAT TM and ETM+ using a mix of histogram threshold and band ratio approaches. Another option is to extract the water–land interface using single-band thresholds such as LANDSAT TM and ETM+ band 5 or band 7 because they are suitable in clear water circumstances. The LANDSAT image data were obtained in GeoTIFF format and were modified to remove the geometric correction. Raster images would be converted to vector images after categorization. The final step is to export the data into shape file format, which can then be processed in a GIS application. Image processing techniques such as cropping, augmentation and correction were used on seven years of LANDSAT data records. The boundary between land and sea becomes clearer when bands 542 are combined, making it easier to scan the shoreline. The historical shorelines data of the Rokan Estuary for the last 14 years can be derived from LANDSAT using image processing and digitization. **Figure 7.29** shows the shoreline data that was extracted from LANDSAT-7 at Rokan Estuary.

7.9.1.2. *Error and uncertainty*

The quantification of errors and uncertainties have been estimated for a long time, and therefore, greater emphasis has been placed on the standards and

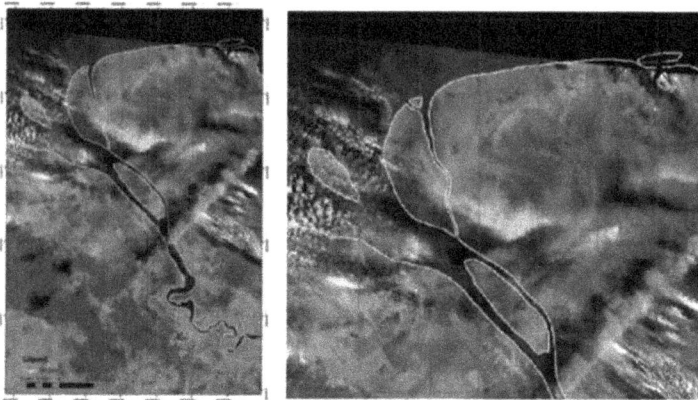

Fig. 7.29.   Rokan Estuary shoreline extracted from LANDSAT – 7 2002 from band 547.

scale of the survey for shoreline monitoring. Improvements in technology and the expanding spectrum of survey methods have definitely resulted in significant reductions in measurement uncertainty. However, when investigating shoreline change over the medium term, that is, on a decadal year scale, the complete range of errors inherent in the measurement and method of shoreline change analysis must be considered. The uncertainty in (1) the original data; (2) the interpretation and derivation of the coastline from various surveys and (3) the statistical models utilized determine the precision of shoreline change measures. The procedures for determining shoreline location are all computed with uncertainty, whether the shoreline position is predefined (i.e., delineated on a map), manually derived (i.e., interpreted visually), numerically extracted (i.e., contour derivation) or automatically delimited through image or data segmentation processes (Jackson and Short, 2020).

Although historical maps may provide a good indication of shoreline position, they are subject to several difficulties due to early geographic projections and the distortions and defects caused by uneven paper shrinkage, splits and creases (Crowell et al., 1991; Underwood and Anders, 1991). Where only printed copies are accessible, older aerial images might suffer from these latter concerns, as well as camera distortions, obliquity (tilt) and relief displacement, all of which lead to relative, within-photo scale and displacement issues (Crowell et al., 1991; Moore, 2000). Where the lens parameters are known or have been calibrated, camera lens distortions can be accounted for. The scale of the photography and the level of observable detail will be determined by the flight elevation. The pixel resolution (i.e., the size of the lowest information unit) of both scanned and digitally acquired aerial images maintains the base level of accuracy of the photograph; the apparent detail of satellite data is similarly controlled by pixel resolution. When maps and pictures require geo-rectification, additional uncertainties and concerns with geographic alignment might arise, which can be evaluated using the root-mean-square error (RMSE) between real and geo-referenced GCP coordinates. Interpretation and digitization introduce new uncertainty. This is influenced to some part by resolution in aerial photography, but it will surely result from the difficulty in interpreting ground features. Although sandy coastal habitats can be easily identified in black-and-white aerial pictures, more distinct characteristics like drift lines are more difficult to spot (Ekebom and Erkkilä, 2003). More information is available with colour photography and, more recently, infrared aerial imagery, allowing for a more exact delineation of coastal features. Uncertainties associated with LiDAR data include some vertical positioning inaccuracy (of the order of 10–15 cm), which can translate to horizontal inaccuracies of up to 3 m when a specific contour (e.g., MHW) is derived (White *et al.*, 2011).

The process for measuring and quantifying uncertainty in estimated shoreline positions is described and defined in several works (Crowell *et al.*, 1991). The square root of the sum of squares of errors associated with each uncertainty source is used to calculate the combined or total inaccuracy in shoreline position. Ruggiero *et al.* (2013) define the total shoreline (observation) position uncertainty based on a study by Hapke *et al.* (2006).

$$U_t^2 = U_g^2 + U_d^2 + U_a^2 + U_w^2 \tag{7.4}$$

where U_g is the geo-referencing error, U_d is the digitizing error, U_a is pixel uncertainty and U_w refers to high-water-line uncertainty. Different data types and shoreline positions derived might therefore accumulate a total uncertainty from a different number and different set of source uncertainties. Estimation of the uncertainties in calculated shoreline trends is specific to the method used to derive the rate of change. For example, the uncertainty in EPR (EPR$_{uc}$) is a simple square root of the summed of uncertainties from the start (U_{ti}) and end (U_{tj}) shorelines divided by the time between ($Y_s - Y_e$).

$$EPR = \frac{\sqrt{U_{ti}^2 + U_{te}^2}}{Ys - Ye} \tag{7.5}$$

when a linear model (e.g., LRR or WLR) is used to determine the trend, the regression statistics can be used to define the regression's strength (R^2), significance (p value) of the regression slope (i.e. rate), and confidence bounds of the regression coefficients **(Fig. 7.30)**. The prediction interval can be used to frame the uncertainty of the prediction when the shoreline trends produced are subsequently used to predict a change at a different point in time (Jackson and Short, 2020) **(Fig. 7.30)**. Beyond the more regular digitizing and comparison of shoreline positions, it is obvious that evaluating uncertainty can be a considerable work in and of itself. It is, nevertheless, necessary for the procedure to acquire confidence in the amount of change and identified trends patterns.

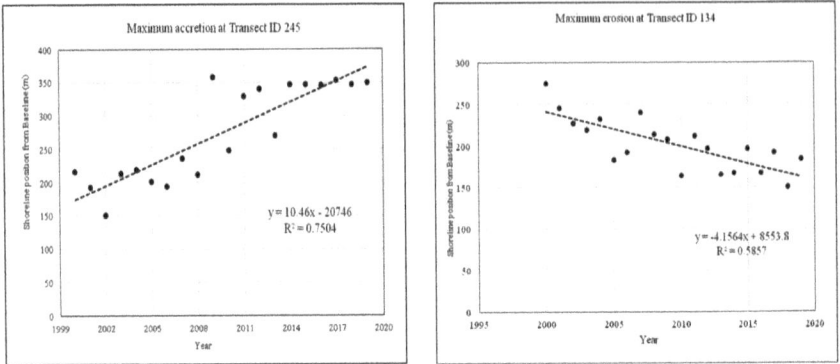

Fig. 7.30. R2 for accretion and erosion transects.
Source: Gracy *et al.* (2020).

7.10. Summary

This chapter briefly explains the various data sources for shoreline mapping and the various indices used in shoreline analysis. In addition, the statistical methods for rate of change calculation in DSAS have been explained. The different methods in shoreline extraction has been explained with a case study. For manual shoreline extraction, the case study of Chennai coast has been explained; for automatic shoreline detection, the case of Bangladesh has been presented while Rokan Estuary has been explained for semi-automatic shoreline delineation. The errors and the associated uncertainties in mapping of shoreline is discussed, along with a formula for total uncertainty calculation.

References

Boak, E.H. and Turner, I.L. (2005). Shoreline definition and detection: A Review. *Journal of Coastal Research*, 21(4): 688–703.

Crowell, M., Leatherman, S.P. and Buckley, M.K. (1991). Historical shoreline change: Error analysis and mapping accuracy. *Journal of Coastal Research*, 7(3): 839–852.

Dhananjayan M., Vasanthakumar S., Sannasiraj S. A. and Murali K. (2022). Historical shoreline analysis and field monitoring at Ennore coastal stretch along the southeast coast of India, *Marine Geodesy*, 45:1, 47–74.

Dolan, R., Hayden, B.P., May, P. and May, S.K. (1980). The reliability of shoreline change measurements from aerial photographs. *Shore and Beach*, 48(4): 22–29.

Ekebom, J. and Erkkilä, A. (2003). Using aerial photography for identification of marine and coastal habitats under the EU's Habitats Directive. *Aquatic Conservation: Marine and Freshwater Ecosystems*, 13(4): 287–304.

Fairley, I., Thomas, T., Phillips, M. and Reeve, D. (2016). Terrestrial Laser Scanner Techniques for Enhancement in Understanding of Coastal Environments. In: Finkl, C., Makowski, C. (Eds.) *Seafloor Mapping Along Continental Shelves*. Coastal Research Library, Vol. 13. Springer, Cham.

Gao, B.C. (1996). NDWI—a normalized difference water index for remote sensing of vegetation liquid water from space. *Remote Sensing of Environment*, 58: 257–266.

Gracy, R.M.M., Sundar, V. and Sannasiraj, S.A. (2020). Analysis of shoreline change between inlets along the coast of Chennai, India. *Marine Georesources & Geotechnology*, 40: 26–35.

Hapke, C.J., Reid, D., Richmond, B.M., Ruggiero, P. and List, J. (2006). National assessment of shoreline change. Part 3: Historical shoreline changes and associated coastal land loss along the sandy shorelines of the California coast. US Geological Survey Open File Report 2006–1219, 72 p.

Himmelstoss, E.A., Henderson, R.E., Kratzmann, M.G. and Farris, A.S. (2018). Digital Shoreline Analysis System (DSAS) version 5.0 user guide: U.S. Geological Survey Open-File Report 2018–1179, 110 p.

Himmelstoss, E.A., Henderson, R.E., Kratzmann, M.G. and Farris, A.S. (2021). Digital Shoreline Analysis System (DSAS) version 5.1 user guide: U.S. Geological Survey Open-File Report 2021–1091, 104 p.

Hossain, M.S., Yasir, M., Wang, P., Ullah, S., Jahan, M., Hui, S. and Zhao, Z. (2021). Automatic shoreline extraction and change detection: A study on the southeast coast of Bangladesh. *Marine Geology*, 441: 106628.

Jackson, D. and Short, A. (2020). *Sandy Beach Morphodynamics*. Elsevier Gezondheidszorg.

Ji, L., Zhang, L. and Wylie, B. (2009). Analysis of dynamic thresholds for the normalized difference water index, *Photogrammetric Engineering and Remote Sensing*, 75(11): 1307–1317.

Kankara, R.S., Chenthamil Selvan, S., Markose, V. J., Rajan, B. and Arockiaraj, S. (2015). Estimation of long and short term shoreline changes along Andhra Pradesh Coast using remote sensing and GIS techniques. *Procedia Engineering*, 116, 855–862.

Komar, P.D. (1998). *Beach Processes and Sedimentation*. Upper Saddle River, New Jersey: Prentice Hall Inc., 544 pp.

McFeeters, S.K. (1996). The use of the Normalized Difference Water Index (NDWI) in the Delineation of Open Water Features. *International Journal of Remote Sensing*, 17(7): 1425–1432.

Moore, L.J. (2000). Shoreline mapping techniques. *Journal of Coastal Research*, 16(1), 111–124.

Morton, R.A. (1991). Accurate shoreline mapping: past, present, and future. *Proceedings of the Coastal Sediments '91* (Seattle, Washington), pp. 997–1010.

Ruggiero, P., Kratzmann, M.G., Himmelstoss, E.A., Reid, D.A.J. and Kaminsky, G. (2013). National assessment of shoreline change: Historical shoreline change along the Pacific Northwest coast. US Geological Survey Open-File Report 2012-1007, 62 p.

Su, L. and Gibeaut, J. (2017). Using UAS hyperspatial RGB imagery for identifying beach zones along the south Texas coast. *Remote Sensing*, 9(2):159.

Sundar, V., Sannasiraj, S.A. and Ramesh Babu, S. (2021). Shoreline changes due to construction of groyne field in north of Chennai Port, India. *Environment Monitoring Assessment*, 193: 830.

Sundar, V., Sannasiraj, S.A. and Ramesh Babu, S. (2022). Sustainable hard and soft measures for coastal protection — Case studies along the Indian Coast. *Marine Georesources & Geotechnology*, 40(5), 600–615.

Sutikno, S., Fatnanta, F., Kusnadi, A. and Murakami, K. (2016). Integrated Remote Sensing and GIS for Calculating Shoreline Change in Rokan Estuary, KnE Engineering, 9 p.

Tyagi, S. and Rai S.C. (2020). Monitoring shoreline changes along Andhra coast of India using remote sensing and geographic information system. *Indian Journal of Geo Marine Sciences*, 49(02), 218–224.

Van der Werff, H.M.A. (2019). Mapping shoreline indicators on a sandy beach with supervised edge detection of soil moisture differences. *International Journal of Applied Earth Observation and Geoinformation*, 74, 231–238.

Underwood, S.G. and Anders, F.J. (1991). Evaluation of the coastal features mapping system for shoreline mapping (No. CERC-91-13). Coastal Engineering Research Center Vicksburg MS, 49 p.

White, S.A., Parrish, C.E., Calder, B.R., Pe'eri, S. and Rzhanov, Y. (2011). LiDAR-derived national shoreline: Empirical and stochastic uncertainty analyses. *Journal of Coastal Research*, SI62, 62–74.

Xu, H. (2006). Modification of normalized difference water index (NDWI) to enhance open water features in remotely sensed imagery. *International Journal of Remote Sensing*, 27(14), 3025–3033.

Chapter 8

Beach Profile Changes Near the Confluence of Estuary and Ocean

Abstract
Beaches undergo gradual changes continuously. The changes over an open beach differ from that over a beach adjoining natural or manmade obstructions. The beach profile changes are of importance in the understanding the coastal processes. This chapter presents the beach profile changes in an open coast and near an estuary quantified through field investigations over a couple of years. The changes in the beach profiles are compared temporally and spatially. The data has been collected on the first spring tide of every month using Real Time Kinematic Global Positioning System (RTK-GPS) from January 2017 to December 2018. The accuracy of data collected using RTK-GPS is 8–15 mm. The study areas considered an open coast in Devaneri, Tamil Nadu, and near an estuary with its mouth being trained by a pair of training walls in Karaikal, Puducherry. Both sites are situated along the southeast coast of Indian peninsula and 300 km apart. The open coast has remained constant with minor changes in the variations in its profile, whereas the one near an estuary undergo major changes seasonally and spatially. It is observed that the presence of coastal structures dictate the erosion/accretion process. Irrespective of the type of coast, northeast monsoon brings forth significant changes along the coast.

8.1. Introduction

The beach profile changes are one of the most studied features in understanding the morphology of a coast, prior to the implementation of any coastal developmental activities. Hence, it is essential to regularly monitor its spatial and temporal evolution. Although several models have been developed to estimate the changes in the beach profiles, the field data helps in cross-referencing and benchmarking the models. An open coast is primarily dominated by the action of waves and other natural factors. Erosion and/or accretion processes along the coast might not follow a particular trend or the drift pattern for certain stretches of the coast, whereas in a few other stretches, the beach may remain stable with just minor changes being observed. The profile shapes of estuarine beaches formed by non–storm-dominant factors such as local wave generation, beach orientation, beach width and slope and sediment grain size (Nordstrom and Jackson, 2013). The presence of hard protection structures dictates the erosion/accretion pattern along a coast (Kannan *et al.*, 2014; Jeyagopal *et al.*, 2020). Beach profiles with nearshore bathymetry are an estimate of erosion/accretion patterns more for artificially nourished beaches in the vicinity of coastal structures (Ananth and Sundar, 1990). The beaches undergo changes as per the seasonal variations. When severe storms or any other extreme coastal hazards impact the coast, the sediments from beaches are usually removed and deposited as an offshore bar. Once the storm recedes, the offshore bar is driven back to its original location by the waves. Continuous monitoring of a shoreline provides great hindsight into its changes, which are sometimes rapid. The data collected from the field is a valuable resource in understanding the behaviour of the beach through quantifying accretion/erosion. The rate of erosion/accretion, due to the normal and extreme weather events and the response of a shoreline to engineering structures can be estimated through regular field surveys, which also serve as ground truth for the validation of numerical models. With a more intense and finer spatial temporal scales, localized impacts on the coast can be easily quantified.

The placing and spacing of profiles along the coast play an important role in the assessment of the coastal processes. In an open and plain coast, profiles can be spaced uniformly and placed with a maximum spacing of 150 m (Timothy *et al.*, 1994). A coast with engineering structures and presence of dunes and vegetation cover needs proper attention while planning the profiles. The ideal time for beach profile data collection in field is during the spring tide, since it facilitates ease of measurement across the maximum possible beach width. The impact of the training wall on either side shoreline varies from 1:1 to 1:10, depending on the perpendicular distance of the structure (Kudale, 2010). Any structure normal to the shoreline along a littoral-drift–dominated coast results in erosion on its down-drift side and accretion on its up-drift side. Along the east coast of India, drift direction varies bi-directionally with season and predominant drift towards north (Suresh *et al.*, 2011; Kunte *et al.*, 2013), hence the shoreline on either side of the training walls needs to be monitored on temporal and spatial scales.

8.2.   Beach Profiling Techniques

The beach profile can be obtained with various types of equipment such as a simple graduated rods and chains (Emery, 1961), standard stadia rod and level or a more accurate auto-tracking geodimeter with a reflecting prism (Birkemeier *et al.*, 1991). Klemas (2011) has presented an overview on the application of LiDAR techniques for topographic and bathymetry survey through three case studies. The standard beach profile survey provides a one-dimensional profile variation representing the difference in relative height between a fixed benchmark and beach level. This exercise enables us to identify the location of beach features such as shoreline, berm, dunes, vegetation cover, etc. It is recommended that each of the measured profile is placed over a permanent structure, such that the original profile can be recreated for the consecutive surveys. Repetitive survey of the same profile must be carried out to achieve a better accuracy, since a slight deviation could lead to misinterpretation of results. The inherent problem with beach profile survey is the assumption of linearity between survey points — this necessitates the surveyor to carefully include all vital features across the beach section. An awareness of these limitations ensures that the measurements are taken in the field to eliminate or reduce factors that are liable to cause potential errors in surveying and in the interpretation of resulting data (Thomas *et al.*, 2018). A comparison of successive beach surveys yields a two-dimensional cross-sectional area quantifying the extent of beach erosion or accretion, whereas a three-dimensional volumetric sedimentation change is obtained by integrating between adjacent cross-sectional areas (Morton *et al.*, 1993). The beach profile survey with a total station, that works on the electronic distance measuring (EDM) principle is more tedious in the field and has a lower spatial resolution compared with airborne LIDAR survey (Cheng *et al.*, 2016). A simple method of measuring the beach profile changes is discussed by Andrade and Ferreira (2006), whereas the methodology for ascertaining the equilibrium of beach profile has been detailed by Aragonés *et al.* (2016). Koroglu *et al.* (2017) has performed experiments to investigate coastal accretion and erosion process using terrestrial laser and concluded that laser scanner can be used for beach profile evaluation exposed to the action of ocean waves.

The three major errors contributing to the deviations in the estimation of beach erosion and deposition is herein discussed. The first error is that all measurements are based relative to a benchmark. If this reference disappears or is lost, it is difficult to compare it with the previous survey or to make it more accurate with subsequent surveys. The second error arises if the successive surveys fail to comply with the identical course of the preceding survey. Lastly, the three-dimensional interpolation of a two-dimensional data may result is significant errors even if the subtle changes in the beach profiles are neglected. The practical limitations associated with conventional beach monitoring are as follows:

- time-intensive survey exercise;
- loss of "permanent" monuments used as reference points where the beach is either rapidly eroding or subjected to substantial wave penetration during storms and
- errors and deviation in the estimation of volumetric changes arising due to inadequate data.

8.3. Beach Profiling using RTK-GPS

Amongst all the survey techniques available, as briefly explained earlier, for acquiring the beach profiles, data collection by RTK-GPS provides quick and best data. It is a satellite navigation technique used to enhance the precision of position data derived from satellite-based positioning systems (Global Navigation Satellite Systems, GNSS) using ground based fixed reference. The base station is fixed reference point near to the survey area. The major set up in the instrument are reference ellipsoid (WGS 1984), local geoid (EGM2008), projection (UTM-44) and exact position of receiver antenna. In a flat beach, points can be taken 10–15 m apart, but in a dune or on an irregular part of beach, maximum points must be collected. The points should be collected on all locations where changes are witnessed. Once the survey is complete, the last point should be taken on the known point of location to avoid any error accumulation. The data collected is a horizontal position (x, y) and vertical (z) levels. These points should be processed and overlaid over previous surveys to check for erosion/accretion in the profiles. The spatial and temporal analysis of data provides the volumetric changes per unit shoreline.

For the case study reported in this chapter, the above-stated procedure was adopted and owing to the recurring seasonal survey the errors involved were minimized. The monthly surveys include a background file in the instrument so as not to disorient from the original profile line. Since all the profiles' starting point is preloaded in the instrument, one need not worry about rapid erosion. The collected data is automatically stored in the instrument, requiring lesser time to complete the surveys. Spatial and temporal analysis of data provides the volumetric changes of highest order. The two case studies carried out are discussed below.

8.4.   Field Studies

8.4.1.   *Case Study 1 (open coast)*

8.4.1.1.   *Study area*

The study area is an open coast along the coastal fishing village in Devaneri located about 50 km south of the major city, Chennai, along the southeast coast of India. The project site (between 12° 38′55″ N to 12° 39′15″ N and 80° 12′14″ E to 80° 12′35″ E) is exposed to wave action from the Bay of Bengal on its east. This fishing village is proximately located next to the major tourist attraction, Mahabalipuram, a UNESCO World Heritage Site comprising 7th- and 8th-century monuments, one which is the Shore Temple, the details of which are as shown in **Fig. 8.1(a).** A total of seven exchanges were considered in this region. Each of the transects extends for about 50–70 m in length. The transects monitored are directly related to the hamlet. The total length of the study area is about 600 m, the mid-transect is positioned at a point 390 m from the northern end and 210 m from the southern end. The transects 1 to 7, viz., T1 to T7 are shown in **Fig. 8.1(b)** are positioned around 100 m beyond the boundary of the village, while the other five transects are placed inside the village boundary. Each transect is 100 m apart from the subsequent ones. A pseudo-baseline is established for spatial and temporal analysis. One or more benchmarks (vertical as well as horizontal) near to site are established by RTK-GPS with respect to Survey of India (SOI) Benchmark. The accuracy of transferring the benchmark is 10–15 mm.

Fig. 8.1(a).   Devaneri location map.

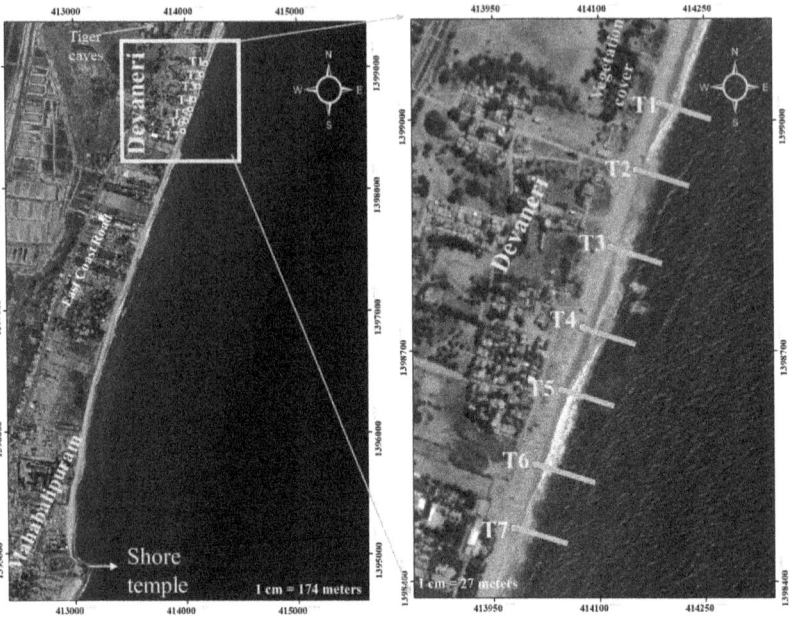

Fig. 8.1(b).   Location of transects (T1 to T7).

8.4.2. *Results and discussion*

Equal spacing of 100 m was adopted between the profile transects as it is an open coast. The above-stated seven transects were monitored from January 2017 to December 2018, with the dynamic portion for beach profile — i.e., the area after the baseline — being taken for comparison. The results on the superposition of the measured beach profiles are presented and discussed herein.

The variations in the beach profiles superposed for each month are shown in **Figs. 8.2** to **8.8.** The seasonal variations of all the transects are compared, and seasonal as well monthly volumetric changes from the beach profile variation over a 1-m length of the stretch were estimated, the results of which are projected in **Figs. 8.9** and **8.10.** In general, the change in volume of sand per meter width of the beach is found to range between 35 m^3 and 65 m^3 for the years 2017 and 2018. A few photographs of the field survey campaign are projected in **Fig. 8.11.** It is observed that, throughout the year, no major accretion/erosion pattern is witnessed, and the change in the quantity of sediments across the survey duration is found

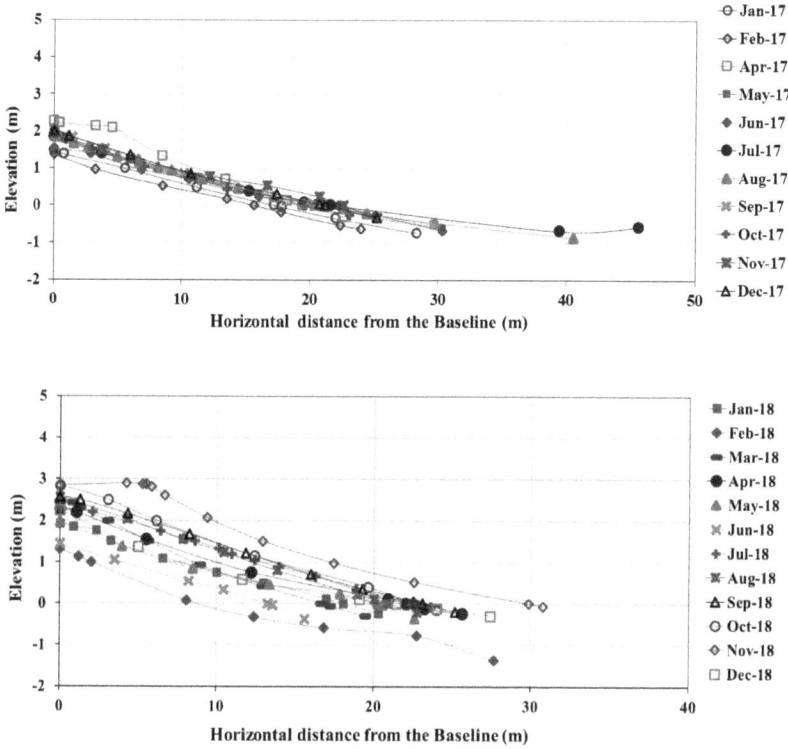

Fig. 8.2.   Transect 1 beach profiles from January 2017 to December 2018.

to be less. The stretch of the Devaneri coast considered is exposed to the southwest monsoon during June to September and the northeast monsoon during October to December, while January to May is the non-monsoon season. Although the transects get accreted during the southwest monsoon season, the dynamics of the impact is found to be comparatively quite small. It is found that, during the end of northeast monsoon and post-monsoon seasons, the profiles tend to erode, and the accretion process commences during the southwest monsoon season. The peak accretion is witnessed during July-August, but the foreshore slopes of the first five transects mostly look similar, and no major changes in the beach width is observed. The spatial changes for all transects are found to be similar as it is an open coast with no man-made structures in the vicinity of the study area.

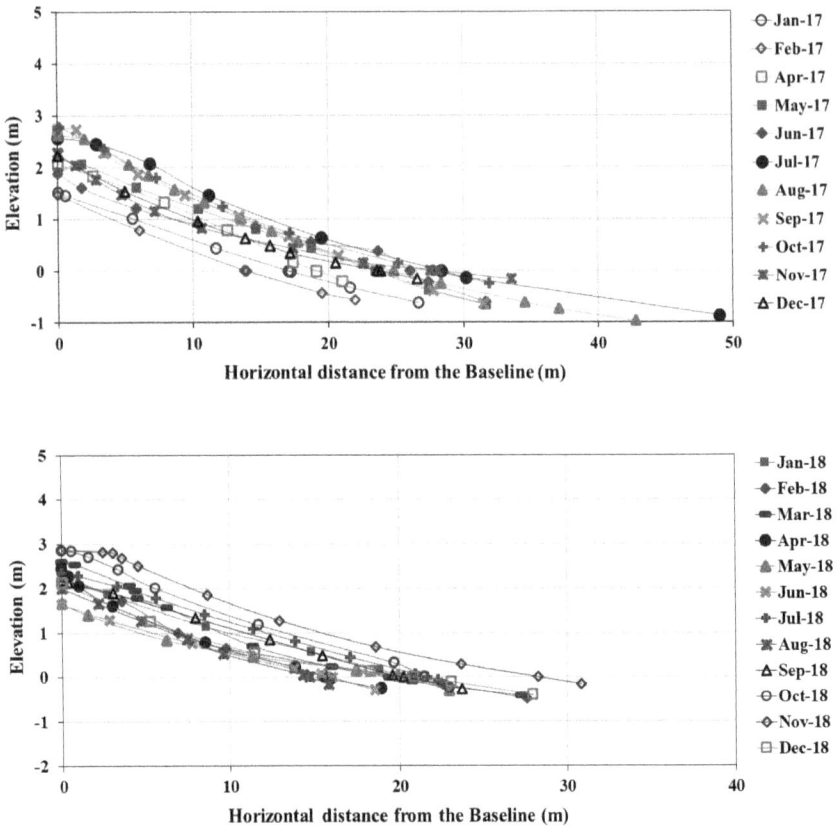

Fig. 8.3. Transect 2 beach profiles from January 2017 to December 2018.

8.4.3. *Case study 2 (near an estuary)*

8.4.3.1. *Study area*

Karaikal is a coastal town with a minor port occupying an area of 160 sq.km., which was formerly part of French India **(Fig. 8.12(a))**. The site falls within the Latitude 10° 54′ 25″ N to 10° 55′08″ N and Longitude 79° 51′ 00″ E to 79° 51′ 23″ E. Arasalar River, a tributary of the River Cauvery, drains into the Bay of Bengal at the tidal inlet. To prevent the sand bar formation near the mouth of the estuary and to train the riverine discharge of the river, a pair of training walls have been constructed on either side of its mouth. These training walls facilitates safe passage for

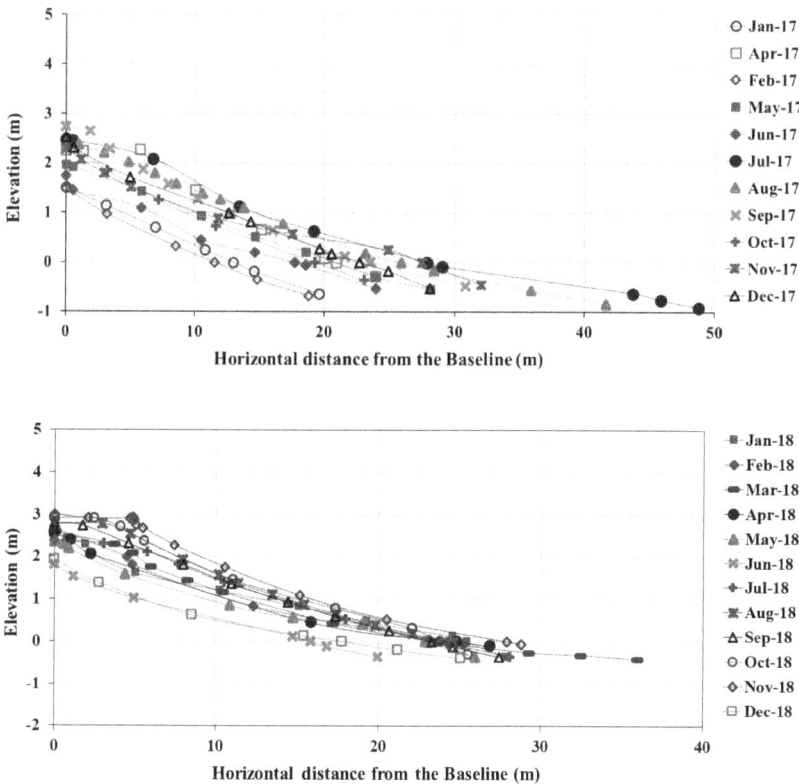

Fig. 8.4.   Transect 3 beach profiles from January 2017 to December 2018.

the fishing vessels, although periodical dredging of shoals in the approach channels is inevitable. The Karaikal Port is a private port located 7 km from the inlet mouth that acts as a hangar for the ships as well as a loading and unloading point of goods. The beach to the north of the training walls is an open beach utilized for recreational purposes by the coastal community. Further north, the coast is protected by a dense vegetation cover, and the beach to the south of training walls comprises of sand dunes covered with creepers which serves as an additional protection cover. Five number of transects on the north and two on the south of the Arasalar river mouth were established for continuous monitoring as shown in **Fig. 8.12(b)**. The beach to the south of training walls is a less popular and has relatively less footfall.

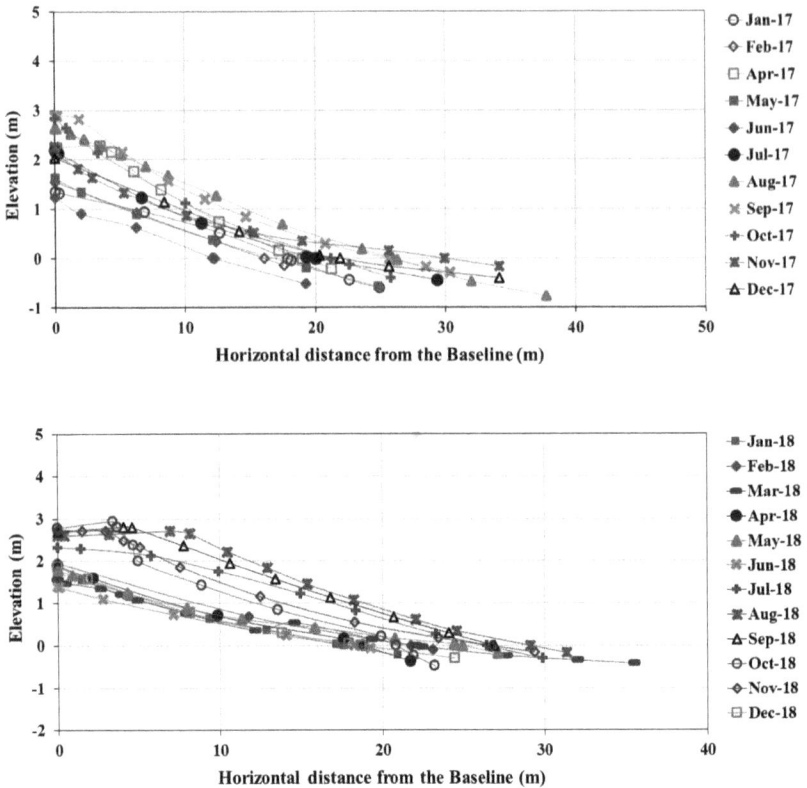

Fig. 8.5. Transect 4 beach profiles from January 2017 to December 2018.

The first two transects are placed on a part of the plain beach, while the third and fourth transects are placed between sand dunes. To the north, the first Transect (T1) is located about 30 m from the northern training wall. The T2 and T3 transects are located 120 m and 280 m from T1, respectively. Transect T4 is located 130 m north from the transect T3. The last Transect T5 is located in front of the vegetation cover and sand dunes, 80 m north from Transect T4. Due to the reduced presence of sand dunes along the south coast and it being a flat beach, just two transects 300 m apart were monitored. Transect T6 is placed over a sand dune, while Transect T7 is placed on a plain turf. The pseudo-baseline is considered for performing spatial and temporal analysis for both north and south of training walls separately. The coastal length of the study area is ~500 m on either side of

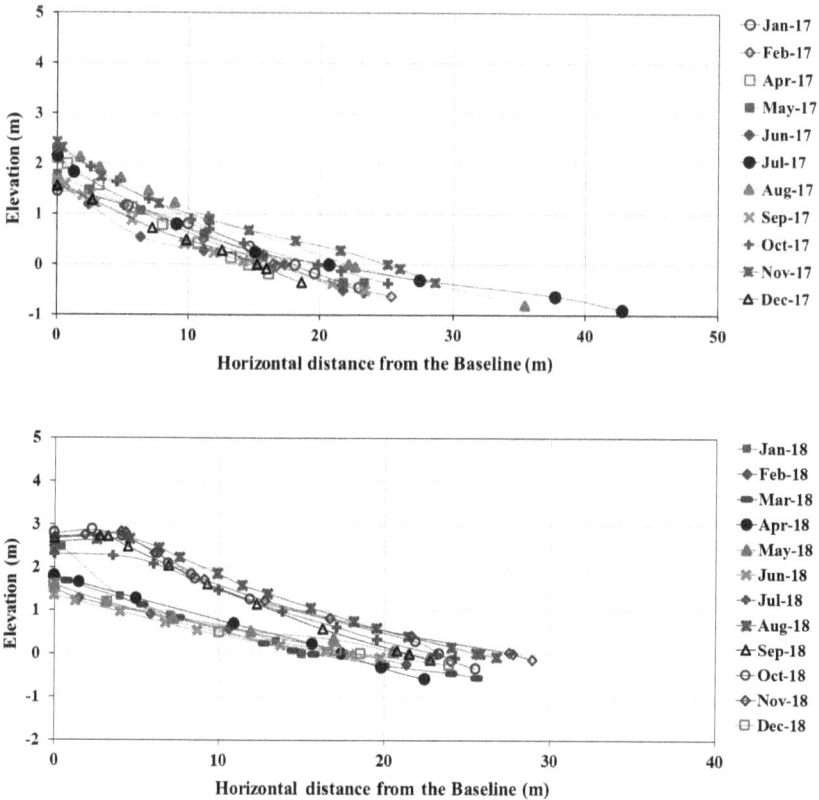

Fig. 8.6. Transect 5 beach profiles from January 2017 to December 2018.

training walls in Karaikal. The baseline considered is 100 m onshore from January 2018 shoreline. The presence of creepers near fourth and fifth transect makes the changes in profiles more interesting. The benchmark for this study area has been transferred from the permanent benchmark fixed by the Survey of India.

Since the study area is located adjacent to the training walls, the direction of monsoon plays a vital role in the determination of accretion/erosion processes. Typical temporal changes in the coast between January 2017 and December 2018 and the corresponding superposed results are projected in **Figs. 8.13–8.19**. The seasonal and monthly volumetric changes per meter length over all profiles shown in **Figs. 8.20 and 8.21**. In general, the change in volume of sand per meter width of the beach is found to range

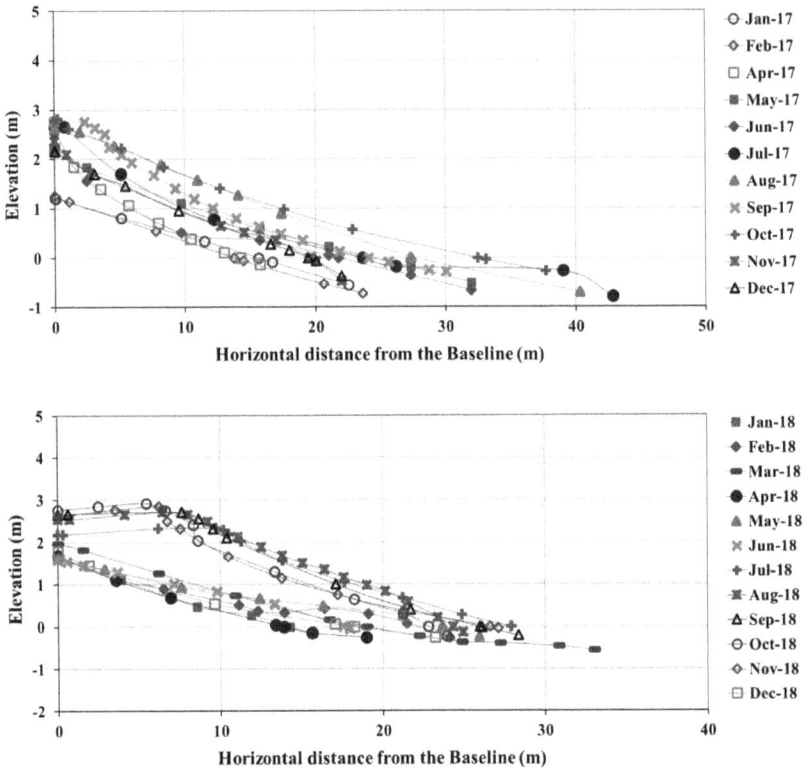

Fig. 8.7.   Transect 6 beach profiles from January 2017 to December 2018.

between $60\,\mathrm{m}^3$ and $100\,\mathrm{m}^3$ in 2017, whereas the said change in volume varies between $100\,\mathrm{m}^3$ and $200\,\mathrm{m}^3$ in 2018, which clearly demonstrates the importance of monitoring the beach profiles along the stretches of coast with manmade structures. During monsoon season, beach to north of the training walls undergo erosion while the coast to south of training walls undergo accretion. Once the post-monsoon started, the processes witness a reversal. Beach to north of training walls undergo accretion while beach to south of training walls undergo erosion. The changes occurring during northeast monsoon is vigorous, and maximum changes occur during this season. Transects 1–4, which are close to the northern training wall, witness rigorous erosional activity during monsoon compared to Transect 5, which is located away from the training wall. Furthermore, Transect 5 is located on the sand

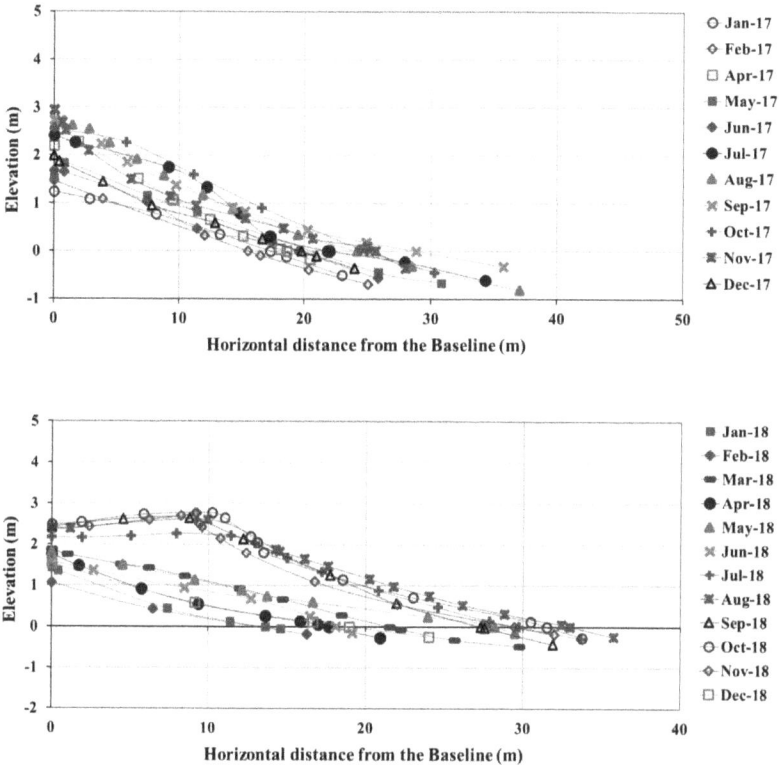

Fig. 8.8. Transect 7 beach profiles from January 2017 to December 2018.

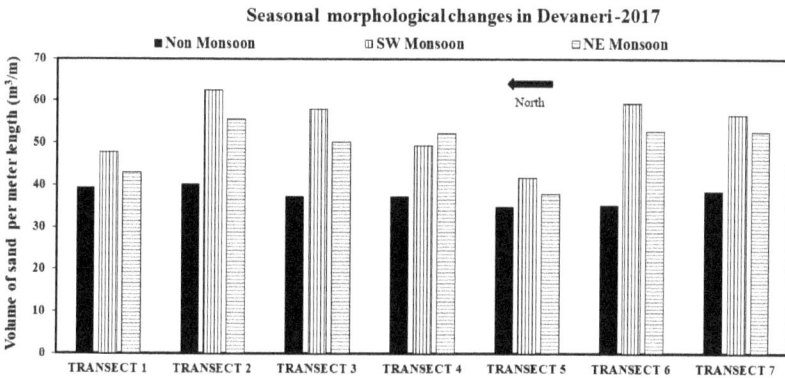

Fig. 8.9(a). Seasonal average volume changes for year 2017 (open coast).

Seasonal morphological changes in Devaneri-2018

■Non Monsoon   ▣SW Monsoon   ▢NE Monsoon

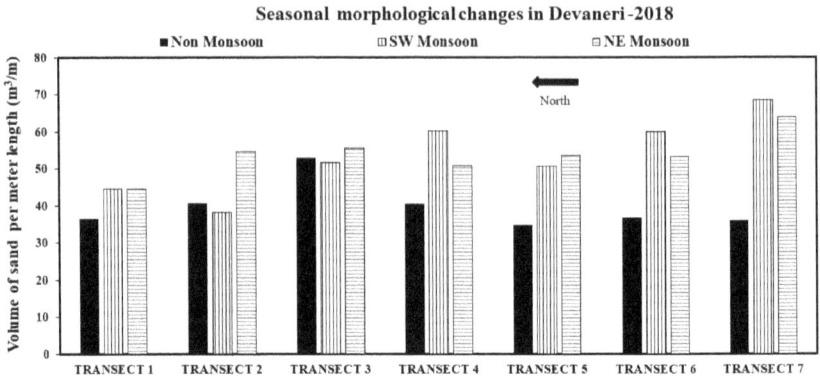

Fig. 8.9(b).   Seasonal average volume changes for year 2018 (open coast).

▣Jan-17 ▨Feb-17 ▤Apr-17 ▦May-17 ▨Jun-17 ≡Jul-17 ▢Aug-17 ▨Sep-17 ▨Oct-17 ▢Nov-17 ■Dec-17

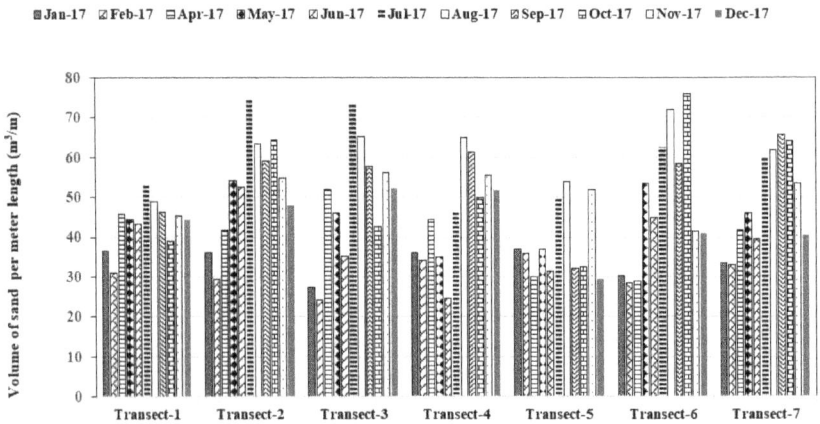

Fig. 8.10(a).   Monthly morphological changes on open coast for year 2017.

dune and near to vegetation. The beach Profiles T1 to T3 on the northern side of training walls, as well as T6 and T7 experiences an increase in beach width. The coastal dynamics during the Northeast monsoon is more pronounced across the east coast of Tamil Nadu, therefore the northern coast in Karaikal experiences net erosion.

8.5.   Summary

The beach profile changes in an open coast as well as near an estuary have been discussed through comprehensive field monitoring. The advantages in using RTK-GPS for beach profile survey over conventional methods has

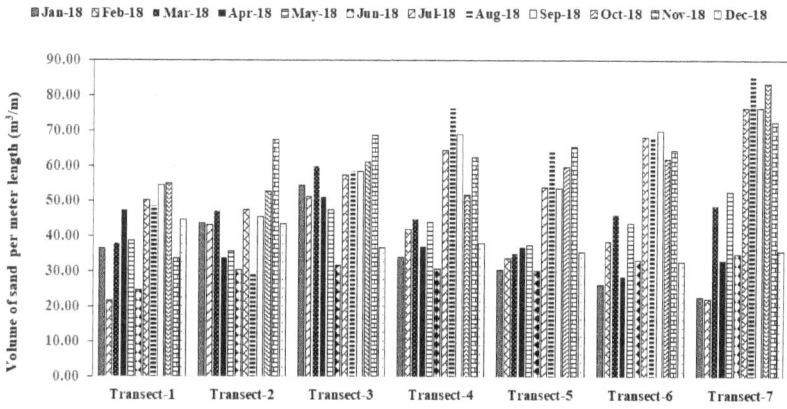

Fig. 8.10(b).   Monthly morphological changes on open coast for year 2018.

Fig. 8.11.   A few photographs taken during field measurement campaign.

Fig. 8.12(a). Karaikal location map.

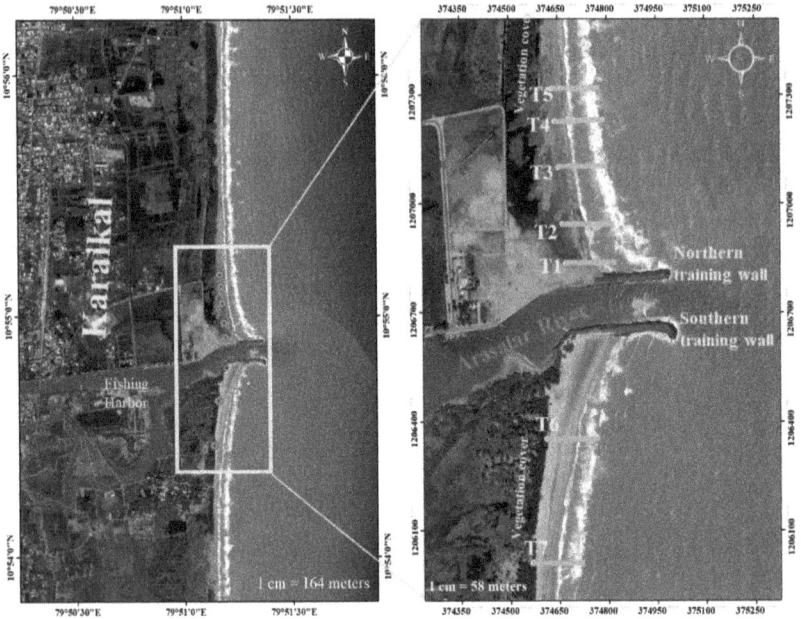

Fig. 8.12(b). Location of Transects T1 to T7.

been discussed. It is observed that open coast beaches do not undergo significant changes with respect to seasons over the two years considered in the study. No major accretion/erosion is witnessed in an open coast.

The presence of hard protection structures in the near-estuary coast dictates the accretion/erosion pattern. Though the beach regains on a yearly scale, the seasonal changes are quite dominant in this region. During the monsoon season, the beach north of training walls starts eroding. Though the duration of the northeast monsoon season is less, the resultant shoreline changes, irrespective of erosion or accretion were intensified.

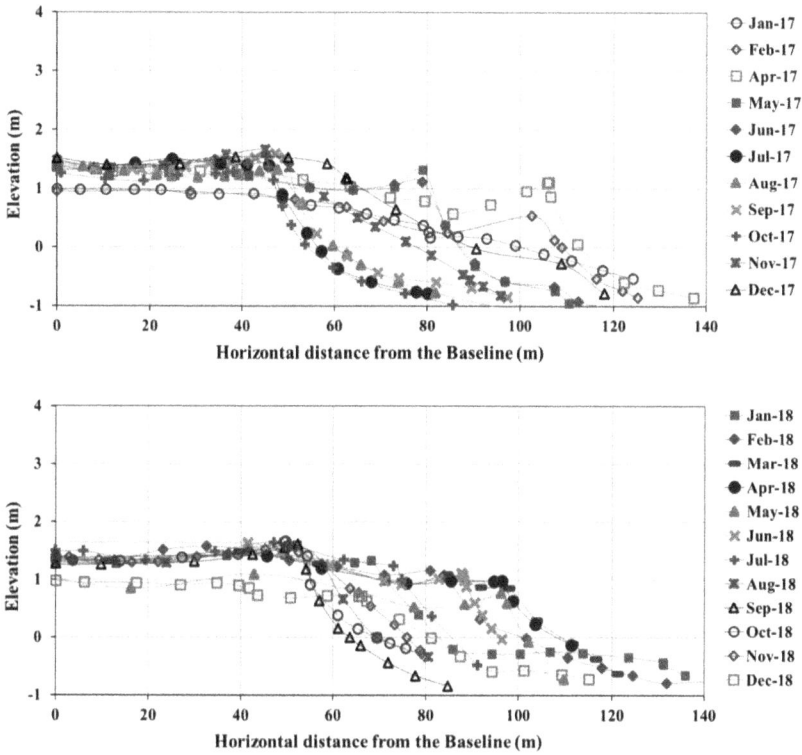

Fig. 8.13.   Transect 1 beach profiles from January 2017 to December 2018 (near estuary).

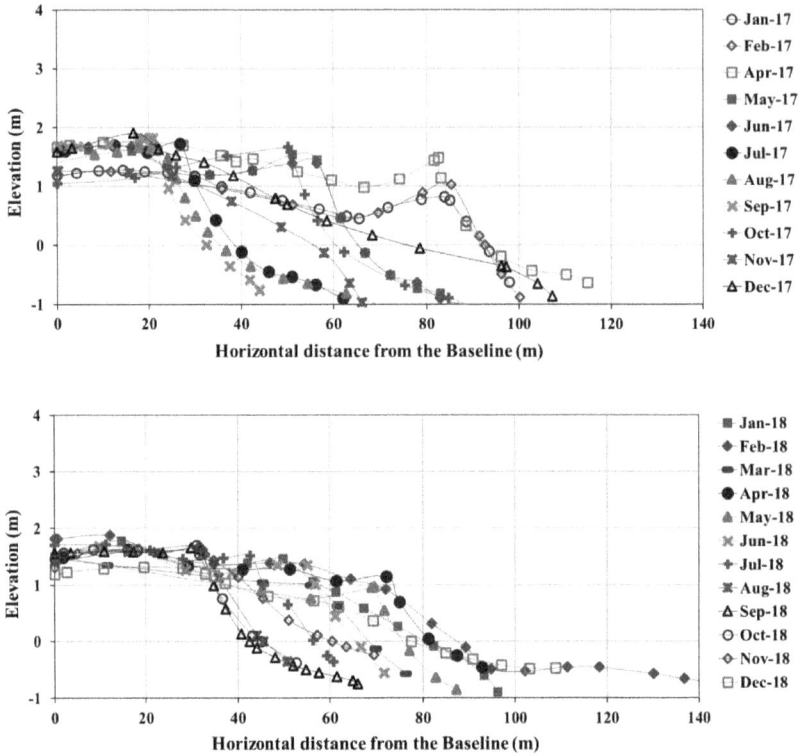

Fig. 8.14.   Transect 2 beach profiles from January 2017 to December 2018 (near estuary).

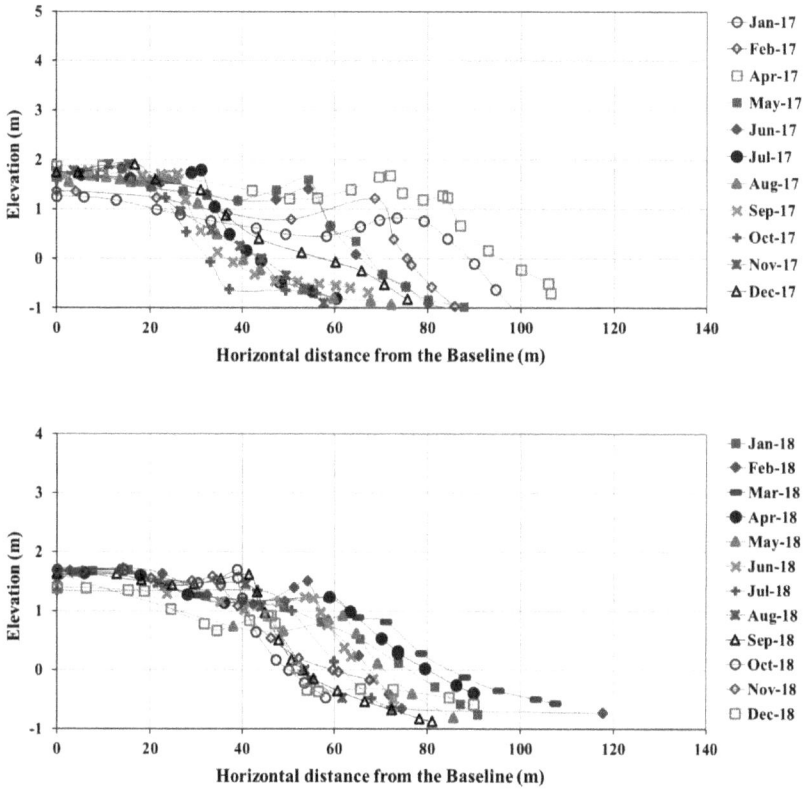

Fig. 8.15.   Transect 3 beach profiles from January 2017 to December 2018 (near estuary).

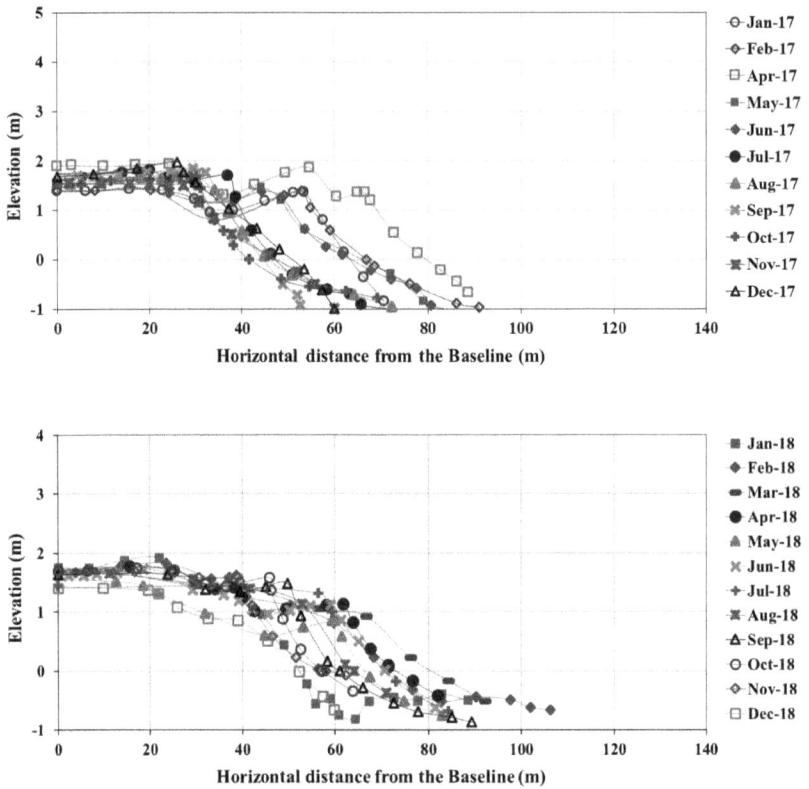

Fig. 8.16. Transect 4 beach profiles from January 2017 to December 2018 (near estuary).

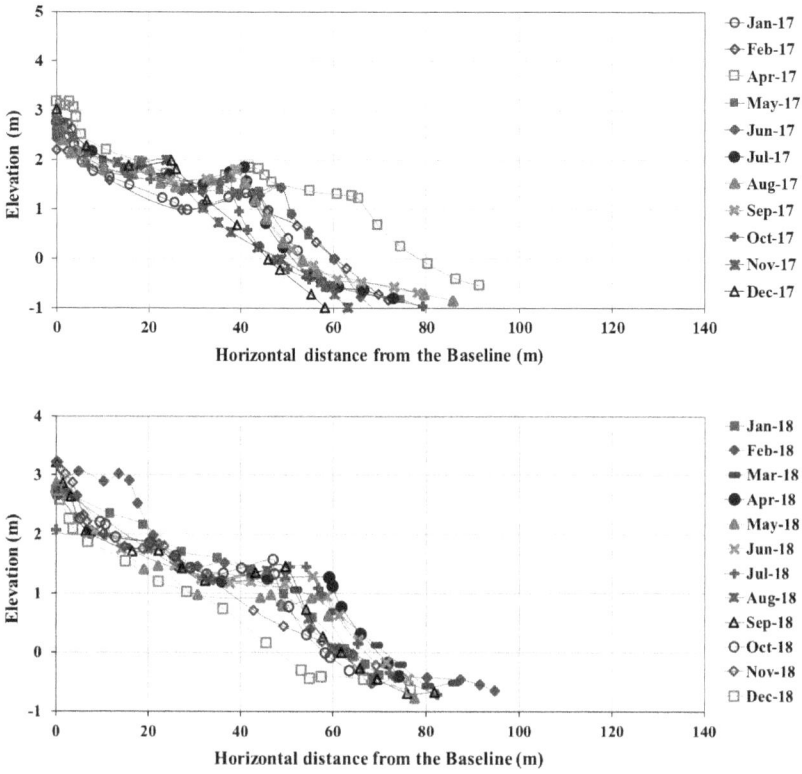

Fig. 8.17.   Transect 5 beach profiles from January 2017 to December 2018 (near estuary).

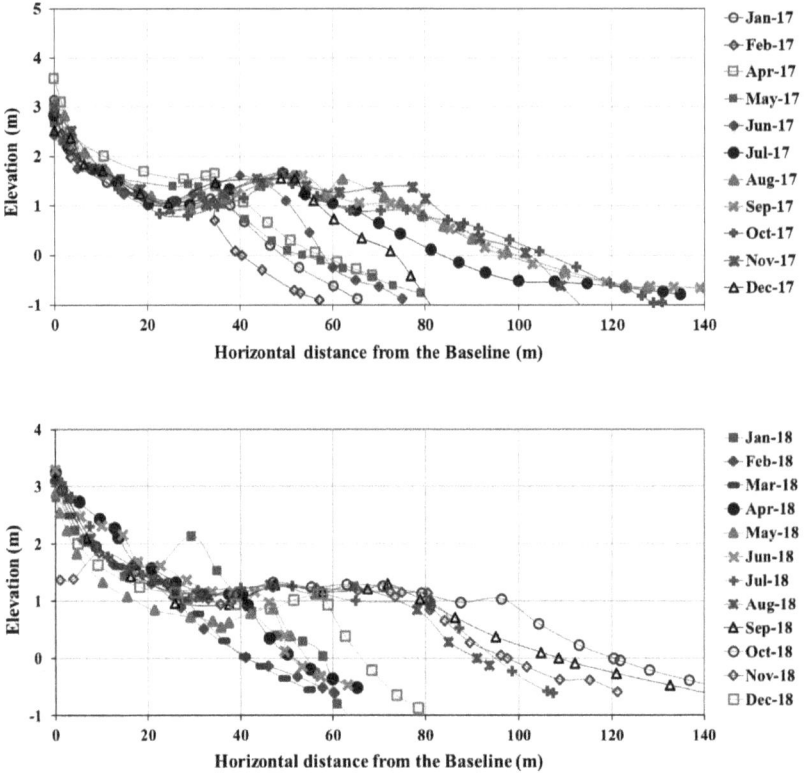

Fig. 8.18. Transect 6 beach profiles from January 2017 to December 2018 (near estuary).

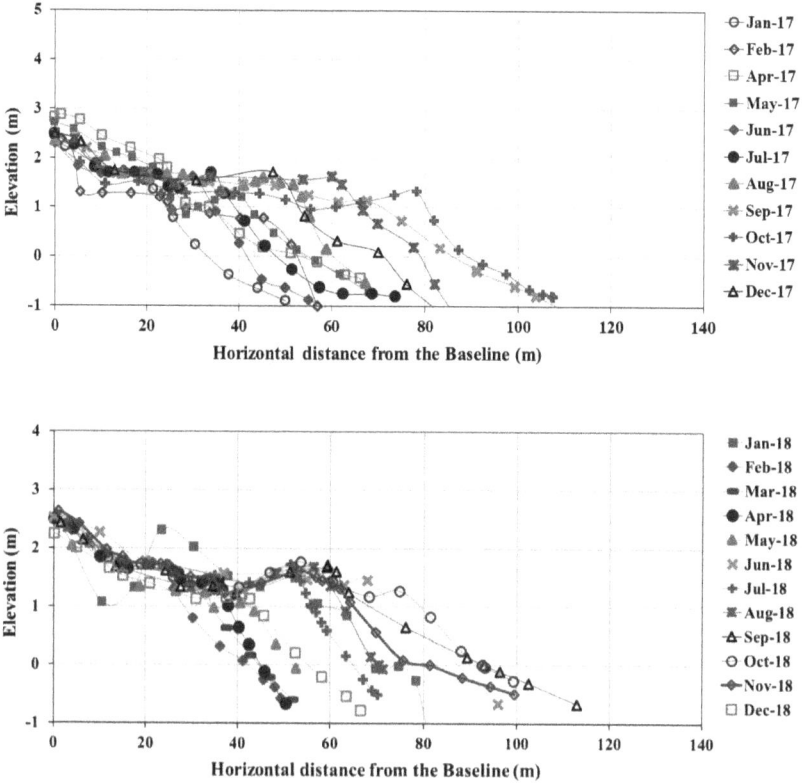

Fig. 8.19. Transect 7 beach profiles from January 2017 to December 2018 (near estuary).

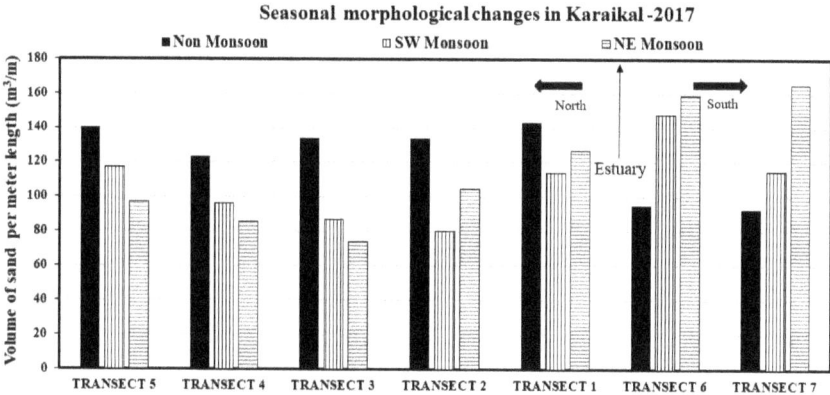

Fig. 8.20(a). Seasonal average volume changes for the year 2017 (near estuary).

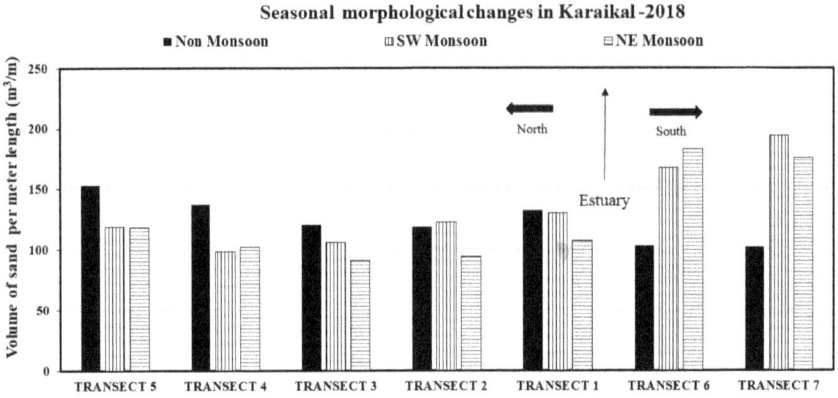

Fig. 8.20(b). Seasonal average volume changes for year 2018 (near estuary).

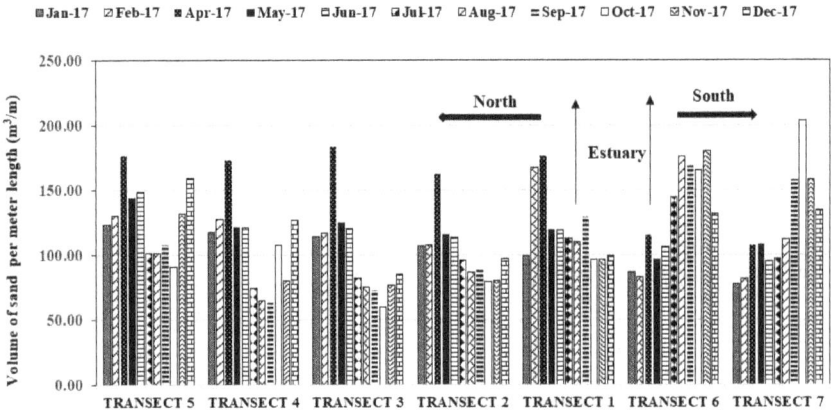

Fig. 8.21(a). Monthly morphological changes in Karaikal coast for the year 2017.

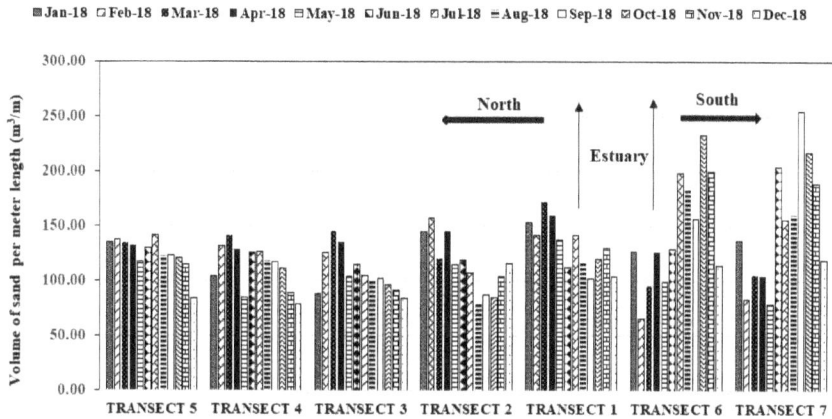

Fig. 8.21(b). Monthly morphological changes in Karaikal coast for the year 2018.

References

Ananth. P.N. and Sundar, V. (1990). Sediment budget for Paradip port, India. *Ocean and Shoreline Management*, 13(1):69–81. https://doi.org/10.1016/09 51-8312(90)90013-8.

Andrade, F. and Ferreira, M.A. (2006). A simple method of measuring beach profiles. *Journal of Coastal Research*, 22(4):995–999.

Aragonés, L., Serra, J.C., Villacampa, Y., Saval, J.M. and Tinoco, H. (2016). New methodology for describing the equilibrium beach profile applied to the Valencia's beaches. *Geomorphology*, 259:1–11. https://doi.org/10.1016/j.geo morph.2015.06.049.

Birkemeier. W.A., Bichner. E.W., McConathy. M.A. and Eiser. C. (1991). Nearshore profile response caused by Hurricane Hugo. *Journal of Coastal Research*, 81(8):113–127.

Cheng, J., Wang, P. and Guo, Q. (2016). Measuring beach profiles along a low-wave energy microtidal coast, West-Central Florida, USA. *Geosciences*, 6: 44. https://doi.org/10.3390/geosciences6040044.

Emery, K.O. (1961). A simple method of measuring beach profiles. *Limnology and Oceanography*, 6:90–93.

Jeyagopal, S., Vasanthakumar, S., Mikkilineni, D., Sundar, V., Sannasiraj, S.A. and Murali, K. (2020). Very severe cyclonic storm impacts to shoreline and beach profiles along the Karaikal coast of India. *ISH Journal of Hydraulic Engineering*. DOI: 10.1080/09715010.2020.1767515.

Kannan, R., Anand, K.V., Sundar, V., Sannasiraj, S.A. and Rangarao, V. (2014). Shoreline changes along the Northern coast of Chennai port, from field measurements. *ISH Journal of Hydraulic Engineering*, 20(1):24–31.

Klemas, V. (2011). Beach profiling and LIDAR bathymetry: An overview with case studies. *Journal of Coastal Research*, 27(6):1019–1028. https://doi.org/10.2112/JCOASTRES-D-11-00017.1.

Koroglu, A., Seker, D.Z., Kabdasli, S.M., Bayram, B., Goktepe, A. and Varol, E. (2017). Evaluation of beach profile changes by using terrestrial laser scanner. *Fresenius Environmental Bulletin*, 26(1):19–28.

Kudale, M. (2010). Impact of Port development on the coastline and the need for protection. *Indian Journal of Geo-Marine Sciences*, 39:597–604.

Kunte, P.D., Alagarsamy, R. and Andursthouse, A.S. (2013). Sediment fluxes and the littoral drift along northeast Andhra Pradesh Coast, India: Estimation by remote sensing. *Environmental Monitoring and Assessment*, 185:5177–5192.

Morton, R.A., Leach, M.P., Paine, J.G., and Cardoza, M.A. (1993). Monitoring beach changes using GPS surveying techniques. *Journal of Coastal Research*, 9(3):702–720.

Nordstrom, K. and Jackson. N. (2013). Removing shore protection structures to facilitate migration of landforms and habitats on the bayside of a barrier spit. *Geomorphology*, 199:179–191.

Suresh, P.K., Sundar, V. and Selvaraja, A. (2011). Numerical modelling and measurement of sediment transport and beach profile changes along south west coast of India. *Journal of Coastal Research*, 27(1):26–34. https://doi.org/10.2112/jcoastres-d-09-00039.1.

Timothy, W.K. and Christopher, J.A. (1994). Beach profile spacing practical guidance for monitoring nourishment projects. *Prof. Coastal Engineering Proceedings*, 24:2100–2114.

Thomas, T., Phillips, M.R., Morgan, A. and Lock, G. (2018). Morphostat: A simple beach profile monitoring tool for coastal zone management, *Ocean & Coastal Management*, 153:17–32. https://doi.org/10.1016/j.ocecoaman.2017.11.016.

Chapter 9

Variation of Sediment Characteristics Along
an Open Coast and Near an Estuary

Abstract
The importance of investigating the sediment characteristics along an open coast and in an estuary are discussed through a comprehensive literature review. The sediment characteristics along an open coast and near an estuary situated on the southeast coast of India are quantified through field investigations conducted over a couple of years. The temporal and spatial variations are discussed. The seasonal variations for both the above-stated environments are also presented and discussed in this chapter.

9.1. Introduction

9.1.1. *Importance*

The sediments present in a coastal environment is the result of weathering actions of rocks and are transported to the coast via air, wind and rivers. The coastal sediments are broadly classified by their size as gravel, silt and clay, which define the coastal features that include erosion and accretion. The sediment characteristics define the nature of the beach, which would vary along an open coast from that in the vicinity of an estuary, which transports sediment of varying characteristics. The phenomena of erosion or deposition of sediments is due to either natural causes or the intervention of man for the purpose of developmental activities.

In the ancient days, insignificant developments had left the estuaries and the coasts in dynamic equilibrium. Since the commencement of the Industrial Revolution, there has a tremendous and progress in infrastructure development, which has been ever increasing. The concentrations of population along the coasts have drastically increased, resulting in instability in the shorelines and causing various problems. The wave and tide-induced motion of sediments is a continuous process. Their quantity and direction of motion are controlled by the flow characteristics as well as the characteristics of the sediments. Combating both sediment deposition and erosion has been a challenge to engineers. The former hampers the free passage of vessels through the approach channel of harbours when silted up or the mouths of rivers get blocked by sand bar formation, thus preventing free exchange of sea and fresh waters. Erosion along the coast has resulted in the loss of land, infrastructures and livelihood of the fishing community, which are just a few of the hazards mentioned here.

An in-depth knowledge of the sediment characteristics is essential as it governs the fate of the sediments in motion being transported either as suspended (entrained in the water) or bed load, which in fact govern the dredging philosophy to be adopted. The porosity and the fall velocity that determines the quantity of sedimentation depend on the sediment characteristics. In a sediment-laden flow, the terminology coarse, fine and cohesive refer to gravels, sand and silt or clay, respectively.

9.1.2.   *Supply of sediments*

As stated earlier, continuous coastal development leads to erosion. The options available to combat coastal erosion include mitigation measures through construction of structures or nourishment of the affected stretch through sediment supply. Human activities that have led to the loss of sediment supply through rivers discharging into the ocean has resulted in several stretches of the affected coast starving for sediment supply. Deforestation and urbanization are of great concern as they lead to erosion of a sediment-laden river basin to a large extent.

The reduction of sediment supply through rivers are mainly due to (i) the creation of reservoirs for the storage of water that act as barriers and trap the sediments; (ii) pumping of water from rivers for irrigation purpose and (iii) failure of monsoon.

9.1.3.   *Coastal features*

The sediment characteristics dictate the coastal features through its deposition and erosion, which are shown in **Fig. 9.1**. Although there are lagoons, spit, salient, tombolo, cusps and bays, as illustrated, beaches are the most common, the conservation of which has been quite demanding in the recent past. The stability of beaches is dependent on the grain size of the sediments deposited responsible for its formation, and its variations in most locations are seasonal.

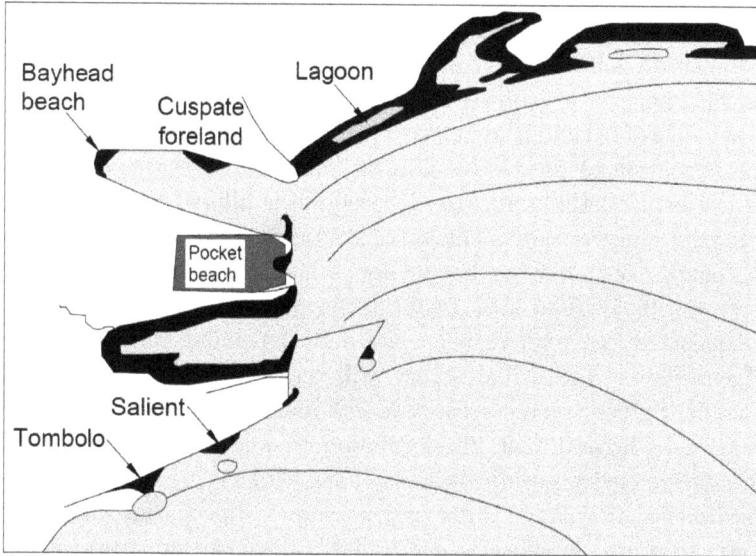

Fig. 9.1. Typical coastal features.

During the fair-weather seasons, the wave energy responsible for the transport of sediments facilitate the building up of the beach, in which case the sediments will be mostly sand. During rough seas, the deposits are usually removed, the quantity of which depends on the sediment characteristics and the wave energy. The removed sand would deposited farther towards the sea as an offshore bar, which again would be removed and re-deposited.

9.1.4. *Sediment transport along the coastal zone*

In the surf zone, breaking wave induces high orbital velocities, which are responsible for the significant suspended sediment concentration. In addition, the wave-generated longshore currents in the surf zone can cause high longshore sediment transport (LST). This is also referred to as "littoral drift" and the movement of sediment particles within the surf zone. If the flow involves sediments, it is no longer said to be a simple fluid flow as it involves two components. Sediment transport is influenced by either waves or currents or a combination of both and characteristics of the sediments. A detailed description of sediment transport within the surf zone, which is dominant and determines the stability of the shoreline, has been discussed in several books and hence is avoided herein.

9.1.5. *Sediment transport along the estuary*

The transition zone between estuaries and the ocean serves as sediment traps. Tide penetration during flood flow enhances the salinity in the river. The exchange of sea and fresh waters bring in considerable nutrients, leading to the formation of productive natural habitats. As the sediment moves along the estuary, the grain size distribution is altered by deposition, re-entrainment and transport. This sorting process in the river deposits much of the coarser sediment on floodplains, which is released only by floods (Lee *et al.*, 1995). The finer material moving in the estuary leads to a high concentration, tend to flocculate and, depending on its weight and the driving force within the estuary, will decide its fate to settle or be in motion. If the flow in the estuary is less and the stretch of the coast is dominated by littoral drift, the formation of sandbars and spits can lead to an adverse environmental impact. If the estuary is laden with a lot of fine sediments, it will naturally find its way to the marine environment. The estuaries are classified as microtidal (tides less than 2 m), mesotidal (2–4 m) and macrotidal (greater than 4 m).

9.2.   Studies in the Past

Along the east coast of India, the littoral drift is predominantly towards the north during the southwest monsoon and towards the south during the northeast monsoon (Ananth and Sundar, 1990). The effect of heterogeneous sediment characteristics on the coastal processes is usually underestimated through conclusions drawn by Holland and Elmore (2008). Merkus (2009) revealed that the grain size of sediments provides a clear indication of the source of the sediments in motion in estuaries and along the beaches. The dynamic sediment transport processes are dictated to a great extent by grain size distribution, as stated by Vandenberghe (2013). Oyedotun (2016) has discussed in detail the sediment characteristics near the confluence of an estuary and beach, through which it was concluded that the processes involve selective sorting, winnowing and transportation of medium to fine sediments. The study related to the changes in the sediment characteristics has been discussed by Pedro Narra *et al.* (2015), Southwell *et al.* (2017), Narayana *et al.* (2007), Preoteasa and Vespremeanu-Stroe (2010) and Das (2009).

The objective of the beach sediment analysis is to measure the individual particle sizes to determine their frequency of distribution and to effectively characterize the sediments (Sathasivam *et al.*, 2015). In an estuary, the

coarse sediments are deposited at the river mouth, and finer sediments re-
deposited at its upper part, as discussed by Narayana *et al.* (2008). Jayakumar *et al.* (2004) stated that the process of beach erosion and accretion is
seasonal which, over a long period, result in the changes in the coastline.
Pardhan *et al.* (2020) investigated the seasonal variation of sediment texture
characteristics at selective beaches along the east coast of India. Sathasivam
et al. (2015) conducted a field study for 11 coastal sites from Kalpakkam to
Colachel, Tamil Nadu (TN), and concluded that the northern part of TN
coast is dominated by fine sand whereas central and southern coasts of TN
were dominated by medium-sized sand.

9.3. Methodology

To assess the sediment characteristics, the measurement in the field involves
seabed sampling to determine the particle size distribution. The sediment samples along the stretch of the coast of the study area were collected monthly. For this study, the beach sediments were collected for
two years continuously in both Devaneri ($12°39'05''$ N, $80°12'29''$ E) and
Karaikal ($10°54'55''$ N, $79°51'10''$ E) sites, both situated along the southeast coast of India, from April 2017 till June 2019, and from July 2020
till March 2021 with a year break in-between. A set of sieves (4.75 mm,
2 mm, 1 mm, 0.6 mm, 0.425 mm, 0.212 mm, 0.150 mm, and 0.075 mm
& pan) $\{IS : 2720\,Part - IV - (1985)\,and\,IS : 2720\,Part - 3 - 1 - (1963)\}$
were used to obtain the classification and distribution of the sediments.

The sieve analysis involved shaking a sediment sample through a set of
sieves that have progressively smaller openings. The collected samples were
dried through a hot air oven at the temperature of $100 \pm 5°C$ for 24 hours.
The samples were taken from the oven and cooled for 24 hours at room
temperature. The lumps were broken into smaller particles before passing
through the sieve. Sample weighing 500 g was taken for sieve analysis. The
sample in the sieve set was placed in the sieve shaker and fastened. After
the completion of a shaking period of 10 minutes, the mass of soil retained
in each of the sieve was weighed using a weigh balance and values were
noted.

The samples were taken along the berm and the shoreline. Each sediment sample of around 0.5 kg were collected after clearing the top loose
sand and digging a pit up to 0.2 m deep. The samples were collected in a
zip-lock pouch with appropriate sample ID. For this chapter, the data were
classified as pre-monsoon, monsoon and post-monsoon for the period of

2017–2018, 2018–2019 and 2010–2021. The average of pre-monsoon, monsoon and post-monsoon data of the three periods were plotted and considered for the discussion.

9.4.   Classification between Coarse and Fine Sediments

The sediments found along the coasts and in estuaries are mostly coarse or fine and is attributed to the mutual interaction of grains in water environment and based on its size. The suspension of coarse grains exhibits independent behaviour from each other — except for mechanical interactions in highly dense suspensions in the event of coarse sediments constituting the seabed, only forces of interlocking and friction are to be considered.

In addition, the shape of the grain size curve has an important effect on the properties of sands and gravels. This can be described with two coefficients, viz., coefficient of curvature C_c and the coefficient of uniformity C_u, defined as follows:

$$C_c = \frac{(D_{30})^2}{(D_{60})(D_{10})} \tag{9.1}$$

$$C_u = \frac{D_{60}}{D_{10}} \tag{9.2}$$

where D_{60} = the grain size at which 60% of the soil is finer, D_{30} = the grain size at which 30% of the soil is finer and D_{10} = the grain size at which 10% of the soil is finer.

If C_c is between 1 and 3, the grain size distribution curve will be smooth, and if C_u exceeds 4 for gravel or 6 for sand, there will be a wide range of sizes. When both criteria are met, the soil is said to be well graded (designated W); otherwise, it is poorly graded (designated P). No practical significance can be attached to the shape of the grain size curve for silts and clays. Attempts to identify one size as a characteristic dimension of the whole sample have been made by Median grain size, $D = D_{50}$. The median grain size D_{50} is commonly used to represent grain size gradation. The grain size Dg corresponding to some other fractions such as D_{90}, D_{65} are also adopted in studies. The median grain sizes for which 50% by weight of the sample is finer or coarser (D_{50}). The diameters D_{90} and D_{65} are the sizes for which 90% and 65% respectively, by weight of the sample is usually taken as,

$$D_g = (D_{84.1}D_{15.9})^{1/2} \tag{9.3}$$

in which $D_{84.1}$ and $D_{15.9}$ are the grain sizes for which 84.1% and 15.9% by weight, respectively, of the sediment is finer. Another classification commonly used by earth scientists is the phi scale.

$$\phi = -\log_2 D_m \tag{9.4}$$

where D_m is grain diameter in mm. The mean value of ϕ for any given sample is usually denoted by M_ϕ. In addition to the characteristic diameter, it is common to specify the nature or spread of the size distribution as follows

$$\text{Skewness} = \frac{\log D_g/D}{\sigma_g} \tag{9.5}$$

$$2^{\text{nd}} \text{ Skewness} = \frac{\log(D_{95}D_5/D^2)^{1/2}}{\sigma_g} \tag{9.6}$$

$$\text{Kurtosis} = \log \frac{(D_{16}D_{95}/D_5 D_{84})^{1/2}}{\sigma_g} \tag{9.7}$$

where σ_g is the standard deviation of the log (grain size) distribution of the sample. It is often found that the distribution of grain sizes is approximately log normal distribution. If this is the case, the standard deviation of the log (grain size) distribution is given by

$$\sigma_g = \log \left(\frac{D_{84.1}}{D_{15.9}} \right)^{1/2} \tag{9.8}$$

For details and for examples for the parameters as discussed above, the reader is suggested to refer to Sundar and Sannasiraj (2019).

9.5. Case Study 1 (Open Coast)

9.5.1. General

The case study for the open coast is Devaneri village **(Fig. 9.2)**, a small fishing hamlet along the east coast near Mahabalipuram, Kanchipuram District, Tamil Nadu. The study area is located 4 km north of Mahabalipuram and 1 km south of Tiger Caves, which are archaeological tourism spots. Mahabalipuram is famous for Shore Temple, which is protected by a seawall. In total, three transects were considered for the observation. The length of the study area along the coast is about 600 m. Transects 1 and 3 are at the boundaries of the village. In addition, Transect 2 is fixed at the centre of the study area. In total, six sediment samples, i.e., two sediment samples

Fig. 9.2.   Location of the study area — Open coast.

along each of three transects, were collected. **Figure 9.3** shows the sample collection at Devaneri site. **Figure 9.4** shows the sample location and **Table 9.1** shows the sediment sample coordinates.

9.5.2.   *Results and Discussion*

The grain size distribution of the sediments for the open coast is projected in **Fig. 9.5(a)**. Based on the pre-monsoon observations (**Fig. 9.5(b)**) along

Fig. 9.3. Devaneri sediment samples collections from berm and shore.

Fig. 9.4. Sample collected location in Google Earth view.

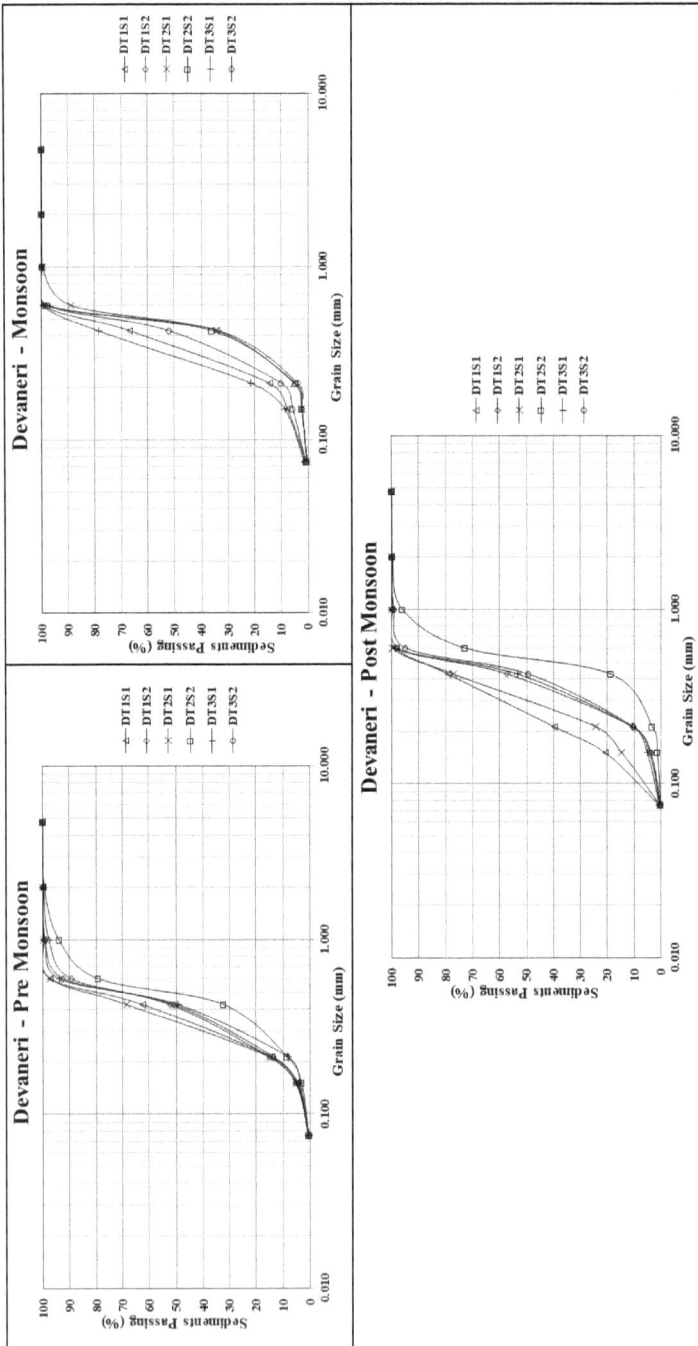

Fig. 9.5(a).   Grain size distribution — Open coast.

Table 9.1. Sediment sample collected location coordinates.

S. No	Sample ID	Remarks	Transects	Latitude	Longitude
1	DT1S1	Berm Sample	T1	12°39′15.12″N	80°12′34.98″E
2	DT1S2	Shore Sample		12°39′14.88″N	80°12′35.34″E
3	DT2S1	Berm Sample	T2	12°39′05.64″N	80°12′31.38″E
4	DT2S2	Shore Sample		12°39′5.52″N	80°12′31.80″E
5	DT3S1	Berm Sample	T3	12°38′57.12″N	80°12′28.08″E
6	DT3S2	Shore Sample		12°38′56.88″N	80°12′28.56″E

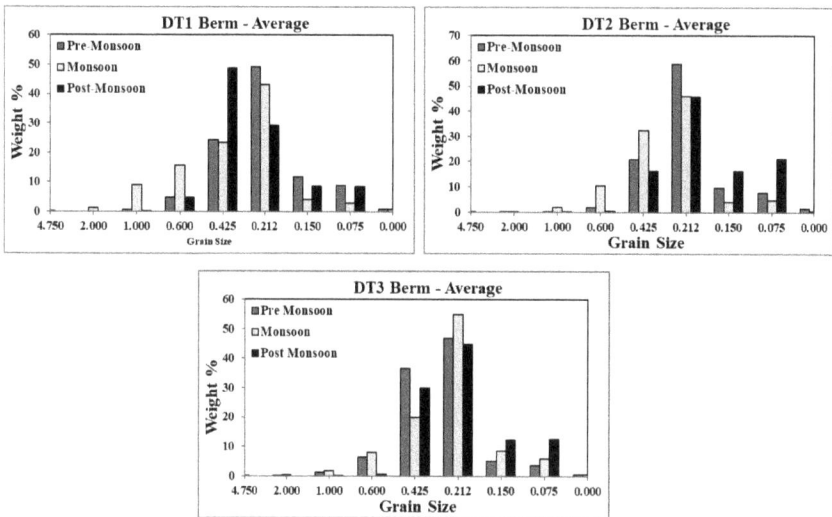

Fig. 9.5(b). Percentage by weight retained w.r.t grain size for the berm samples.

the berm sediment sample of three transects in the open coast, approximately 50% of the sediments are found to be of 0.212 mm grain size, falling under the category of fine sand. During monsoon, more than 40% of sediments are fine sand, with grain size of 0.212 mm. During the post-monsoon period, the sediments in the Transect 1 contain approximately 50% of 0.425 mm grains, which is coarse sand, the Transects 2 and 3 show more than 40% of 0.212 mm grains, i.e., fine sand.

Based on the pre-monsoon observations (**Fig. 9.6**) of the shore sediment sample of three transects, in the first and the third transects, we can see

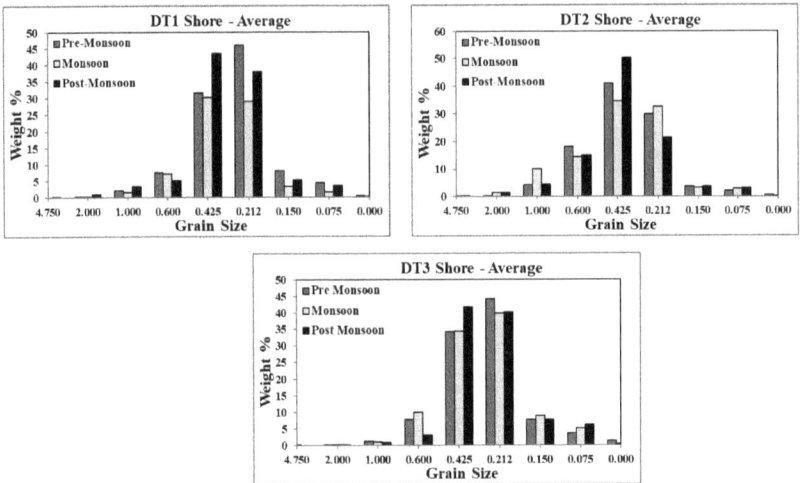

Fig. 9.6.   Percentage of weight retained w.r.t grain size for the shore samples.

approximately 50% of the sediments are of 0.212 mm grain size, whereas in the second transect, more than 50% of the sediments are of 0.425 mm sized coarse sand. In the first and second transects during the monsoon, it is seen that the 0.212-mm- and 0.425-mm-sized sands are distributed equally, whereas in the third transect, 45% of the sediments are of 0.212 mm size fine sand. However, during the post-monsoon, approximately 50% of the sediments are of 0.212 mm coarse sand in all three transects.

The analysis on the grain size showed that the median sediment size D_{50} (**Fig. 9.7**) of berm sediments ranges from 0.33 to 0.40 mm during the pre-monsoon period and ranges from 0.36 to 0.49 mm during the monsoon period, whereas during the post-monsoon period, D_{50} ranges from 0.27 to 0.41 mm. However, the analysis on the grain size showed that the median sediment size D_{50} of shore sediments ranges from 0.39 to 0.47 mm during the pre-monsoon period and 0.38 to 0.54 mm during the monsoon period, whereas during post-monsoon period, D_{50} ranges from 0.40 to 0.50 mm.

9.6.   Case Study 2 (Near Estuary)

9.6.1.   *General*

The Karaikal coast of Union Territory of Puducherry is taken as the case study for the estuary (**Fig. 9.8**). The study area along Karaikal is

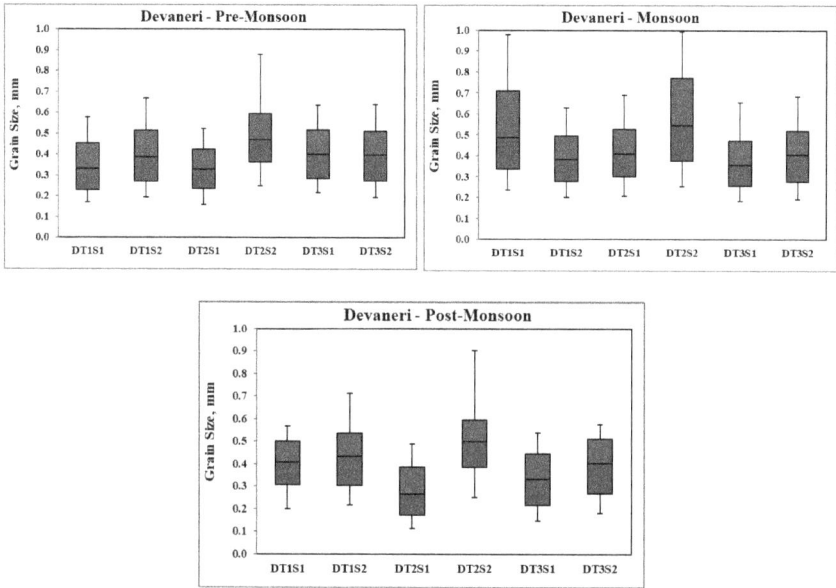

Fig. 9.7. D_{50} of the sediment samples.

dominated by sandy beaches and tidal flats. In Karaikal, a tributary of the Cauvery River called Arasalar River empties into the Bay of Bengal. A pair of training walls (breakwaters) have been constructed on the river mouth. The Karaikal fishing harbour is located at the southern bank of the Arasalar River. These breakwaters enable safe movement for the fishing vessels. The beach near the northern breakwater is a leisure spot for the Karaikal people and it has a beach width of about 800 m. As we move towards the north for about 600 m along the coast, it can be noticed that the coast is protected by a dense vegetation. The beach of the southern breakwater is less visited by the people and covered with more vegetation. The length of the study area along the northern coast is about 500 m and about 1 km along the southern coast. In the northern beach, three transects are fixed, in which the first transect is close to the north Breakwater and Transects 2 and 3 were about 250 m and 500 m from the Transect 1, respectively. On the southern coast, the first transect is near the southern breakwater and Transect 2 is about 1 km from Transect 1. Therefore, a total of 10 sediment samples were collected along the Karaikal coast. Views of the sample collection and their locations are shown in **Figs. 9.9** and **9.10**, respectively. The sample location and sediment sample coordinates are shown in **Table 9.2.**

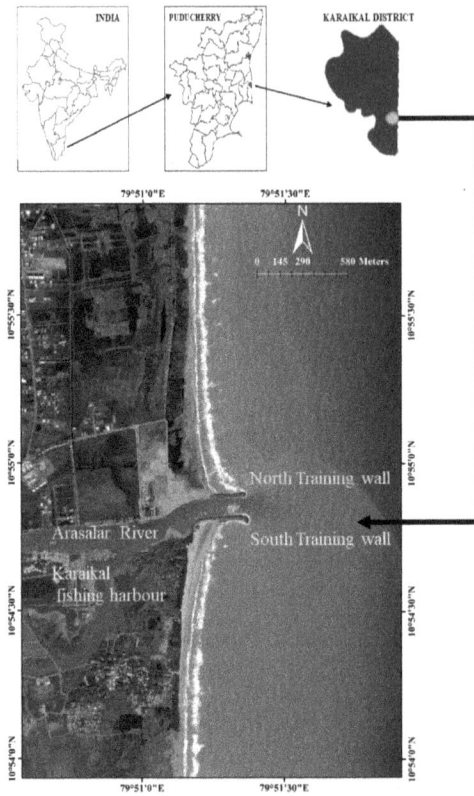

Fig. 9.8.　Study area — Near estuary.

Fig. 9.9.　Karaikal samples collections from berm and shore.

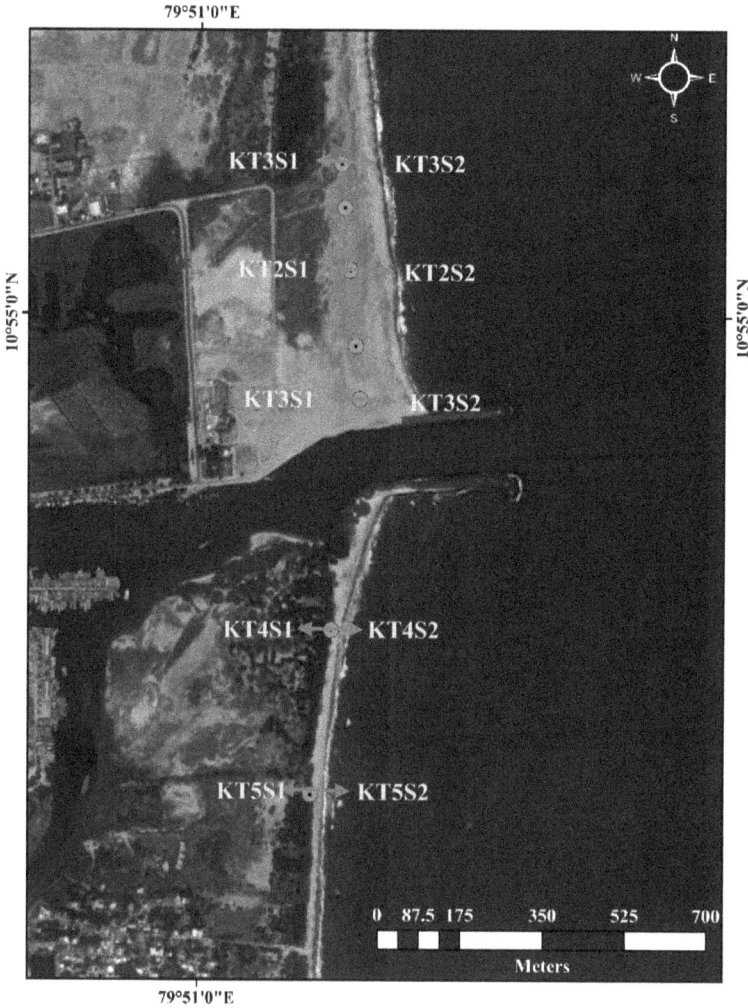

Fig. 9.10. Sample collected location in Google Earth view.

9.6.2. *Results and discussion*

In Karaikal, in the vicinity of the confluence of the estuary and the sea, the grain size distribution of the sediments for the open coast is projected in **Fig. 9.11(a)**. The pre-monsoon observation (**Fig. 9.11(b)**) of berm sediments along the north transects comprised 0.212-mm-diameter grains for more than 50% of sediments, whereas along the south transects, the berm sediments of the first transect, which was close to the breakwater, shows

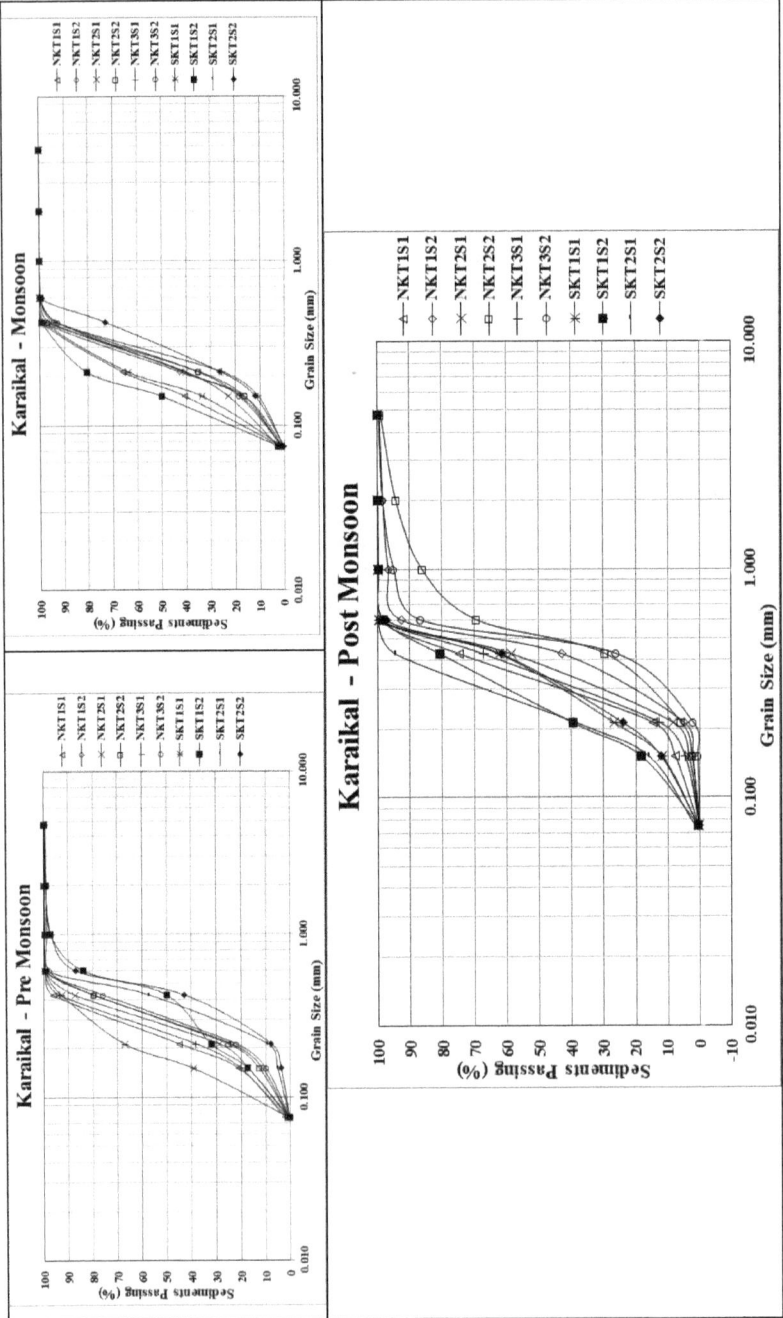

Fig. 9.11(a).   Grain size distribution — confluence of ocean and estuary.

Table 9.2. Sediment sample collected location coordinates.

S. No	Sample ID	Remarks	Transects	Latitude	Longitude
1	KT1S1	Berm Sample	T1	10°54′54.84″N	79°51′14.34″E
2	KT1S2	Shore Sample		10°54′54.90″N	79°51′14.52″E
3	KT2S1	Berm Sample	T2	10°55′03.18″N	79°51′12.36″E
4	KT2S2	Shore Sample		10°55′03.24″N	79°51′12.54″E
5	KT3S1	Berm Sample	T3	10°55′08.70″N	79°51′11.82″E
6	KT3S2	Shore Sample		10°55′08.70″N	79°51′12.12″E
7	KST1S1	Berm Sample	T4	10°54′38.33″N	79°51′11.33″E
8	KST1S2	Shore Sample		10°54′38.33″N	79°51′11.69″E
9	KST2S1	Berm Sample	T5	10°54′27.71″N	79°51′09.69″E
10	KST2S2	Shore Sample		10°54′27.68″N	79°51′09.95″E

equally distributed sediments of size 0.212 mm, 0.150 mm and 0.75 mm of about 30% each of sediments, and most of the grains were only fine sand and slit sediments. The berm sediments of second transect of the south shows that more than 50% of sediments were fine sand measuring 0.212 mm grain size.

The monsoon observation in all the five transects shows that more than 50% of the sample consists of sediment size ranging between 0.21 mm and 2 mm (medium to fine sand). On comparing the first south transect observation during the monsoon, the size of the sand grains is almost similar to that of the grains during the pre-monsoon observations. As for the post-monsoon observation, more than 50% of the sand grains are found to be of 0.212 mm size, i.e., fine sand.

The pre-monsoon observation of shore sediments (**Fig. 9.12**) shows that along the north transects, more than 50% of the sediments were fine sand of size 0.212 mm, whereas along the south transects, more than 40% of the sediments are fine sand of size 0.212 mm. The monsoon observation of shore sediments shows that in the north transects, more than 60% of the sediments were fine sand of size 0.212 mm, whereas in the first south transect, 40% of the sediments are fine sand, sized 0.212 mm and the second south transect contains more than 50% of the sediments fine sand sized 0.212 mm. The post-monsoon observation of the north transect shows that, in first transect, more than 40% and 30% of sediments contains 0.212 mm and 0.425 mm of fine sand and medium sand, respectively, whereas Transects 2 and 3 show approximately equally distributed sediment of 0.212 mm and 0.425 mm grain size. In the post-monsoon observation of the south transect, the first transect shows that more than 50% of the sediments were of

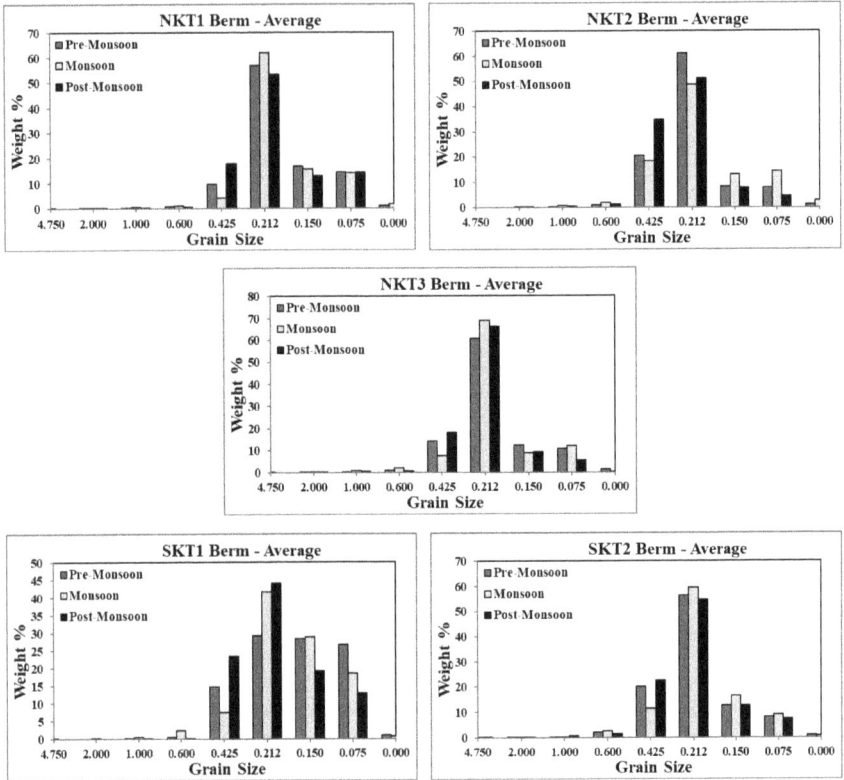

Fig. 9.11(b). Percentage of weight retained w.r.t grain size for the berm samples.

fine sand sized 0.212 mm and the second transect shows that more than 40% and 30% of sediments contains 0.212 mm and 0.425 mm of fine sand and medium sand, respectively.

The analysis on the grain size showed that the median sediment size D_{50} (**Fig. 9.13**) of berm sediments ranges from 0.23 to 0.33 mm during the pre-monsoon period and ranges from 0.24 to 0.31 mm during the monsoon period, whereas during the post-monsoon period, D_{50} ranges from 0.30 to 0.37 mm. However, the analysis on the grain size showed that the median sediment size D_{50} of shore sediments ranges from 0.31 to 0.41 mm during the pre-monsoon period and from 0.21 to 0.31 mm during the monsoon period, whereas during post-monsoon period, D_{50} ranges from 0.43 to 0.57 mm.

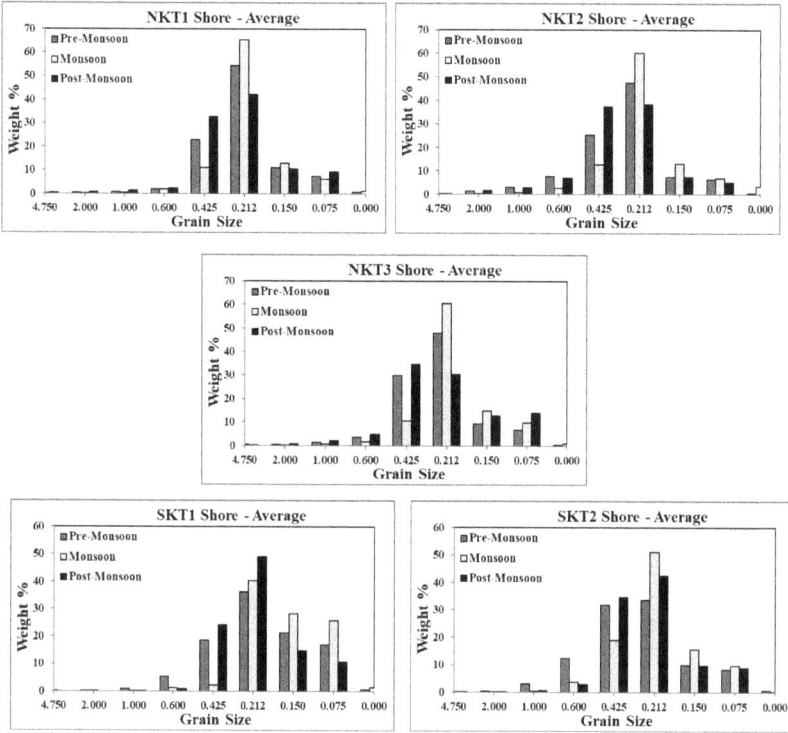

Fig. 9.12. Percentage by weight retained w.r.t grain size for the shore samples.

9.7. Summary

Based on the observations from the cases and the discussions over the obtained results, we can conclude that in the Devaneri (open coast) case, 50% of the grains are fine sand during both the pre-monsoon and post-monsoon periods whereas in the monsoon period, 50% of the grains are both fine sand and medium sand. However, in the Karaikal (estuary) case, 50% of the grains are seemed to be fine sand in all year round (i.e., pre-monsoon, monsoon and post-monsoon periods). In general, it can be noted that the slope of the beach varies with respect to the grain size. In the case of open coast, due to the equal distribution of fine and medium sand, the steep slope can be observed along the Devaneri coast. In the case of near estuary, due to the presence of a pair of training walls, the finer sediments will get settled during LST and the above study also reveals that the Karaikal coast is composed of 50% of finer sand.

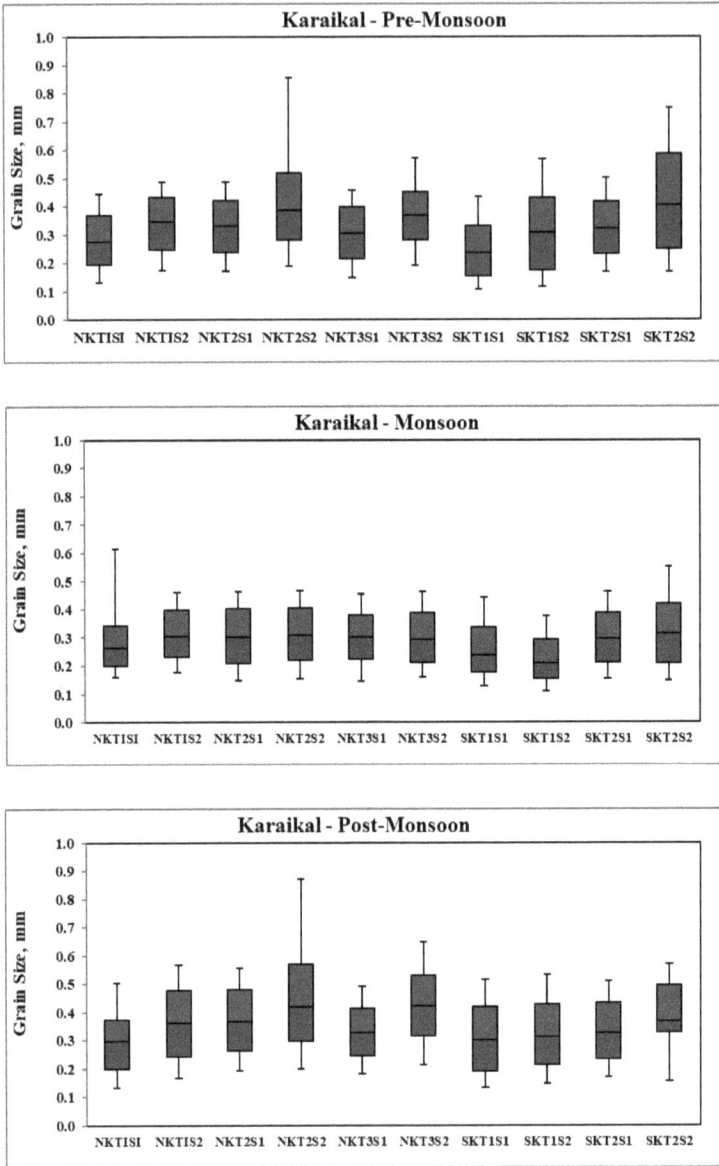

Fig. 9.13.   The variation of D_{50} of the sediment samples.

References

Das, G.K. (2009). Grain size analysis of some beach sands from the Indian Coast. *Geographical Review of India*, 71(1): 10–18.

Holland, K.T. and Elmore, P.A. (2008). A review of heterogenous sediments in coastal environments. *Earth Science Review*, 89: 116–134.

Jayakumar, S., Raju, S.N.N., and Gowthaman, R. (2004). Beach dynamics of an open coast on the west coast of India. 3rd *INCHOE-2004*, Goa, India.

Lee, E.M., Clark, A.R., Doornkamp, J.C., Boardman, J., Hooke, J., Lewin, J., Pethick, J., Brunsden, D., Jones, D. and Newson, M. (1995). The occurrence and significance of erosion, deposition and flooding in Great Britain. HMSO, London.

Merkus, H.G. (2009). *Particle Size Measurements: Fundamentals, Practice, Quality*. Springer, New York.

Narayana, A.C., Tatavarti, R., Shinu, N. and Subeer, A. (2007). Tsunami of December 26, 2004 on the Southwest coast of India: Post-tsunami geomorphic and sediment characteristics. *Journal of Marine Geology*, 242:155–168.

Narayana, A.C., Jago, C.F., Manojkumar, P., and Tatavarti R. (2008), Nearshore sediment characteristics and formation of mudbanks along the Kerala coast, southwest India. *Journal of Estuarine, Coastal and Shelf Science*, 78:341–352.

Oyedotun. T.D.T. (2016). Sediment characterization in an estuary — beach system. *Journal of Coastal Zone Management*, 19:3. DOI: 10.4172/2473-3350.1000433.

Narra, P., Coelho, C. and da Fonseca, J.E.R. (2015). Sediment grain size variation along a cross-shore profile-representative D_{50}. *Journal of Coastal Conservation*. DOI: 10.1007/s11852-015-0392-x.

Peroteasa, L. and Vespremeanu-Stroe, A. (2010). Grain size analysis of the beach-dune sediments and the geomorphological significance. *Revista de Geomorfologie*, 12:73–79.

Pradhan, U.K., Sahoo, R.K., Pradhan, S., Mohany, P.K. and Mishra, P. (2020). Textural analysis of coastal sediments along east coast of India. *Journal of Geological Society of India*, 95:67–74.

Sathasivam, Kankara, R.S., ChenthamilSelvan, S., Muthuswamy, M., Samykannu, A., and Bhoopathi. R. (2015). Textural characterization of coastal sediments along the Tamil Nadu Coast, East Coast of India. *Procedia Engineering* 116:794–801.

Southwell, M.W., Veenstra, J.J., Adams, D., Elizabeth, C.D., Scarlett, V. and Payne, K.B. (2017). Changes in sediment characteristics upon oyster reef restoration, NE, Florida, USA. *Journal of Coast Zone Management*, 20:442.

Sundar, V. and Sannasiraj, S.A. (2019). *Coastal Engineering-Theory and Practice, Advanced Series on Ocean Engineering*, Vol. 47, World Scientific, Singapore.

Vandenberghe, J. (2013). Grain size of fine-grained windblown sediment: A powerful proxy for process identification. *Earth-Science Review*, 121:18–30.

Section C

Modelling, Calibration and Validation

Chapter 10

Wind-Wave Prediction and Design Wave Climate

Abstract

The most dynamic part of the ocean environment is the wind-wave. It has the maximum energy content among all the ambient environmental forces in the ocean. The prediction of such wave climate is of prime importance for various scientific and engineering applications. It is possible to estimate the integrated wave parameters through analytical equations in an unbounded domain with constant wind fields but it requires numerical models for the prediction over complex domain and varying wind conditions. It is always required to establish the operational wave climate for several day-to-day activities and extreme wave climate for the estimate of the design wave. In this chapter, a brief introduction to the historical understanding on the prediction of wave climate to the state-of-the art numerical wave models is presented. The extraction of design wave climate is highlighted.

10.1. Wave Prediction: Wind-Waves

The prediction of sea states is of great importance for ship routine and offshore activities.

Often the design requirements are derived from the design wave height and wave period. The extent of maximum wave height and wave period achieved during an extreme event dictate the lifetime design of the structures as well as the number of design components involved. The design requirements vastly diverge based on the type of structures under study, which involves numerous parameters. The design of offshore structures needs the extreme wave climate while an operational wave climate is required to be established for the prediction of tranquillity condition inside a harbour.

In early days, wave prediction was based on the observations of an experienced sailor. Recently, remote sensing images provide valuable wave information and are now widely adopted for the observation of the ocean. However, remote sensing data needs a numerical model to assimilate data in the domain of interest from its limited tracks over the region. Numerical models can also independently be applied for the prediction of wind waves. In this chapter, wind-wave prediction is addressed from the simple approximation method to the complex numerical prediction that needs relatively considerable effort. Both the extreme methods of prediction are still being commonly adopted from the conceptual design stage to the detailed engineering design.

10.2. Historical Wave Assessment

It is possible to estimate wave parameters, significant wave height, H_s and the mean zero crossing period, T_z from local geography and wind information in addition to historical representation of wave records. The oceanography maps provide H and T values over 50- or 100-year storm. The information about different sea states, H_s and T_z, over say a one-year period, can be best be represented by the wave scatter diagram. It indicates also the number of occurrences and hence will be useful for fatigue studies.

The wave rose diagram as projected in **Fig. 10.1** provide information about percentage exceedance of H_s from various directions, θ_m. The wave height exceedance diagram is used to model the extreme conditions of storm for fatigue or foundation settlement problems.

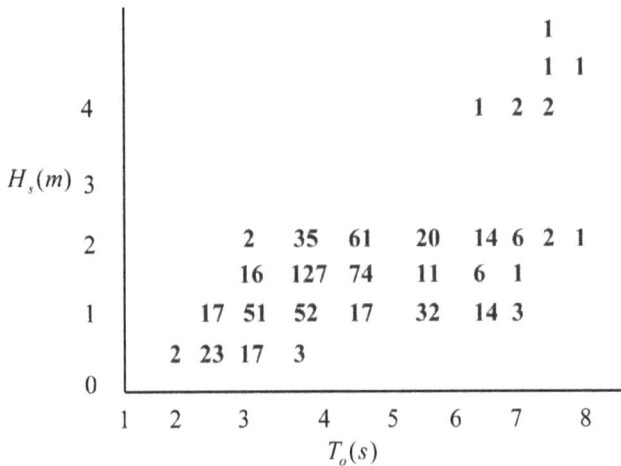

Fig. 10.1. Wave scatter diagram.

10.3.   Principle of Wind-Wave Generation

10.3.1.   *General*

The most common reason for the formation of waves is the wind that transfers the energy when it blows across the ocean. Higher-speed winds will transfer more energy to the water, resulting in stronger waves. Several theories have been evolved to address the wind-wave generation, which is a complex process. As the wind blows over the water surface, wrinkles are initially formed on the surface due to pressure and shear effects, and hence, small ripples are formed. Continued transfer of wind energy on the water surface accelerates the growth of the waves. The growth of waves also depends on the duration of the wind blowing over the sea surface. The magnitude and direction for a particular location is presented as wind rose diagram, a typical diagram which projects the magnitude of the wind speed in knots represented as contours (concentric circles) and its magnitude from each direction, as shown is **Fig. 10.2**. The strength of the wind and wave is usually calculated in accordance with the Beaufort wind force or Beaufort scale (**Table 10.1**), devised by Commander Francis Beaufort in 1805 for observing and classifying wind force at sea. Even today, the Beaufort scale serves as a useful means to caution fishermen venturing into the ocean.

10.3.1.1.   *Pressure effect*

From Bernoulli's theorem, the pressure increases if the velocity drops and vice versa.

(1) Above the crest, the wind velocity increases and hence pressure decreases.
(2) Above the trough, the wind velocity decreases and hence pressure increases.

10.3.1.2.   *Shear effect*

Water surface is stretched due to wind shear. Friction forces between air and water induce shear forces and push water particles to make small hills and hence a down-hill on its upstream.

Due to pressure and shear effects, ripples are formed on the water surface. Three main factors governing wave growth from the ripples stage are:

(1) wind speed,
(2) wind duration and
(3) fetch (length over which wind blows).

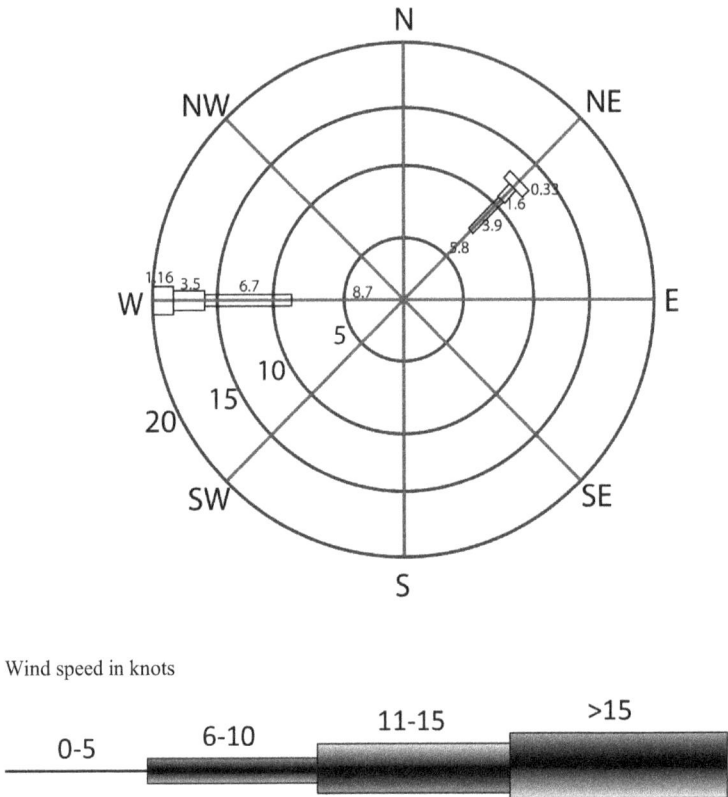

Wind speed in knots

Fig. 10.2. Wind rose diagram.

Table 10.1. Beaufort wind force scale.

Beaufort number	Description	Wind Speed (m/s)	Wave height (m)	Sea conditions
0	Calm	0.0–0.2	0	Calm undisturbed sea
1	Light air	0.3–1.5	0.1	Ripples without form
2	Light breeze	1.6–3.3	0.2	Small wavelets
3	Gentle breeze	3.4–5.4	0.3–1.0	Large wavelets
4	Moderate breeze	5.5–7.9	1.0–1.5	Small waves
5	Fresh breeze	8.0–10.7	1.5–2.5	Moderate waves
6	Strong breeze	10.8–13.8	2.5–3.5	Larger waves
7	Near Gale	13.9–17.1	3.5–5.0	Sea heaps up
8	Gale	17.2–20.7	5.0–6.5	Moderately high waves
9	Strong gale	20.8–24.4	6.5–8.0	High waves
10	Storm	24.5–28.4	8.0–10	Very high waves
11	Violent storm	28.5–32.6	10–13	Exceptionally high waves
12	Hurricane	≥ 32.7	≥ 14	Phenomenally high waves

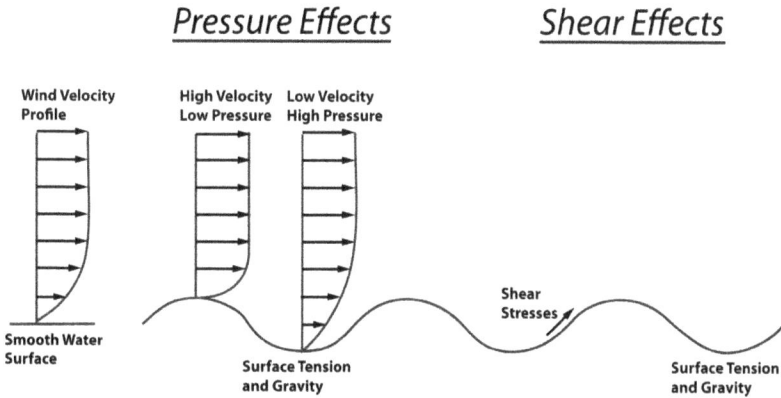

Fig. 10.3. Shear and pressure differences along a wave.

Ripples grow if the wind continuously blows over the surface of the ocean. Let us assume wind of constant mean velocity blows over the ocean. Initially, high-frequency short waves formed. After the wave breaks, it is transformed to low-frequency components, and the nonlinear wave–wave interaction shifts the energy to low frequency. The above-mentioned effects are illustrated in **Fig. 10.3**.

Fact to ponder: Water is a frictionless medium!

So, energy continuously transfers from high frequency to low frequency until the phase velocity of wave (C) is equal to the wind velocity (U), i.e., $C = U$. Generally, we can take $C_g = U$.

Beyond, $C_g > U$, wind does not supply energy to waves directly. That means, waves reach an equilibrium at certain extent. This equilibrated sea state is called as *fully developed sea*. The growth of the waves due to the wind blowing over its surface is pictorially represented in **Fig. 10.4**.

Two physical constraints may restrict the above process:

(1) Before the wave attains its equilibrium, if the wind stopped blowing, the sea state can be called "underdeveloped" from its potential. This sea is called *duration limited*.
(2) On the other hand, if there is a constraint in the space for wave propagation for the wave to grow, then also waves would stop growing. This sea is called the *fetch-limited* sea.

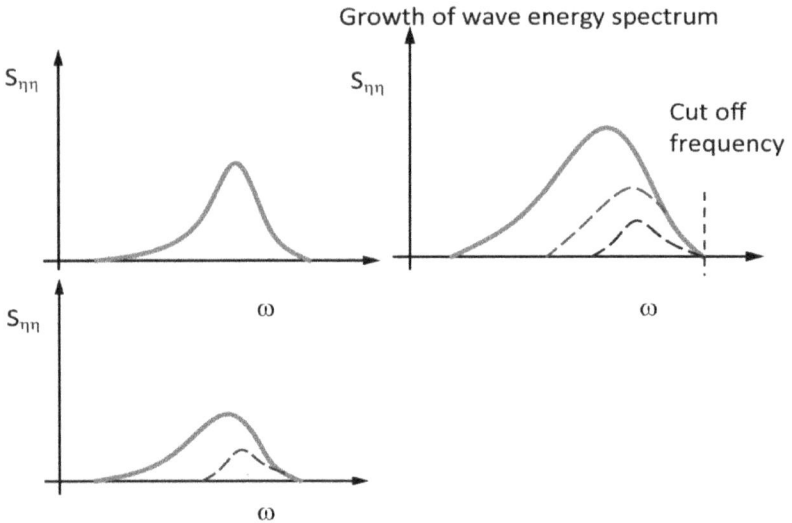

Fig. 10.4.   Spectral wave growth.

Hence, wind is the exciting force in this case and gravity is the restoring force. Hence, the sea is developed as a vertically oscillating component influenced by gravitational constant, g. Hence, the wind-waves are also called as gravity waves.

From the above discussion, Fully developed sea $= f(U, g)$
Duration-limited sea $= f(U, t_d, g)$
Fetch-limited sea $= f(U, L_F, g)$

where t_d is the duration of wind and L_F is the fetch.

10.3.1.3.   *Fully developed sea state*

It is known that any sea state has an equilibrium for the given wind climate, therefore, it is obvious to seek a solution to arrive at the wave characteristics, such as significant wave height and mean wave period, directly. Several theories predict reasonably well, explaining that the energy from the wind will be equal to the energy of the breaking waves making an equilibrium sea state.

Spectral model such as Pierson–Moscowitz (PM) spectrum is a unidirectional spectrum for a fully developed sea.

10.3.2.   *Wave climate and sources*

The general state of a sea condition at a particular location is referred to as the wave climate. The design wave climate comprise parameters such as wave height, wave period and wave direction. In the operation and design of coastal structures, wave climate plays a major role, as underestimation of wave condition in the location of structures will lead to their failure. The operation of the coastal structures is governed by the normal wave condition while the stability of the coastal structure depends on the storm wave conditions.

The waves with a a short duration of 20 minutes are classified as short-term statistics of an individual wave. The wave climate of about one year are long-term statistics while storm wave data of 30–50 years are termed as extreme wave data, which is applicable for the design of marine structures. Wave data sources are as follows:

(1) instrumentally measured data,
(2) visually observed data and
(3) hindcast storm wave data.

Wave rider buoy is most common instrument used for measuring wave data. The quality and duration of wave data will differ based on the site conditions.

10.3.3.   *Wave spectral model: PM spectrum*

For a fully developed sea, following the PM formulation (Pierson and Moskowitz, 1964), the estimates for significant wave height ($H_{mo} = H_s$) and peak wave frequency (f_p) are derived as follows:

$$H_{mo} = \frac{0.21\,U^2}{g} \tag{10.1}$$

$$f_p = \frac{0.87g}{2\pi U} \tag{10.2}$$

where U is the wind speed in m/s, and g is the gravitational constant. The above formulae were derived for wind speeds between 10 and 20 m/s, and it was assumed that the sea was neither fetch-limited nor duration-limited.

10.3.4. *SMB wave prediction curves*

Utilizing a limited amount of field data for calibration, a wave prediction procedure was developed by Sverdrup and Munk (1947) based on the wave energy growth concepts. Bretschneider (1952, 1958) improved this procedure by calibration with a vast field data. The term SMB method is coined after the three authors. Consider a dimensional analysis of the basic wave prediction relationship,

$$H_s, T_s = f(U, L_F, t_d, g) \tag{10.3}$$

Depending on whether the wave generation is fetch- or duration-limited, the fetch or the duration term on the right side would control the estimation.

$$\frac{gH_s}{U^2} = 0.283 \tanh\left[0.0125\left(\frac{gX}{U^2}\right)^{0.42}\right] \tag{10.4}$$

$$\frac{gT_s}{2\pi U} = 1.2 \tanh\left[0.077\left(\frac{gX}{U^2}\right)^{0.25}\right] \tag{10.5}$$

$$\frac{gt}{U} = K \exp\left[\left\{A\left(\ln\left(\frac{gX}{U^2}\right)^2\right) - B\ln\left(\frac{gX}{U^2}\right) + C\right\}^{0.5} + D\ln\left(\frac{gX}{U^2}\right)\right] \tag{10.6}$$

where $K = 6.5882$; $A = 0.0161$; $B = 0.3692$; $C = 2.2024$; $D = 0.8798$.

The above relation has been presented in the form of empirical equations and dimensional plots and is shown in **Fig. 10.5** (U.S. Army Coastal Engineering Research Centre, 1977).

For a fetch-limited wave condition, the solid lines can be used to predict the significant wave height and period. For a duration-limited wave condition, the dashed line can be used. Note that the parameters, fetch, duration, significant wave height and wave period were non-dimensionalized in terms of wind speed. The curves tend to become asymptotic to each other and horizontal lines on the right-hand edge. This limit is the fully developed sea condition.

10.3.5. *SPM: Deep water wave prediction*

A parametric model based on JONSWAP studies has estimated wave characteristics under fetch-limited and duration-limited wind conditions.

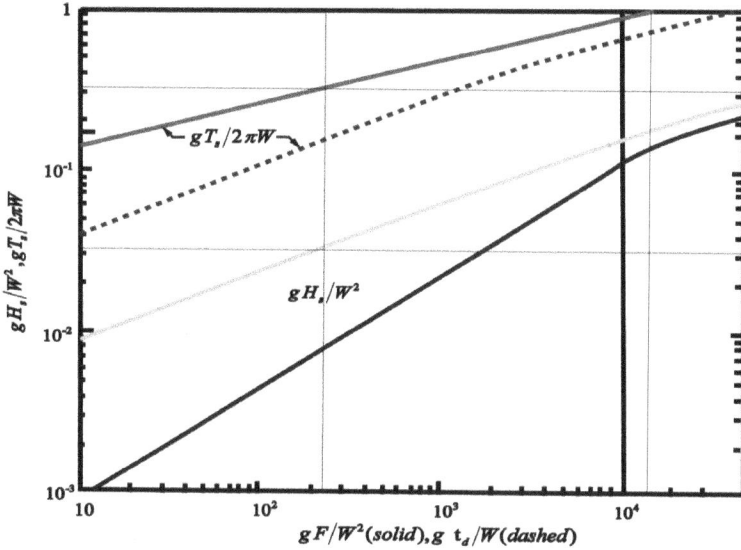

Fig. 10.5. SMB wave prediction curves.

Note: W is the wind speed, t_d is the duration of wind and F is the fetch.

Source: Bretschneider (1952, 1958).

For fetch-limited condition:

$$\frac{gH_{m0}}{U_A^2} = 0.0016 \left(\frac{gL_F}{U_A^2}\right)^{1/2} \tag{10.7}$$

$$\frac{gT_p}{U_A} = 0.286 \left(\frac{gL_F}{U_A^2}\right)^{1/3} \tag{10.8}$$

For duration-limited condition:

$$\frac{gt_d}{U_A} = 68.8 \left(\frac{gL_F}{U_A^2}\right)^{2/3} \tag{10.9}$$

Here, the wind is adjusted to U_A from U_{10} as, $U_A = 0.71U_{10}^{1.23}$. It is to be noted that the above expressions have empirical coefficients and hence, it is sensitive to units. Wind speed is given in terms of m/s.

10.4. Extreme Sea States

The maximum significant wave height, $H_{\max}$ generated by a tropical cyclone in deep waters is given as,

$$H_{\max} = 0.2(P_n - P_c) \tag{10.10}$$

where, $(P_n - P_c)$ is the pressure drop from the environment to the cyclone centre in hPa (Hsu, 1991). However, the wave height attenuates quickly outside this region, and in most cases, it may not be of any interest.

10.5. Wind-Wave Modelling

10.5.1. *General*

All the above efforts in predicting wave heights are based on the fact that the domain is either unbounded or in the defined boundedness. However, in most of the field situations, one needs to predict (forecast) wave climate for operational reasons, such as to establish a navigational route (this is a million-dollar industry); to establish a time window for the erection of jacket structure; or to evaluate the number of operational days in an offshore jetty. Such a near-accurate estimate could not be made with the historical methodologies.

Assisting the above objectives, world wars had established the sea route to invade countries. This led to a huge investment in operational wave prediction models by the naval fleets across the globe. It had led to sudden surge in the funding to the research developments of theoretical modelling aspects of wind-wave modelling after the World Wars I and II. The wind velocities and pressure variations over the sea-surface are the primary factors driving the wind waves as seen in **Fig. 10.6**. The resultant shear and pressure effects over a typical wind-generated wave is depicted in **Fig. 10.7**.

Jeffrey's theory (1924, 1925) laid a foundation to the concept of momentum transfer from wind to wave even in 1920's. It addressed whether variation in pressure, as discussed earlier, can result in a flux of energy from the wind to the waves.

$$\frac{\partial E}{\partial t} = \frac{1}{\rho_w g} \overline{\rho_a \frac{\partial \eta}{\partial t}} \tag{10.11}$$

The pressure component that is correlated with $\frac{\partial \eta}{\partial t}$ (*slope*) yields to an energy flux to the wave (i.e., the component of pressure in quadrature with the water surface). So, there should be a phase shift between $\eta \& p$ for positive energy flux, higher pressure on the windward side of the wave.

From Potential flow formulations,

$$\frac{p}{\rho_a g} = -ae^{-kz}\left(1 - \frac{U}{C}\right)^2 \sin(kx - \omega t) - z \tag{10.12}$$

Here, p is in out of phase with η. From Eq. (10.11), this leads to no flux of energy from the air to the water. This is counter to our intuition. Also, Eq. (10.12) shows that p exponentially decreases with z above water surface.

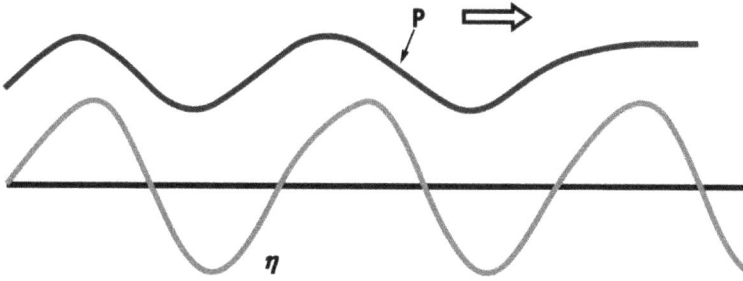

Fig. 10.6. Air pressure variation above the wave surface.

10.5.2. *Jeffrey's sheltering theory*

The energy transfer by form drags associated with flow separation (on lee-ward side) is expressed by Jeffrey's sheltering theory as,

$$p = S\rho_a (U_\infty - C)^2 \frac{\partial \eta}{\partial x} \tag{10.13}$$

where S = sheltering coefficient < 1, and
   U_∞ = wind speed (no boundary layer is considered).
   From Eqs. (10.11) and (10.13),

$$\frac{\partial E}{\partial t} = \frac{1}{2\rho_w g} S\rho_a (U_\infty - C)^2 (ak)^2 C \tag{10.14}$$

Note, $E\alpha a^2$ (phase speed)

$$\alpha (U_\infty - C) \quad \text{(So, no energy flux if } U_\infty = C)$$

where α is the wave slope. So no energy flux would happen if $U_\infty = C$, and, $E \propto ak$, i.e., wave slope (due to flow separation assumption). Here, S depends on U_{min}^3, i.e., to keep minimum wind for the wave to sustain against the losses.

After nearly three decades from the initial work of Jeffrey, the work of Philips (1957, 1960) and Miles (1957) achieved milestones in trans-ferring the understanding of energy transfer mechanism to mathematical formulation.

The key to Philips' (1957, 1960) study is the pressure effect. The main cause of the wave growth is hypothesized as a resonant interaction between forward-moving pressure fluctuations and free waves propagating at the same speed as the pressure fluctuations.

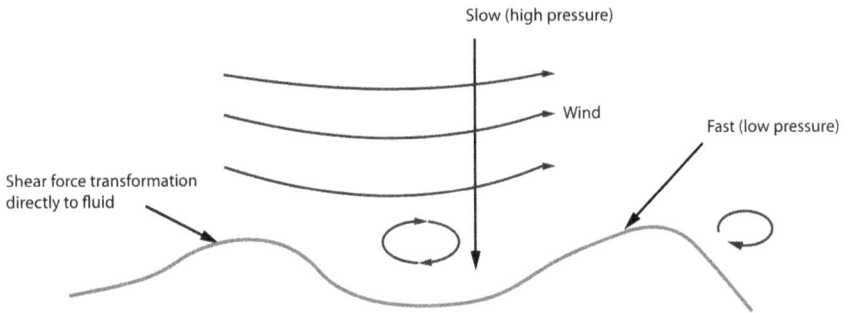

Fig. 10.7.   Pressure and shear effects on a wind-generated wave.

Miles (1957) hypothesized on the air flow patterns above the free surface that develops due to the wavy surface. It specifies the development of a secondary air circulation around an axis parallel to the wave crest by the wind velocity profile. This is based on the fact that if one closely follows the wind velocity profile, as discussed earlier just above the free surface, below one point, the wind velocity becomes lesser than the wave phase velocity. Below this point, air flow is reversed relative to the forward moving wave profile. However, above this point, air flow direction is same as wave propagating direction. This result in a relative flow circulation in a vertical plane. This causes an out of phase pressure distribution on the wave surface with the surface displacement, η. If we refer back potential flow formulation Eq. (10.12), there is a momentum transfer to the wave at particular wave components.

Many other theories have evolved but Philips and Miles' theories dominate others in addressing the initial knowledge on the development of the wind waves. One of the significant findings from other theories which is absorbed in the theoretical modelling is that the ripples on the free surface create more friction and hence frictional forces can be enhanced if the wind blows over the surface compared to over a very calm sea.

10.6.   Wave Evolution Modelling

Let us define a mathematical model for the description of the evolution of wind waves from the physical processes which influences this evolution. Now, let us attempt to describe the evolution of wind waves.

From Jeffery's theory (1924, 1925), it can be understood that the flux of energy is averaged over a period of time, and hence modelling the rate of change of wave energy is an ideal modelling parameter compared to the wave profile and its velocity components.

In 1950s, the governing equation for wave propagation was formulated.

$\frac{dE}{dt} = 0$ for no wind condition, i.e., only wave propagation is considered.

If there is wind, we can include the input wind energy as a source function.

$\frac{dE}{dt} = S_{in}$, where S_{in} is the input source function (addition of energy).

If there is only addition of energy without the provision of any extraction, numerically the energy spectrum will grow without attaining the equilibrium state. Hence, the energy extraction process through wave breaking and friction are modelling through a negative source function, called dissipation source.

$\frac{dE}{dt} = S_{in} + S_{dis}$, where, S_{dis} is the dissipation source function (sink).

By observing the above source functions, S_{in} will pump the energy on the higher frequency band up to the limit of $U = C_g$, where the resonant interaction ceases to transfer energy to the water surface. Moreover, the dissipation function again takes away the energy from higher-frequency band relative to the S_{in}. Overall, for a given wind speed, the energy spectrum will not grow towards lower-frequency bands with the above definitions. The waves grow, and there is transfer of energy from high- to low-frequency bands due to nonlinear wave–wave interaction.

Hence, another source function modelling nonlinear wave–wave interaction has been included to complete the wind-wave modelling equation.

$$\frac{dE}{dt} = S_{in} + S_{dis} + S_{nl} \tag{10.15}$$

Even with the present-day capabilities of supercomputers, it is uneconomical to fully solve the nonlinear wave–wave interaction term. Hence, only resonant interaction terms by four wave–wave interaction has been modelled by Hasselmann and Hasselmann (1985). This reduced order has been included in the recent state-of-the art models, WAM and WaveWatch III, as Discrete Interaction Approximation (DIA) term.

For nearshore wave prediction, an additional nonlinear wave–wave interaction term, called triad interaction, has to be included to consider the three-wave resonant interaction in coastal waters.

Figure 10.8 shows the energy distribution contribution of various source function with reference to the given spectrum.

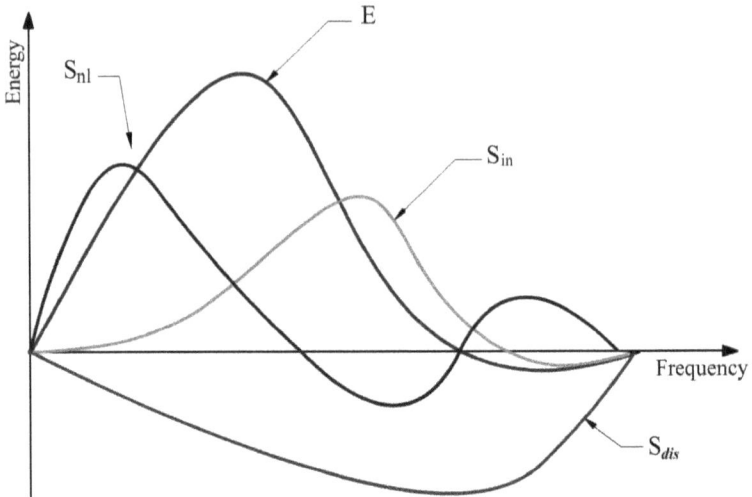

Fig. 10.8.   Distribution of energy by various source functions.

10.7.   Numerical Wind-Wave Model

10.7.1.   *Background*

The spectral approach for developing a numerical wave modelling has been in existence since the early 1960s. The solution for the wave energy transport equation was arrived at with the concept of spectral energy balance. Linear formulation and prescribed spectral shapes were the basis for the first-generation wave models. This linear formulation of the wind-wave momentum transfer over-predicted the wave climate. In the second-generation wave models, nonlinear terms for wave–wave interactions were included. Owing to the computational restrains involved in solving for the integration of nonlinear energy transfer terms, the complete formulation was not achieved. The complex wave–wave interaction terms were solved by the method of discrete interaction approximation (Hasselmann and Hasselmann, 1985), which laid foundation for the third-generation models. The wind input source term was defined by Janssen *et al.* (1984). Over two decades post the implementation of the third-generation models, gradual improvements are being made.

In the third-generation WAM model, the wave transport equation is solved explicitly without making assumptions on the shape of the resultant spectrum with the estimation of the evolution of the ocean wave energy spectrum (WAMDI, 1988; Komen *et al.*, 1994). Hasselmann (1963) proposed an equation for the energy balance of the wave spectrum, which is the basis for the exact theory of wave spectrum dynamics.

$$\frac{\partial F(f, \theta; x, t)}{\partial t} + v \cdot \nabla_x F(f, \theta; x, t) = S \qquad (10.16)$$

where $F(f, \theta; x, t)$ is the wave energy spectrum in terms of frequency f and propagation direction θ at the position vector, x and at time t; v is the group velocity. The second term on the left-hand side is the divergence of the convective energy flux, $v \cdot \nabla_x F(f, \theta; x, t)$, and S is the net source function which considers all physical processes which contribute to the evolution of the wave spectrum. The source function is represented as superposition of source terms due to wind input, non-linear wave–wave interaction, dissipation due to wave breaking, and bottom friction.

$$S = S_{in} + S_{nl} + S_{ds} + S_{bot} \qquad (10.17)$$

The combination of the source terms depicts our understanding of the physical processes of wind waves. Spectral shapes corresponding to the measured wind wave spectra are formed from the inputs of wind fields, non-linear interaction and bottom friction. Except for the non-linear source term, which uses the discrete interaction approximation that simulates an exact non-linear transfer process formulated by the four-wave resonant interaction Boltzmann equation and characterizes the third-generation model, all the other source terms are individually parameterised to be proportional to the action density spectrum, F.

The wind input source function was adopted from Snyder *et al.* (1981) and Komen *et al.* (1984). The nonlinear source function S_{nl} is represented by the discrete interaction operator parameterisation proposed by Hasselmann *et al.* (1985),

$$S_{nl}^{di}(k_4) = \sum_{\gamma=1,2} A_\gamma \omega_4 \left[n_1^\gamma n_2^\gamma (n_3^\gamma + n_4^\gamma) - n_3^\gamma n_4^\gamma (n_1^\gamma + n_2^\gamma) \right] \qquad (10.18)$$

where A_γ are coupling coefficients and the action densities

$$n_i^\gamma = F(k_i^\gamma)/\omega_i^\gamma, \quad i = 1, 2, 3; \ \gamma = 1, 2 \qquad (10.19)$$

are evaluated at discrete wave numbers $k_i^\gamma = T_i^\gamma k_4$. The discrete wave numbers are related to the reference wave number k_4 through fixed linear transformations, T_i^γ. These discrete interactions have been tested for fetch- and duration-limited wave growth. In the finite depth hind cast studies, the depth-dependent angular refraction term is generally ignored.

10.7.2. *Wave propagation over constant depth bathymetry*

The wave generation over a basin of constant water depth was considered. A water depth of 250 m was assumed over a region of 20° × 20°. The grid resolution was 1/12° × 1/12°. In this case, the model was run for different combinations of wind and current fields. These were:

(1) Constant wind blowing over the entire region in the absence of current field;
(2) Constant wind in addition to in-line current field;
(3) Constant wind over opposing current field.

A constant northerly wind of speed 10 m/s was assumed to blow over the entire region. An initial wave was set up with the same wind condition. The simulation was then carried out for 48 hours and the steady state was reached. In the second case, with the above wind field, a constant in-line current of 5 m/s was assumed to be present. In the last case, a constant opposing current field of 5 m/s was assumed. The current direction in the second condition was the same as the wind direction while, in the last condition, it was 180° out-of-phase with the wind direction.

The simulated wave field was analyzed for estimated spectral parameters such as significant wave height, H_s and peak wave period, T_p. **Figure 10.9** shows the evolution of wave field in the virtual constant depth basin under the action of constant wind field over period of 48 hours. The wave parameters such as significant wave height and mean wave period approached asymptotic values at the end of the propagation period. The estimates were compared to the values from the analytically derived equations for the constant wind field. The comparison of wave characteristics from WAM with the wave spectral model and SMB prediction curves is presented in **Table 10.2** for the constant wind condition. **Table 10.3** presents the variation in wave conditions in the presence of in-line and opposing current field.

The spectra for the three cases are shown in **Fig. 10.10**. It can be seen that the in-line current field reduces the wave height and shifts the peak frequency toward higher harmonics. The opposing current field, however, made the waves steep by focusing on the narrow band of frequencies. The frequency components were shifted towards lower harmonics.

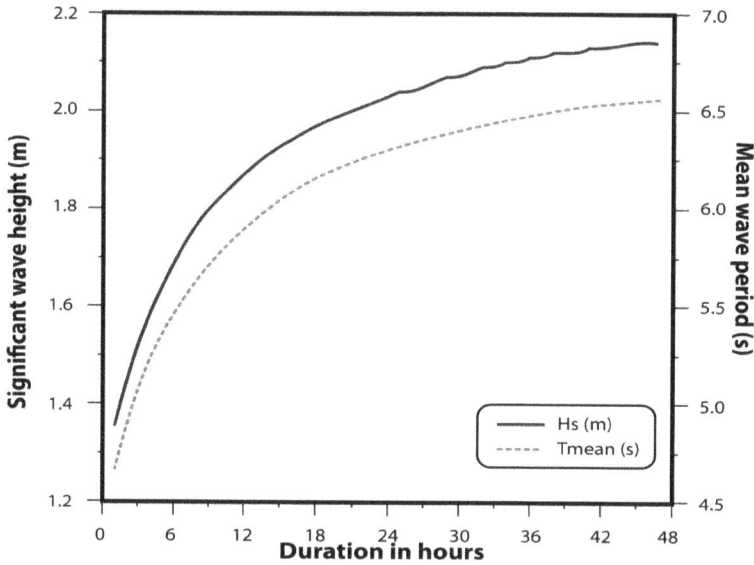

Fig. 10.9.   Evolution of wave components with time in a constant uni-directional wind field blowing over the entire region.

Table 10.2.   Variation of simulated wave estimates under the action of constant wind and current fields.

S.No.	External Forcing	WAM	Wave Spectral Model	SMB
1.	Significant Wave height	2.14	2.14	2.14
2.	Peak wave period	7.44	7.4	7.69

Table 10.3.   Comparison of simulated and analytically derived wave estimates in a constant northerly wind of $10 \, \text{m/s}$.

S.No.	External Forcing	H_s (m)	T_p (s)
1.	Wind	2.14	7.44
2.	Wind + Inline current	2.07	5.09
3.	Wind + Opposing current	2.10	13.19

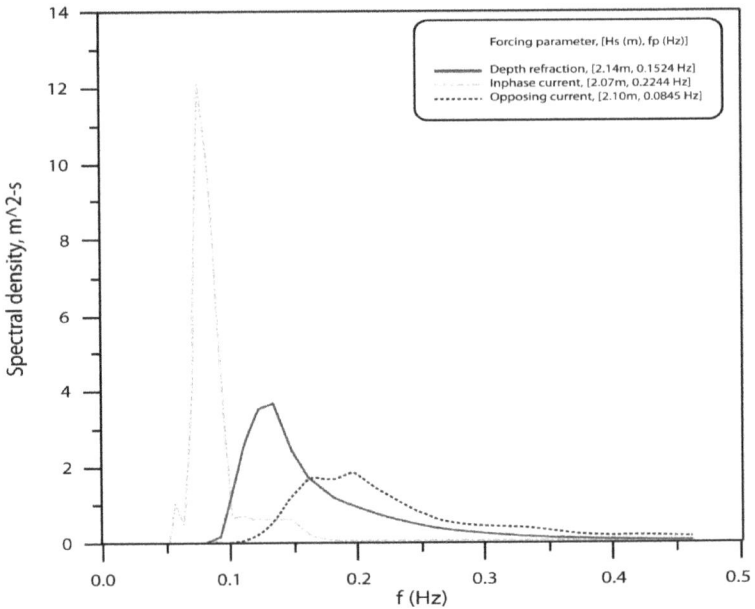

Fig. 10.10. Variation of generated wave spectra under different wind and current fields.

10.7.3.   *Real-field wave prediction*

The bathymetry of required resolution over the domain and the wind field at required time steps are essential for better wave prediction. Given the constraint of the numerical modelling capabilities, the wave prediction accuracy depends on the precision of wind vectors. In this section, the main focus is the simulation of wave climate along southeast coast of India during operational conditions and during severe cyclonic sea state. The model domain considered is 0°–25°N and 75°E–95°E covering Bay of Bengal that is shown in **Fig. 10.11**. The 2-minute resolution ETOPO2 bathymetry is obtained from National Geophysical Data Center (NGDC), a division of National Oceanic and Atmospheric Administration (NOAA). In the present study, a model grid resolution of $0.1° \times 0.1°$ is considered. The wind data is obtained from National Centers for Environmental Prediction (NCEP), NOAA, with a grid resolution of $0.5° \times 0.5°$. A typical wind vector over the domain at 0600 hours on 29 December 2011 is projected in **Fig. 10.12**. The simulation period has been chosen in accordance with the field measurement program during a normal sea state as well as during the occurrence of cyclones.

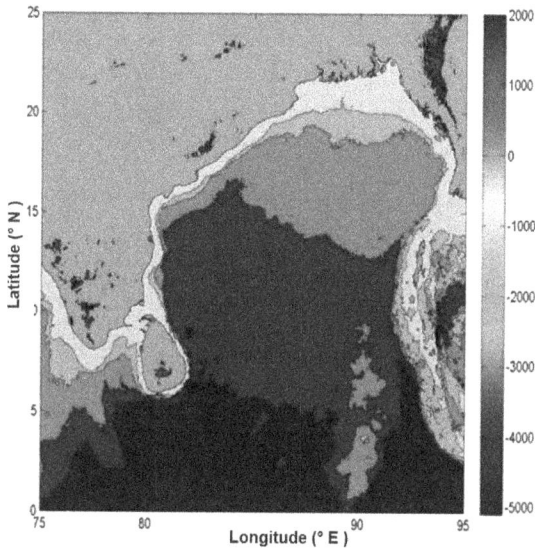

Fig. 10.11. The domain used in WAM for simulation.

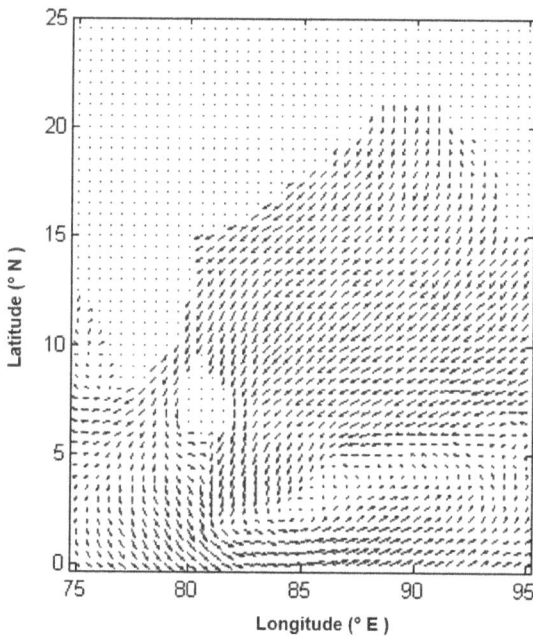

Fig. 10.12. Typical wind vectors (12:00 hrs, 22 December 2011) obtained from ENCEP.

The idea is that this would also facilitate the validation of the predictions of wave characteristics with the third-generation wave model, WAM during extreme events. It is to be noted that the offshore wave boundary has been derived from the large Indian Ocean domain up to 50°S latitude.

Functionality of WAM and output of the model:

The following wave propagation processes are implemented in the model:

- spherical propagation.;
- deep and shallow water;
- depth refraction;
- dissipation due to white-capping and depth limitation;
- wave generation by wind;
- nonlinear wave–wave interaction.

The model provides the following output quantities:

- significant wave height;
- mean wave direction;
- mean frequency.

10.7.4.   *Results and discussion*

10.7.4.1.   *Wave characteristics*

The wave characteristics such as significant wave height, mean wave period and mean wave direction at a deep-water location off Chennai coast have been extracted, which are sampled every six hours. Fundamentally, the wave fields generated follows the wind pattern blown over the ocean surface. A close relation of the spatial variability amongst the wave characteristics can be observed, i.e., a maximum value of significant wave height corresponds with the maximum of wind speed. **Figure 10.13** represents the typical wave climate exceedance for an offshore location and **Fig. 10.14** represents the wave rose diagram of the significant wave height along with the direction. A typical contour plot of H_s obtained from WAM over the entire domain at 0600 hours on 29 December 2011 is projected in **Fig. 10.15**.

The study of a phenomenon at its extreme conditions requires the use of statistical methods specially designed for this purpose, such as the Extreme Value Analysis. Here the extreme wave analysis has been carried out based on Generalized Extreme Value (GEV) distribution method using significant wave height arrived from the WAM model covering a period of 38 years (1981–2018) for Chennai, which is presented in **Fig. 10.16**. A similar WAM

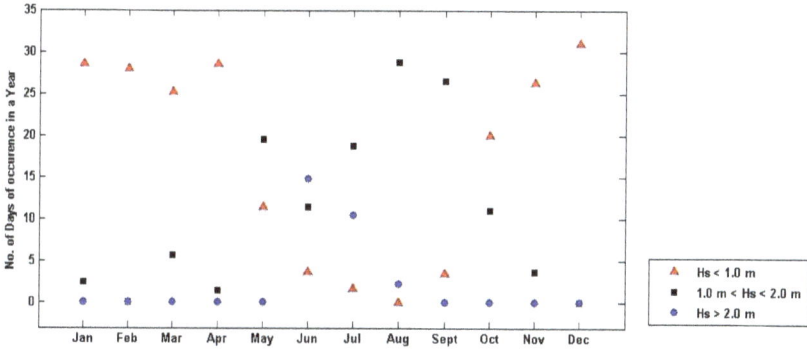

Fig. 10.13. Wave climate exceedance at offshore location.

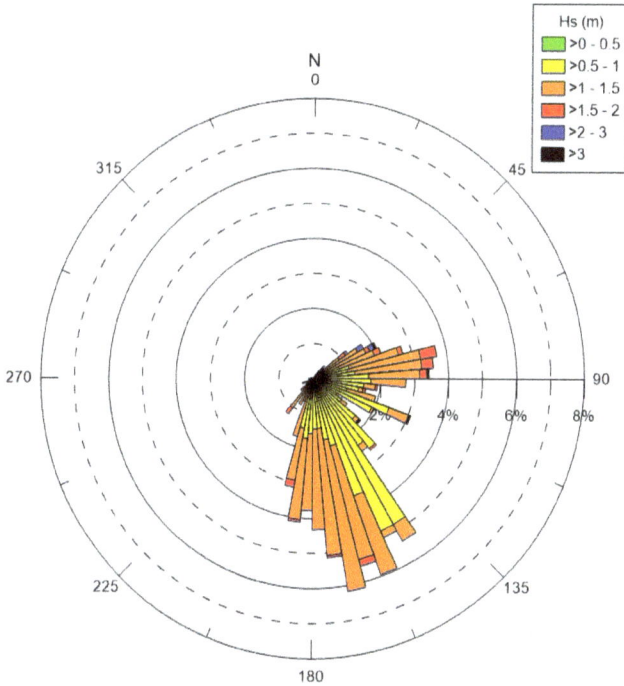

Fig. 10.14. Wave rose diagram representing the significant wave height (m) along the particular direction for an annual year.

model for Kerala is presented in **Fig. 10.17** and details for extreme wave heights on Kochi is shown in **Table 10.4**.

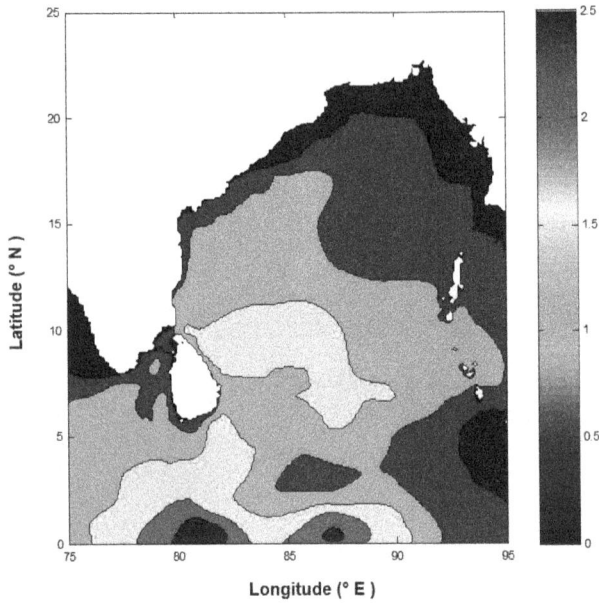

Fig. 10.15. Typical significant wave height contour (12:00 hrs, 22 December 2011) obtained from WAM.

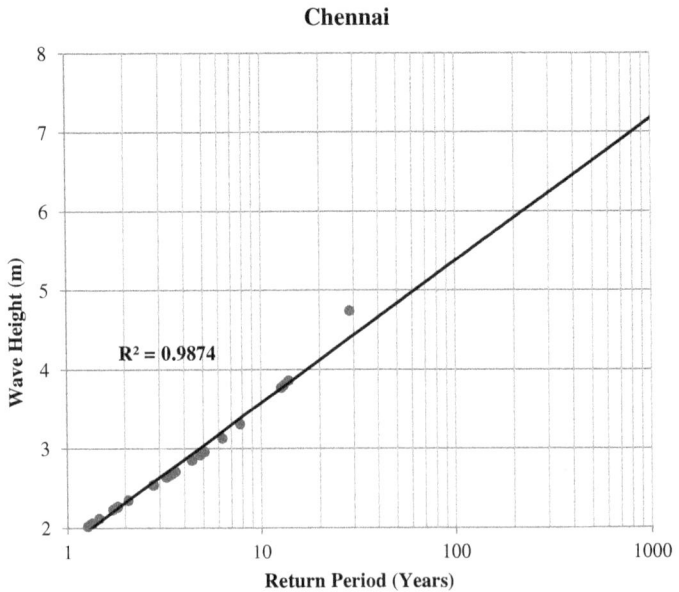

Fig. 10.16. Extreme value for the location of Chennai.

Table 10.4. Extreme wave heights for different return periods of Kochi.

Location	Latitude	Longitude	Return Period (Years)				
			30	50	100	400	1000
Kochi	10.0N	76.0E	4	4.2	4.5	5.1	5.6

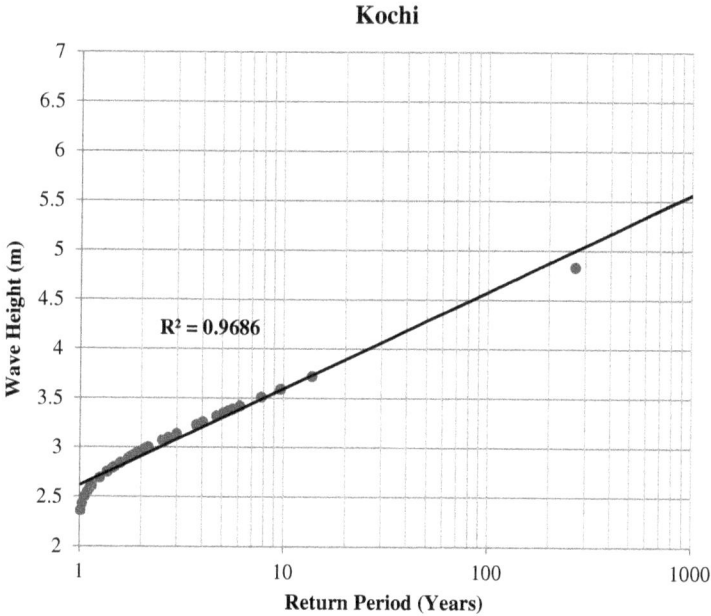

Fig. 10.17. Extreme wave analysis at Kochi.

10.7.4.2. *Validation with field measurements*

The measured wave characteristics, H_s, T_m and θ_m at an offshore location off Chennai coast covering Cyclone Thane are compared with that simulated through WAM as presented in **Fig. 10.18**. The H_s and θ_m values under both normal and cyclonic duration near the study area are found to be in good agreement. The predicted T_m deviates while the cyclone builds up the wave climate, i.e., just after 27 December 2011. Further, the correlation coefficients of computed and measured wave characteristics

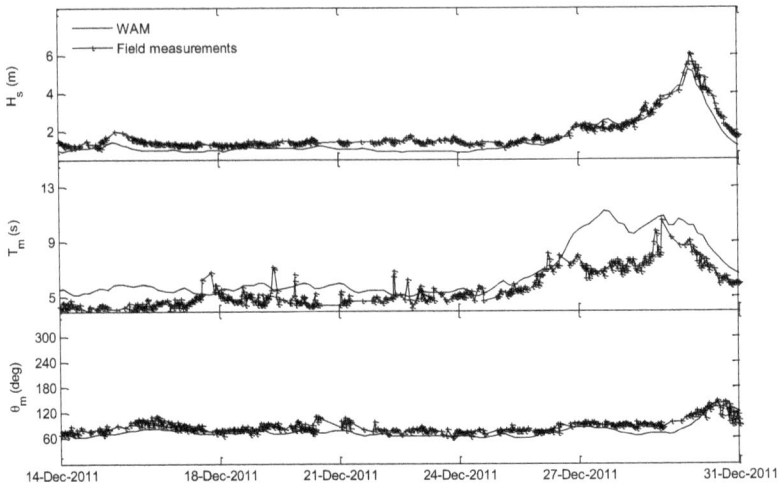

Fig. 10.18. Comparison of results from WAM with field measurements in 20 m water depth at off Chennai coast during second phase data acquisition covering Cyclone Thane (using wave rider buoy).

have been estimated. The correlation coefficient for H_s, T_m and θ_m are found to be 0.89, 0.79 and 0.92, respectively.

The significant wave height, H_s, from buoy measurements at 21 m water depth off Kochi coast during June and July 2016 are compared with the results obtained from WAM simulation in **Figs. 10.19(a) and (b)**.

The wave characteristics H_s and θ_m from measurements off Chennai coast have been successfully simulated for the year 2011 and are projected in **Figs. 10.20–10.32**.

10.7.4.3. *Extreme wave analysis*

The study of a phenomenon at its extreme conditions requires the use of statistical methods specially designed for this purpose like the Extreme Value Analysis. Here the extreme wave analysis has been carried out based on the Generalized Extreme Value (GEV) distribution method using significant wave height arrived from the WAM model covering a period of 38 years (1981–2018).

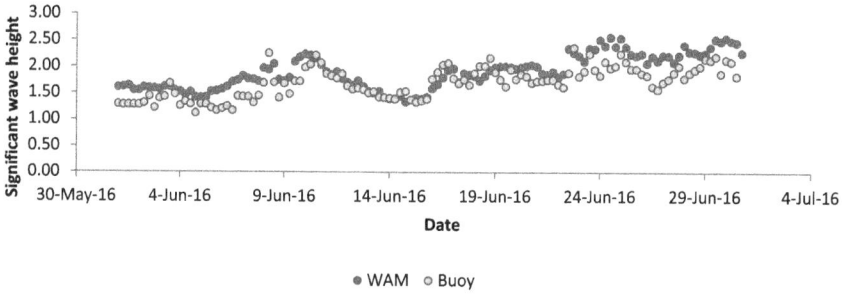

Fig. 10.19(a). Significant wave height off Kochi coast during the month of June 2016 with buoy data.

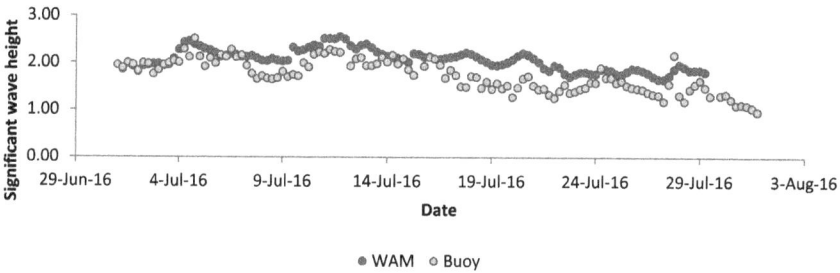

Fig. 10.19(b). (b) Significant wave height off Kochi coast during the month of July 2016 with buoy data.

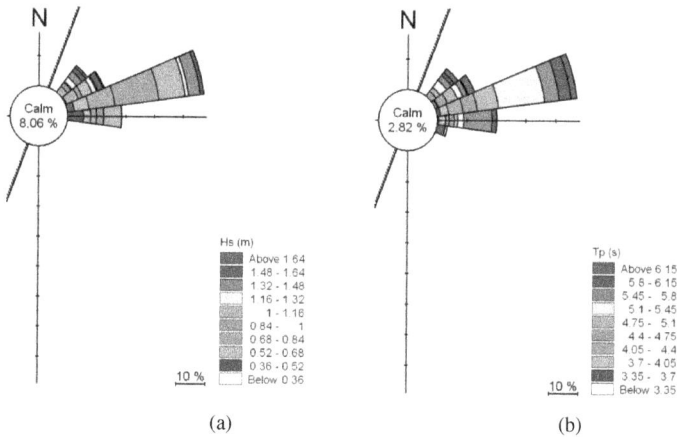

Fig. 10.20. Wave rose diagram for the month of January 2011: (a) wave height and (b) wave period.

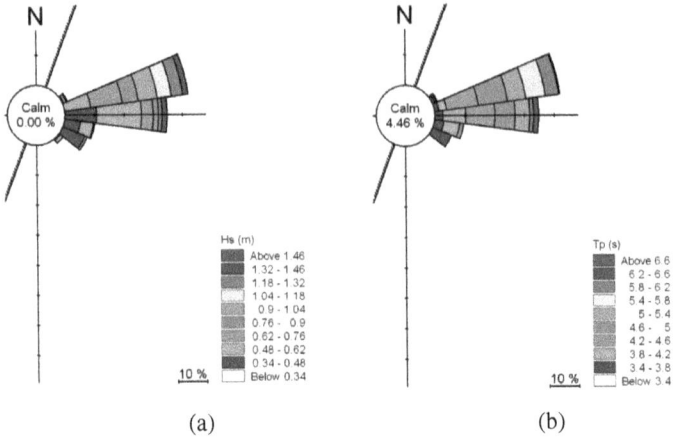

Fig. 10.21. Wave rose diagram for the month of February 2011: (a) wave height and (b) wave period.

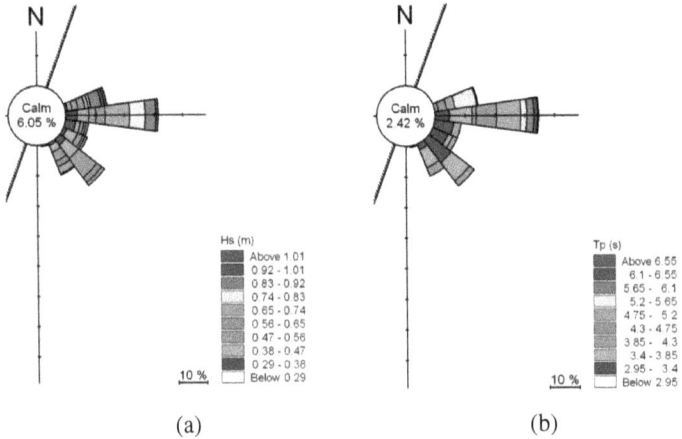

Fig. 10.22. Wave rose diagram for the month of March 2011: (a) wave height and (b) wave period.

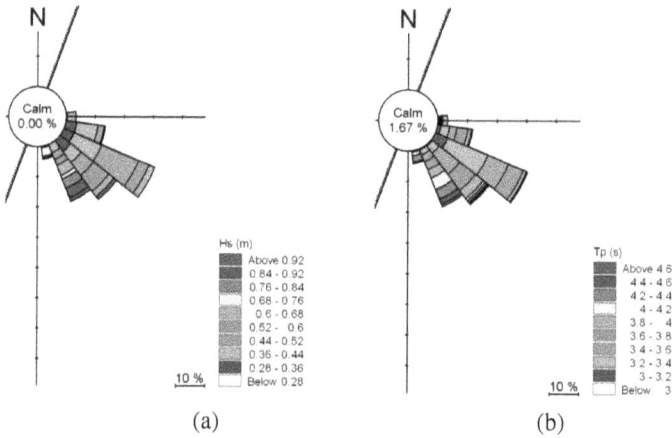

Fig. 10.23. Wave rose diagram for the month of April 2011: (a) wave height and (b) wave period.

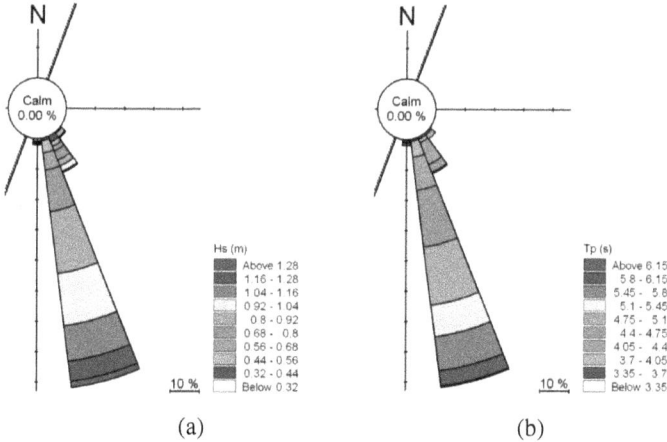

Fig. 10.24. Wave rose diagram for the month of May 2011: (a) wave height and (b) wave period.

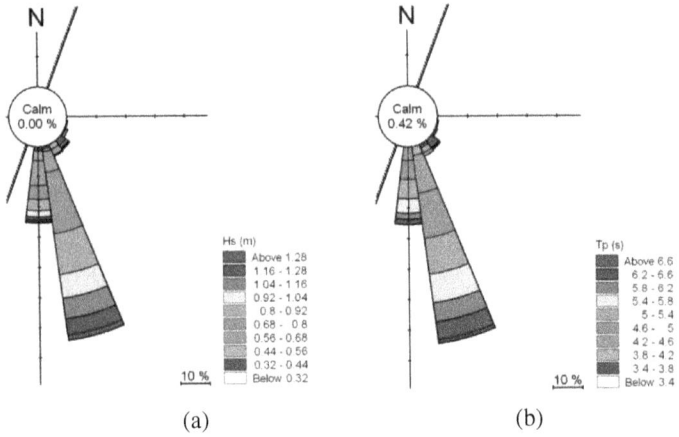

(a)   (b)

Fig. 10.25.   Wave rose diagram for the month of June 2011: (a) wave height and (b) wave period.

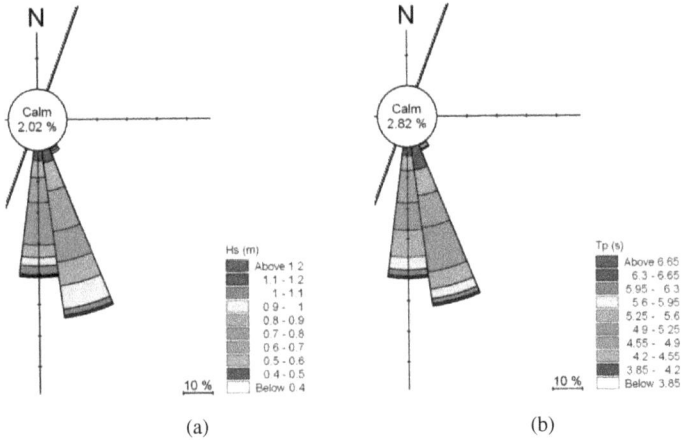

(a)   (b)

Fig. 10.26.   Wave rose diagram for the month of July 2011: (a) wave height and (b) wave period.

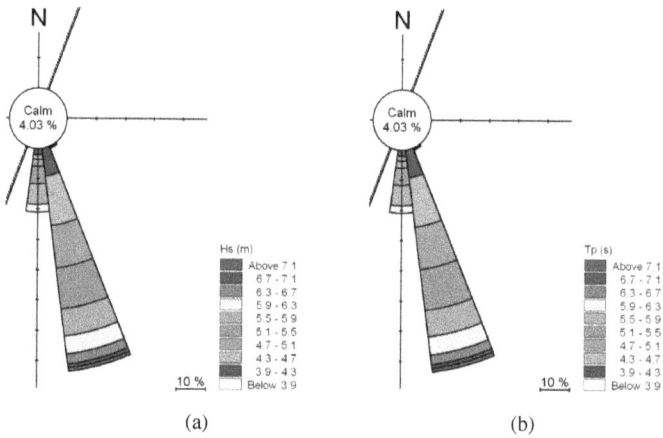

Fig. 10.27. Wave rose diagram for the month of August 2011: (a) wave height and (b) wave period.

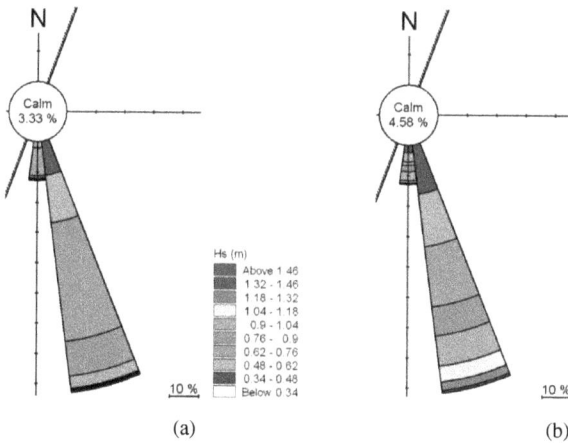

Fig. 10.28. Wave rose diagram for the month of September 2011: (a) wave height and (b) wave period.

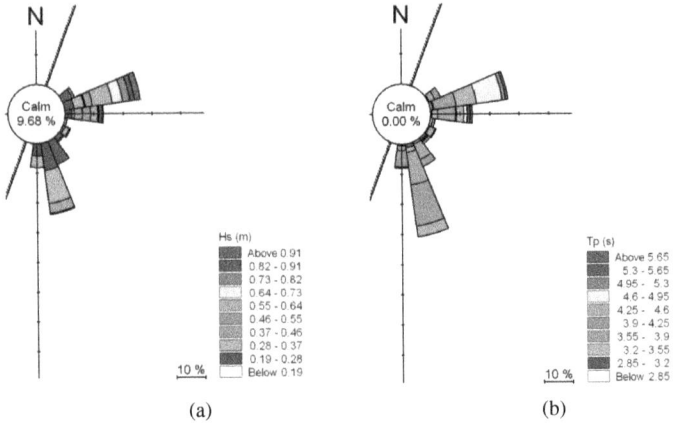

Fig. 10.29. Wave rose diagram for the month of Oct, 2011(a) wave height and (b) wave period.

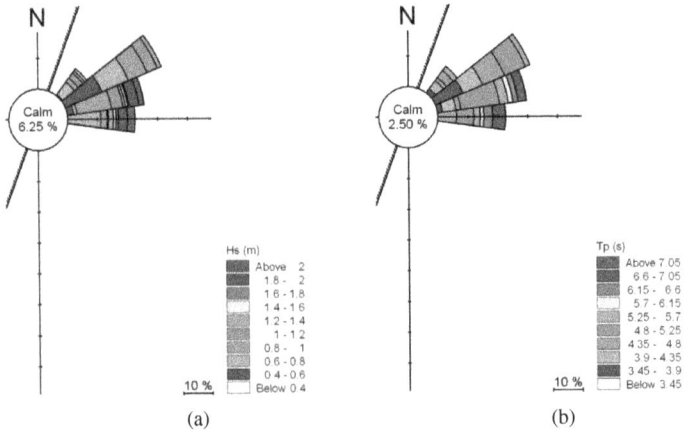

Fig. 10.30. Wave rose diagram for the month of November 2011: (a) wave height and (b) wave period.

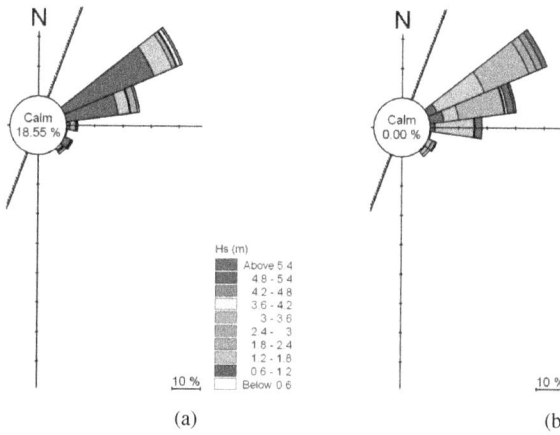

Fig. 10.31. Wave rose diagram for the month of December 2011: (a) wave height and (b) wave period.

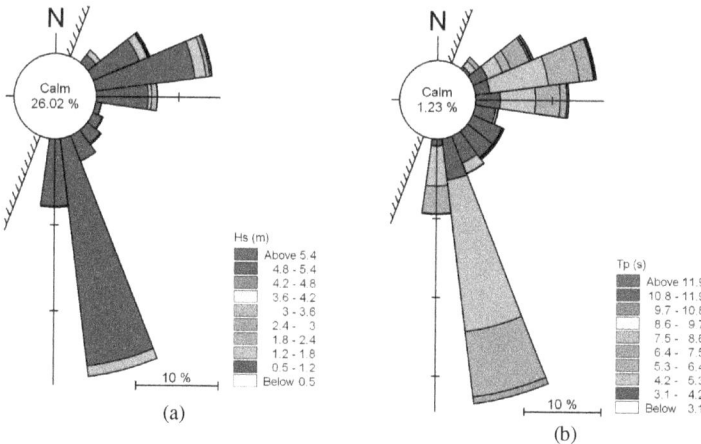

Fig. 10.32. (a) Annual occurrence of significant wave height with respect to wave direction. (b) Annual occurrence of mean wave period with respect to wave direction.

10.8.   Summary

Wave prediction is of prime importance for many offshore activities, from determining optimal ship routing to designing of maritime structures. With the advancement in numerical modelling and computational facilities, one can forecast or hindcast wind-waves with reasonable accuracy in any complex domain even in a cyclonic sea state. Further, ocean wave climate could be analytically estimated with reasonable accuracy even under even fetch-limitation or duration-limitation conditions. The extreme wave analysis further provides a good estimate for design wave climate. This chapter discussed on the above aspects in detail which would help the readers to predict the wave characteristics with reasonable estimate.

References

Beji, S. and Battjes, J.A. (1993). Experimental investigation of wave propagation over a bar. *Coastal Engineering*, 19, 726–750.

Benoit, M., Marcos, F. and Becq, F. (1996). Development of a third-generation shallow-water wave model with unstructured spatial meshing. *Proceedings of the 25th International Conference on Coastal Engineering*, ASCE, Orlando, pp. 465–478.

Booij, N., Ris, R.C. and Holthuijsen, L.H. (1999). A third-generation wave model for coastal regions: 1. model description and validation. *Journal of Geophysical Research*, 104, 7649–7666.

Bowden, K.F. (1983). *Physical Oceanography of Coastal Waters*. Ellis Horwood Limited, Chichester, 67 pp.

Bretschneider, C.L. (1952). Revised wave forecasting relationship. *Proceedings of the 2nd Conference on Coastal Engineering*, Council on Wave Research, University of California, Berkeley, pp. 1–5.

Bretschneider, C.L. (1958). Revisions in wave forecasting: Deep and shallow water. *Proceedings of the 6th Conference on Coastal Engineering*, Council on Wave Research, University of California, Berkeley, pp. 1–18.

BS 6349:1984, Part I. Wave prediction charts.

Cavaleri, L. and Malanotte-Rizzoli, P. (1981). Wind wave prediction in shallow water theory and applications. *Journal of Geophysical Research*, 86, 10961–10973.

CERC (2002). Coastal Engineering Manual.

Coastal Engineering Research Center (US) (1984). Shore protection manual. Department of the Army, Waterways Experiment Station, Corps of Engineers, Coastal Engineering Research Center.

Collins, J.I. (1972). Prediction of shallow water spectra. *Journal of Geophysical Research*, 77(15), 2693–2707.

Dingemans, M.W. (1997). *Water Wave Propagation Over Uneven Bottoms*. Part 1 – Linear wave propagation. *Advanced Series on Ocean Engineering*, 13, World Scientific, Singapore.

Dodd, N. (1998). A numerical model of wave run-up, overtopping and regeneration. *ASCE Journal of Waterways, Ports, Coastal and Ocean Engineering*, 124(2), 73–81.

Eldeberky, Y. (1996). Non linear transformation of wave spectra in the nearshore zone. Ph.D. Thesis, Delft University of Technology, The Netherlands.

Freilich, M.H. and Guza, R.T. (1984). Nonlinear effects on shoaling surface gravity waves. *Philosophical Transactions of the Royal Society of London, Series A*, A311, 1–41.

Giarrusso, C.C. and Dodd, N. (2000). ANEMONE: OTTO-1d – A User manual. Report TR87, HR Wallingford.

Golding, B. (1983). A wave prediction system for real time sea state forecasting. *Quarterly Journal of the Royal Meteorological Society*, 109, 393–416.

Gunther, H., Hasselmann, S. and Janssen, P.A.E.M. (1992). The WAM model cycle 4 – Technical Report No. 4. Modellberatungsgruppe, Hamburg, Germany.

Hasselmann, K. (1963). On the non-linear energy transfer in a gravity-wave spectrum. Part 3. Evaluation of the energy flux and swell-sea interaction for a Neumann spectrum. *Journal of Fluid Mechanics*, 15(3), 385–398.

Hasselmann S. and Hasselmann K. (1985). Computations and parameterizations of the non-linear transfer in a gravity-wave spectrum. Part 1: A new method for efficient computations of the exact non-linear transfer integral. *Journal of Physical Oceanography*, 15, 1369–1377.

Hasselmann, K., Barnett, T.P., Bouws, E., Carlson, H., Cartwright, D.E., Enke, K., Ewing, J.A., Gienapp, H., Hasselmann, D.E., Kruseman, P., Meerburg, A., Muller, P., Olbers, D.J., Richter, K., Sell, W., and Walden, H. (1973). Measurements of wind-wave growth and swell decay during the joint North Sea project (JONSWAP). *Deut. Hydrograph. Z.*, 8, 1–95.

Hasselmann, S. and Hasselmann, K. (1985). Computations and parameterizations of the non-linear energy transfer in a gravity wave spectrum. Part II: Parameterizations of the non-linear transfer for application in wave models, *Journal of Physical Oceanography*, 15(11), 1378–1391.

Holthuijsen, L.H. and De Boer, S. (1988). Wave forecasting for moving and stationary targets. In: Schrefler, B.A. and Zienkiewicz, O.C. (Eds.), *Procs. Intl. Conference on Computer Modelling in Ocean Engng.*, held Venice, pp. 231–234.

Janssen, P.A.E.M., Komen, G.J. and de Voogt, W.J.P. (1984). An operational coupled hybrid wave prediction model. *Journal of Geophysical Research*, 89, 3635–3654.

Kantha, L.H., Blumberg, A.L. and Mellor, G.L. (1990). Computing phase speeds at open boundary, *Journal of Hydraulic Engineering, ASCE*, 116(4), 592–597.

Khandekar, M.L. (1989). *Operational Analysis and Prediction of Ocean Wind Waves*. Springer-Verlag, New York.

Komen G. J., Cavaleri, L., Donelan, M., Hasselmann, K., Hasselmann, S. and Janssen, P.A.E.M. (1994). *Dynamics and Modelling of Ocean Waves*. Cambridge University Press.

Komen G. J., Hasselmann, S. and Hasselmann, K. (1984). On the existence of a fully developed windsea spectrum. *Journal of Physical Oceanography*, 14, 1271–1285.

Leenderste, J.J. (1967). Aspects of a computational model for long period water wave propagation, The Rand Corporation, Rept. RH-5299-RP, Santa Monica, CA.

Madsen, O.S. and Sorensen, O.R. (1992). A new form of the Boussinesq equations with improved linear dispersion characteristics. A slowly-varying bathymetry. *Coastal Engineering*, 18, 183–205.

Massel, S.R. (1996). *Ocean Surface Waves: Their Physics and Prediction. Advanced Series on Ocean Engineering*, Volume 11. World Scientific Publishing Co. Ltd., Singapore.

Miles, J.W. (1981). Hamiltonian formulations for surface waves. *Applied Science Research*, 37, 103–110.

Pierson, W.J. and Moskowitz, L. (1964). A proposed spectral form for fully developed wind seas based on the similarity theory of S.A. Kitaigorodskii. *Journal of Geophysical Research*, 69, 5181–5190.

Ris, R.C., Booij, N., Holthuijsen, L.H., Padilla-Hernandez, R., and Haagsma, I.G. (1998). *User manual SWAN cycle 2 version 30.75*, Delft University of Technology, Department of Civil Engineering, Delft, The Netherlands. Available at: http://www.wldelft.nl/soft/swan/

Ris, R.C., Holthuijsen, L.H. and Booij, N. (1999). A third-generation wave model for coastal regions: 2. Verification. *Journal of Geophysical Research*, 104, 7667–7681.

Snyder R.J., Dobson, F.W., Elliott, J.A., and Long, R.B. (1981). Array measurements of atmospheric pressure fluctuations above surface gravity waves. *Journal of Fluid Mechanics*, 102, 1–59.

Sverdrup, H.U. and Munk, W.H. (1947). Wind, sea and swell: Theory of relations for forecasting. Publication 601, US Navy Hydrographic Office, Washington, DC.

The WAMDI Group (1988). The WAM Model - A third-generation ocean wave prediction model. *Journal of Physical Oceanography*, 18, 1775–1810.

Tolman, H.L. and Chalikov, D. (1996). Source terms in a third-generation wind-wave model. *Journal of Physical Oceanography*, 26, 2497–2518.

U.S. Army Coastal Engineering Research Center (1977). *Shore Protection Manual*, 3rd Edition, US Government Printing Office, Washington, DC.

Wittmann, P.A. and Clancy, R.M. (1993). Implementation and validation of a third-generation wave model at Fleet Numerical Oceanography Center. In O.T. Moagoon and J.M. Hemsley (Eds.), *Ocean Wave Measurements and Analysis, Proceedings of the Second International Symposium*, 25–28 July, New Orleans, ASCE, pp. 406–419.

Young, I.R. (1999). *Wind Generated Ocean Waves. Elsevier Ocean Engineering Book Series*, 2, 1–288.

Chapter 11

Wave-Induced Sediment Transport
Through Measured Wave Characteristics

Abstract
Longshore sediment transport (LST) plays a vital role in dictating the stability of a shoreline. The estimation of LST rate poses problems due to assigning the value for the driving parameters and the method adopted. In this chapter, the driving parameters, i.e., the wave characteristics, predicted from the field-measured wind data for a year were considered for the estimation of LST. The predicted wave data was initially validated with the field measurements, acquired by deploying a bottom-mounted directional wave recorder. LST were evaluated with five different empirical formulae, the details of which are discussed in this chapter.

11.1.   General

The coastal zone, which is a stretch between land and the ocean, is of paramount importance as it progressively undergoes development and is a high-productivity zone. The abundant resources available in these stretches of land are vastly exploited for the development of industries, mining, tourism and other similar activities. The stability of the shoreline is often questionable due to the above-stated man-made activities or due to its exposure to natural hazards, like tsunamis, storm surges and flooding. The movement of sediments along the coast is controlled by several driving parameters, and the wave-induced currents largely dictate the zones of erosion or deposition. While sediment deposition across tidal inlets and harbour approach channels poses a challenge in maintaining the required depths for the vessels in motion for maritime trade purpose, the erosion along the coast is a perennial problem as valuable stretches of the coast are being swallowed by the waves, which during storms become unmanageable. To provide suitable coastal protection measures or to facilitate a dredging strategy, a comprehensive knowledge on the prediction of longshore sediment transport (LST) through a combination of applying empirical methods and measurements is warranted. The assessment of the direction and magnitude of the sediment transport over the past decades has been based on the utilization of ship-observed or visually observed wave data or that derived from the wind and through empirical formulae that are well entrenched in literature. This chapter deals with the prediction of sediment transport through predicted characteristics of waves after its validation through measurements.

11.2. Littoral Transport and Process

Littoral transport is the sediments in motion mostly within the surf zone driven by the wave-induced longshore current (termed as LST) or the cross-shore currents (termed as cross-shore sediment transport). LST could be calculated using instantaneous sediment concentration and instantaneous velocity.

A wave propagating towards the shore feels the seabed, when the water depth is less than half the deep-water wave length, from which location the seabed and the water particle velocity under that wave begin to interact. When the water particle velocity is higher than the critical velocity to lift sediment of particular characteristics, it will be set in to suspension, and this suspended material will oscillate with the orbital velocity. The wave will develop nonlinearity when approaching the shore, which undergoes a change in orbital velocity by adding a translatory motion to its oscillatory motion. Thus, the suspended sediments will approach the shore along with the wave. This forms the part of sediment transport as suspended transport. Further, slightly heavier sediments move under the wave action along the bed in the form of ripples parallel to the wave motion, and sand moves from one side of the crest to the other with the passage of each wave. This is termed as bed load sediment transport. As the waves continue to propagate towards the shore, it undergoes shoaling and eventually break in the near-shore by depositing sediments whilst exceeding the orbital motion over critical velocities.

When the waves steepen and finally break, the orbital velocities continue to increase. Finally, the orbital motion will exceed the critical velocity of sheet flow. At this point, the ripples are flattened. A layer of sediment now moves over the bottom, resulting in high suspended sediment concentrations. With this breaking event, a coastal current will be generated, and based on the net direction, alongshore or cross-shore sediment transport occurs.

11.3.   Analytical and Numerical Works Comparison with Field Measurements

The equilibrium beach profile through detailed investigations along the coast of California and North Sea off Denmark is attained through an equation as claimed by Brunn (1954). The equilibrium profile (depth) was reported to be proportional to the offshore distance (y) through a relation $Ay2/3$, where A is a constant arrived through field observations.

Exon (1975) reported that the presence of engineering structures reduced the size of the bar field along the western Baltic Sea. Most of the LST estimations in vogue till date are based on the works of Komar (1976). The estimates of the littoral transport rate by Chandramohan and Nayak (1991) over the entire coast of India serve as a basis on which the coastal process is understood, it facilitates planning for the coastal protection measures. Wellen *et al.* (1998) evaluated the performance of a number of widely used transport equations of CERC (1984), Damgaard and Soulsby (1996), Kamphuis *et al.* (1986), Kamphuis (1991), for the prediction of LST and compared with field-measured LST. They found that most of the equations tend to over-predict the expected transport by a factor of 1.5 to 4.

The presence of natural or man-made barriers intercept the free passage of the littoral drift that results in the instability of the shoreline. Sundar *et al.* (1992) investigated the velocity field inside a groin field along with the effect of swash oscillation and concluded that the magnitude of swash oscillation can be order of about 30%–35% of the measure of wave surface elevation. The estimation of alongshore sediment transport within the surf zone close to a temporary groin along a beach of Florida through a measurement campaign was reported by Wang and Kraus (1999). The variation of the measured quantity with the predicted values from the formulation of CERC was reported to be of an order between 0.6 and 1.6. Nordstorm *et al.* (2003) measured the alongshore sediment transport rates by conducting experiments using tracers on a micro-tidal estuarine beach. The measured values were compared with the sediment transport formulae of CERC, Inman and Bagnold (1963) and Kamphuis (1991). It was reported that most of the formulae under-predicted the transport by a factor of 3 to 6.

Sanil Kumar *et al.* (2004) measured the sediment transport rate along central west coast of India by conducting field experiments using traps for one year. The measured quantity of sediment was compared with the values obtained using widely used formulae like CERC (1984) and Walton and

Bruno (1989). The behaviour of waves in the nearshore zone in the presence of groin field was studied for their directional characteristics using Maximum Entropy Method (MEM) by Sundar *et al.* (1994). It was concluded that the directional wave spectrum tends to become wider while the waves approach the shore, and inside the groin field, the angle of wave approach from the shore normal will increase due to the presence of cross reflection, which effect the alongshore current pattern.

Ramana Murthy *et al.* (2008) evaluated the performance of beach fill at Ennore along the coast of Chennai, India. Beach profiles and sediment sample collections were carried out at regular intervals, and beach fill morphology was studied in detail. Based on the results, details of the beach fill to be maintained were arrived at. Suresh *et al.* (2011) estimated the LST along the east and west coasts of Tamil Nadu in South India, with the Van-Rjin (2001) formula. The results indicated that the sediment transport rate varied between 60,000 and 80,000 m^3/m width for the east coast of Tamil Nadu. Anand *et al.* (2011) reported that the results from a preliminary exercise of the measurement and analysis of the wave and flow fields in a groin field north of the Chennai port along the southeast coast of India and reported that the sediment transport prediction from the numerical model was found to be **15%** less than predicted by the **CERC** and **KOMAR** models and **30%** higher than **KAMPHUSIS** and **sediment distribution model**.

The LST rates using the simulated wave characteristics were estimated along Karaikal on the southeast coast of India, the details of which are discussed in the subsequent sections.

11.4.   Study Area

This study area, Karaikal, is situated in the union territory of Puducherry on the east coast of India (10.9254°N, 79.8380°E) as shown in **Fig. 11.1**. This district encompasses an area of 157 km^2 (MoEFCC, 2019). The coastline is a flat terrain, its alignment is almost straight oriented along the North-South direction, with a moderate beach width of about 500–700 m width along with the presence of backshore and gentle foreshore. Extreme events like cyclone can directly damage the coast due the configuration of coast (Loveson *et al.*, 2014). Mohapatra (2015) attempted to classify hazard due to cyclones through a hazard criterion based on frequency and intensity of cyclone, wind strength and probable maximum storm surge. Karaikal was identified as one of the coastal hazard zones exposed to cyclones,

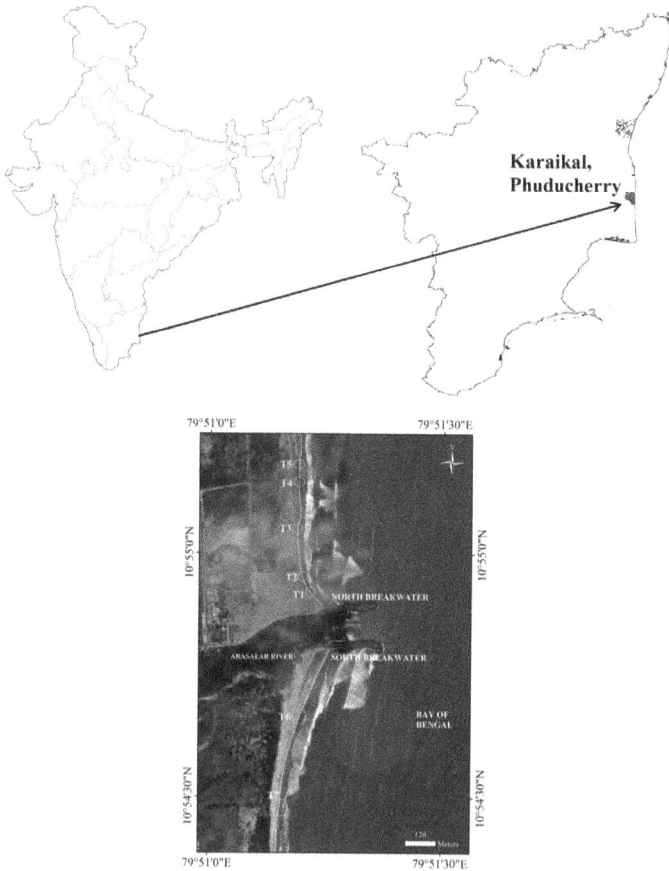

Fig. 11.1.   Study area location 1: Karaikal.

falling under the category highly prone. Therefore, continuous monitoring/ maintenance and estimation of sediment transport along this coast is necessary for planning coastal protection measures.

11.5.   Wave Characteristics

For the prediction of wave characteristics from the wind data, the third-generation wave model; WAve Model (WAM) (Komen *et al.*, 1994; Holthuijsen, 2007) is used. Sundar and Suresh (2011), implemented the wave data generated from WAM along the tip of the west coast of India for the year 2007 to predict the sediment transport rate using the formulation of Van Rijn (2001), and the obtained results were closer to field-measured sediment transport rate. Vimala *et al.* (2014) compared the WAM results with the wave rider buoy data for both west and east coasts of India and concluded that the prediction of WAM results along the west coast was better than that in the east coast. The field measurement of waves was carried out with a bottom-mounted Directional Wave Recorder (DWR) deployed at a depth of 10 m from 5 May to 27 October 2018. The wave data were recorded continuously at a sampling frequency of 2 Hz and 2,048 samples of wave burst duration with an interval of 60 minutes. A high-accuracy piezo-resistive pressure sensor is fitted with DWR to measure pressure in shallow waters less than 90 m having an accuracy of ±0.01% and a resolution of 0.001%. A flux gate compass is incorporated within the DWR with a range of 0°–360° with an accuracy ±1° and resolution 0.1°. To analyze the pressure and current oscillations generated by the wave action, this type of DWR uses the Linear Wave Theory for computation of the wave parameters. The data were simulated for a period of one year from October 2017–September 2018. The stimulated results were validated with the field-measured wave data.

11.6. Numerical Modelling of Wave Characteristics

WAM, a third-generation wave model based on the energy transport equation (Komen *et al.*, 1994) has been used for the assessing wave characteristics. The source function is represented as superposition of wind input, nonlinear wave–wave interaction and dissipation due to wave breaking and bottom friction.

The governing equation for the WAM is given in Eq. (11.1).

$$\frac{\partial F(f,\alpha;x,t)}{\partial t} + \upsilon\nabla_x F(f,\alpha;x,t) = S \qquad (11.1)$$

where $F(f,\alpha;x,t)$ is the wave energy spectrum in terms of frequency, f and propagation direction, α at the position vector x at time t, whereas υ is the group velocity. The second term on the left-hand side is the divergence of the convective energy flux $(\upsilon\nabla_x F)$. The net source function S considers all physical processes that contribute to the evolution of the wave spectrum. In the present study, 0.1° × 0.1° grid resolution is adopted, covering the Bay of Bengal domain of 0°–25°N and 75°E–95°E (**Fig. 11.2**).

The ETOPO2 bathymetry data from National Geophysical Data Centre, NOAA (NGDC) and the wind data from National Centre for Environmental Prediction (NCEP), NOAA, with a grid resolution of 0.5° × 0.5°

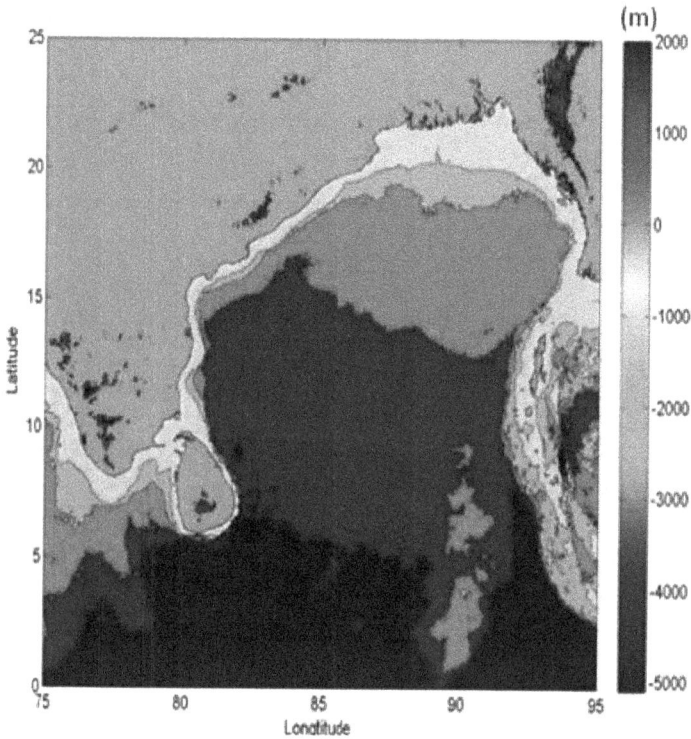

Fig. 11.2.   The domain used in the WAM model.

at six-hour interval have been used as the forcing condition for the model. Initially, the model has been set up to execute for one year from October 2017 to September 2018. The model results are validated with the measured wave data. The wave characteristics thus derived were applied for evaluating the sediment transport rates off the study area.

The simulation of waves along the Karaikal coast has been carried out for a period of one year from October 2017 to September 18. Since the comparison between measured and simulated were found similar over the entire measurement campaign, as mentioned earlier, to obtain clarity in the variations, the results are reported only for a limited duration. The wave characteristics viz., significant wave height, H_s, the peak period, T_p, and mean wave direction, θ_m, are extracted from the prediction. The predicted H_s, T_p and θ_m are compared with the field-measured wave data in **Figs. 11.3(a)**, **(b)** and **(c)**, respectively. The agreement is observed to be reasonable, with

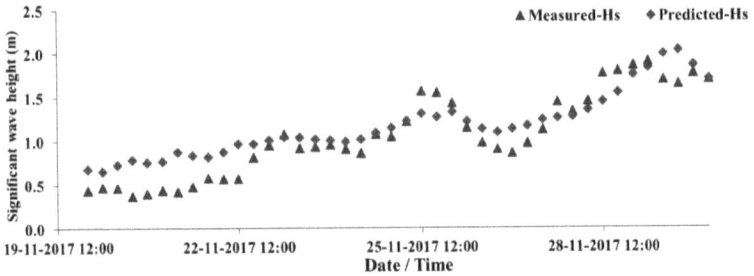

Fig. 11.3(a).   Comparison of measurement H_s with model prediction.

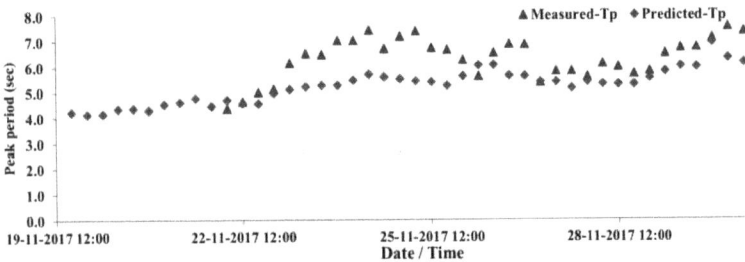

Fig. 11.3(b).   Comparison of measurement T_p with model prediction.

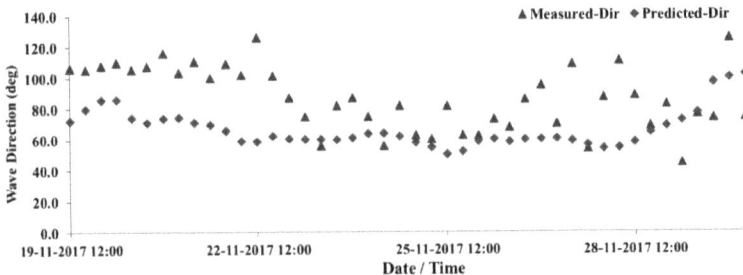

Fig. 11.3(c).   Comparison of measurement θ_m with model prediction.

the largest deviation being observed for θ_m. The measured wave rose diagram for the period of measurement is depicted in **Fig. 11.3(d)**. The wave direction off the coast of Karaikal varies from northeast to southeast. Predominantly, the wave approaches from the southeast direction with a higher frequency of occurrence. This would dictate the predominantly northerly sediment transport during most part of the year. However, the energy concentration in the waves is higher in the northeast monsoon period, during

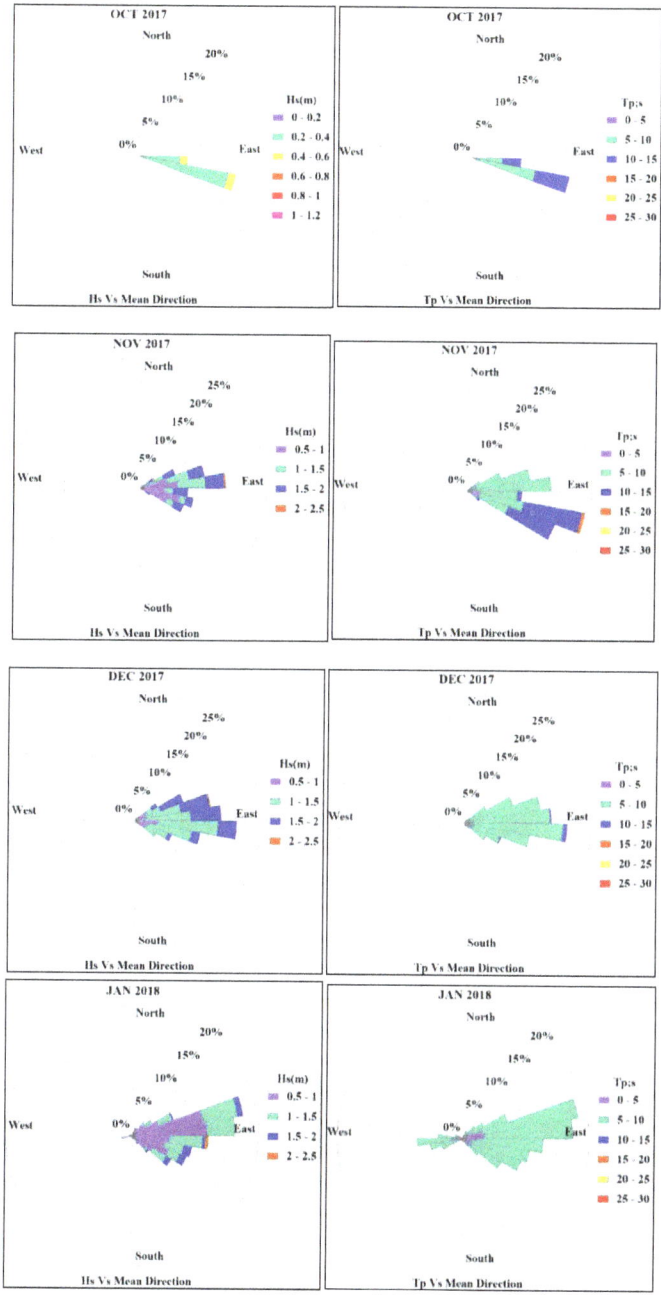

Fig. 11.3(d). Wave rose diagram for the period of measurement.

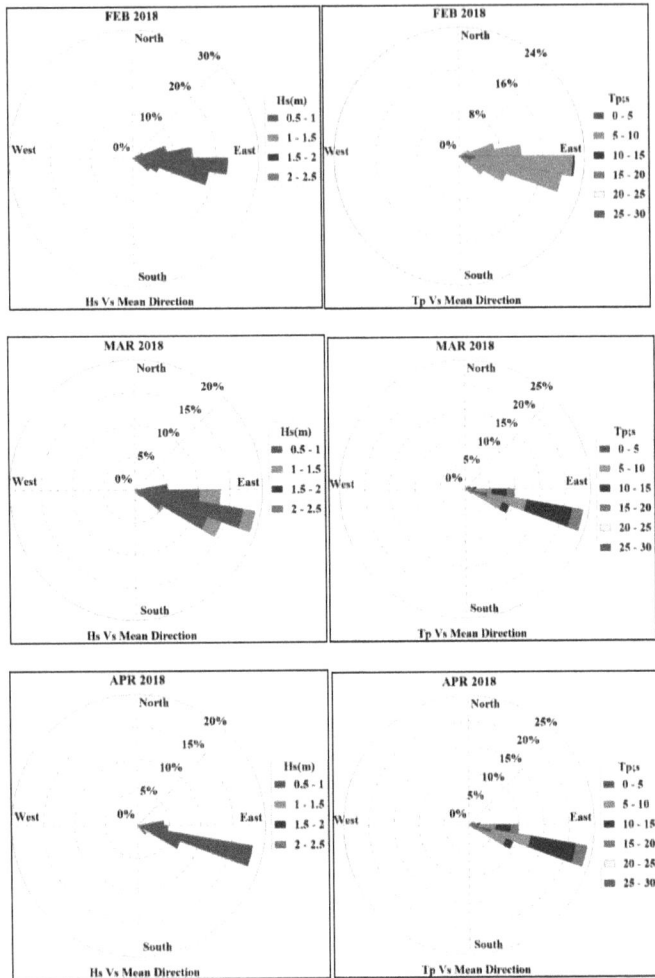

Fig. 11.3(d). (*Continued*)

which the waves approach from east to northeast direction. The sudden impact on the coast is higher during this period due to monsoon waves with cross-shore sediment transport. The coast would regain after the monsoon, hence the impact due to northeasterly waves is relatively less.

The basic statistics of monthly predicted data are depicted in **Table 11.1**. These parameters were used to obtain the nearshore breaker wave characteristics. From the aforesaid table, the lowest minimum H_s of

Table 11.1.   Monthly predicted wave characteristics off Karaikal.

Month/ year	H_s (m)			T_p (sec)			direction (deg) w.r.t north		
	min	max	mean	min	max	mean	Min	max	mean
Oct-17	0.33	1.04	0.61	3.33	7.12	5.00	14	311	146
Nov-17	0.39	2.03	1.05	4.37	7.02	5.46	6	120	60
Dec-17	0.45	1.58	1.03	4.29	7.79	5.44	43	100	62
Jan-18	0.64	1.56	0.99	4.44	6.0	5.08	46	85	63
Feb-18	0.50	1.73	0.86	3.89	6.57	5.19	60	111	83
Mar-18	0.50	1.04	0.69	4.38	5.72	4.94	73	153	114
Apr-18	0.41	0.98	0.67	3.34	6.58	4.65	40	184	139
May-18	0.46	1.19	0.67	3.45	6.17	4.47	128	234	174
Jun-18	0.42	1.14	0.76	3.95	6.07	4.64	160	231	195
Jul-18	0.31	1.07	0.64	3.58	8.11	5.22	34	337	179
Aug-18	0.42	0.88	0.64	4.06	8.69	5.77	134	230	174
Sep-18	0.32	1.22	0.64	3.90	7.92	4.91	125	228	175

0.3 m was observed during July; and the highest mean and maximum H_s of 1.0 m and 2.0 m were observed during November, respectively. The month of October recorded the lowest T_p of 3.3 s; and the highest mean and maximum of 5.8 s and 8.7 s, respectively, were recorded during August. The wave approach angle with respect to geographic North during southwest monsoon was found to vary on an average from 150° to 250°; and, during the non-monsoon, it varies from about 50° to 200°. During the northeast monsoon, the variation in the wave direction is wide. The seasonal distribution of wave characteristics shows that the maximum significant wave height of 2 m was observed during the northeast monsoon season, and H_s was found to be less during the southwest monsoon. H_s is found to vary from 0.4 m to 1.7 m during the non-monsoon season. The maximum T_p is about 8.7 s during the southwest monsoon season.

11.7.   Longshore Sedimentation Formulae

For this study, five of the well-established sedimentation formulae viz., Komar (1976), Komar Distribution (1977), CERC (1984), Kamphuis (1991) and Van Rijn (2001) were adopted for the estimation of LST rate. Herein, only the Kamphuis and Van Rijn formulae consider the effect of beach slope and particle grain size, whereas the other three formulations are based only on the breaking characteristics of the wave, with varying dependent and independent variables.

11.7.1. *Komar (1976)*

Longitudinally transported directions and volume of sands (m^3/day) were calculated using Eq (11.1) proposed by Komar.

$$Q = 3.4(EC)_b \sin \alpha \cos \alpha \qquad (11.1)$$

where Q is the LST rate, E is the wave energy density at breaking, C is Celerity, with suffix "b" referring to C at breaking and α is the angle between wave crest and shoreline.

11.7.2. *Komar Distribution (1977)*

Komar (1976, 1977) utilized the Bagnold (1963) model to analyze the distribution of the littoral drift, which was then "calibrated" to yield the total transport, I_1, as shown in Eq. (11.2).

$$I_1 = 0.77\,P_1$$
$$Q = k\,P_1 \qquad (11.2)$$

where I_l is the immersed Weight transport rate, k is the dimensionless coefficient and P_l is the wave power. (Refer to *Coastal Engineering Manual* (2006) for a complete discussion on the P_l parameter as a function of the available input parameters describing the wave characteristics.)

11.7.3. *CERC (1984)*

In the Coastal Engineering Research Centre (CERC) (Shore Protection Manual, 1984), the total LST rate (Q) is calculated using Eq. (11.3).

$$Q = H_{sb}^2 C_b a_1 \sin(2\theta_b) \qquad (11.3)$$

wherein H_{sb} is the significant wave height at breaking, C_b is the group celerity at breaking, a_1 is the dimensionless parameter and θ_b is the breaker angle (deg).

11.7.4. *Kamphuis (1991)*

Kamphuis (1991) presented LST formula as seen in Eq. (11.4), where nominal particle size D_{50} and the beach slope within surf zone were considered. It is a major function of beach slope, and the effect of particle size is minimum. The below expression shows that the rate of sediment transport is directly proportional to significant wave height.

$$Q = 6.4 \times 10^4 H_{sb}^2 T_p^{1.5} (\tan \beta)^{0.75} D_{50}^{-0.25} (\sin 2\theta_b)^{0.6} \qquad (11.4)$$

where H_{sb} is the significant wave eight at breaking, β is the beach slope, T_p is the peak period, D_{50} is the particle size (mm) and θ_b is the breaker angle (deg)

11.7.5.   Van Rijn (2001)

Van Rijn (2001) proposed a formula as seen in Eq. (11.5) for the calculation of LST rate, including factors for normalizing the swell, particle grain size and the beach slope.

$$Q = 40 \, K_{swell} K_{grain} K_{slope} (H_{sb})^3 \sin(2\theta_b) \qquad (11.5)$$

where K_{swell} is the swell correction factor, K_{grain} is the particle size correction factor and K_{slope} is the slope correction factor.

11.8.   Estimation of Sediment Transport

The monthly net LST rates estimated using different formulae (CERC, 1984; Kamphuis, 1991; Komar, 1976; Komar, 1977; Van Rijn, 2001) are shown in **Fig. 11.4**. The positive and negative y-axes denote the direction of the drift towards the north and south, respectively. It is observed that during the northeast monsoon, (October–December 2017), except during October 17, the sediment transport is heading southwards and found to be in the range between 0.22×10^5 and $0.49 \times 10^5 \, \mathrm{m^3/month}$ (October 17), whereas the southward rate ranges between 0.29×10^5 and $0.95 \times 10^5 \, \mathrm{m^3/month}$ (November–December 2017).

Like the seasonal changes in temperature and humidity, the offshore and nearshore wave climate also changes according to the season and the changes in wind circulation. Annually, the seasons are broadly classified as non-monsoon (January–May), northeast monsoon (October–December) and southwest monsoon (June–September). During the non-monsoon period (March–May 2018), the sediment transport is heading northwards and found to be in the range between 0.12×10^5 and $0.29 \times 10^5 \, \mathrm{m^3/month}$ (March–May 2018). Whereas the southward transport ranges between 0.074×10^5 and $0.83 \times 10^5 \, \mathrm{m^3/month}$ during January 2018 and February 2018. During the southwest monsoon period, (June 2018 to September 2018), the net sediment transport is towards north with its magnitude in the range between 0.16×10^5 and $0.58 \times 10^5 \, \mathrm{m^3/month}$. The highest magnitude

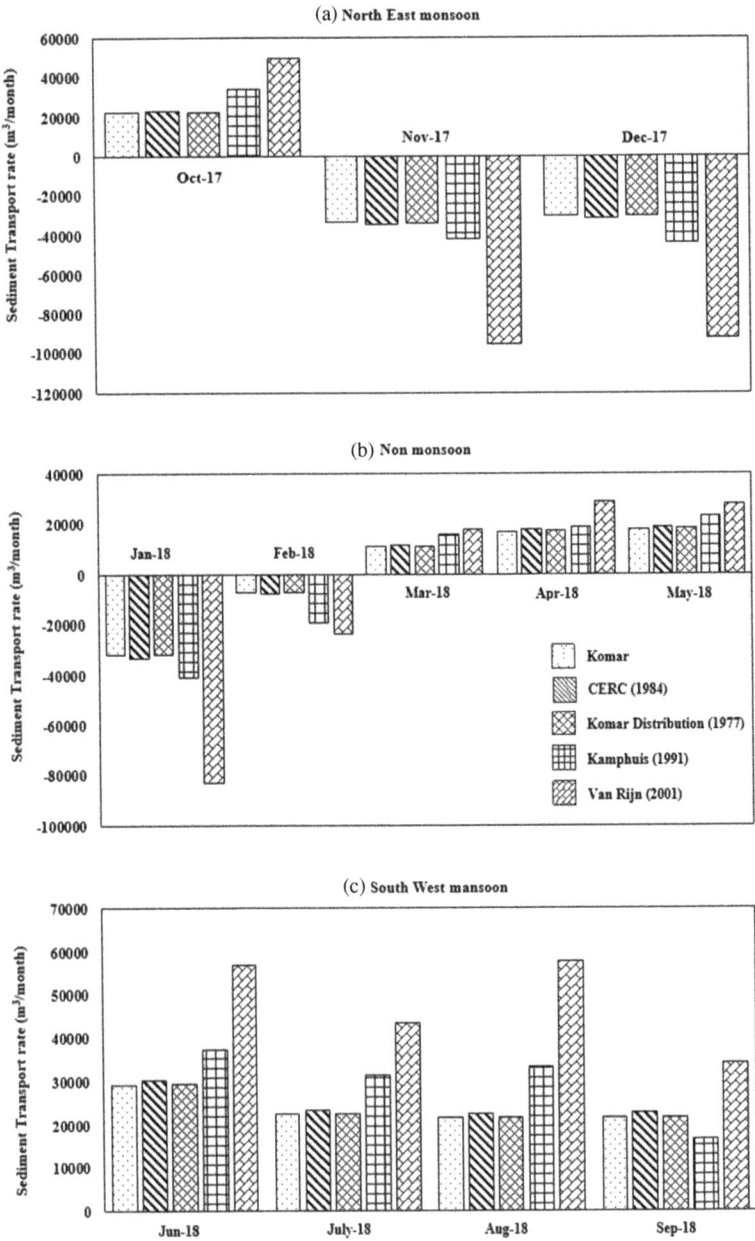

Fig. 11.4. Seasonal distribution of sediment transport rate (a) northeast monsoon, (b) non-monsoon and (c) southwest monsoon.

of LST rate is observed along the coast during southwest monsoon (June 2018 to September 2018) season, predominantly directed towards north, whereas, during the northeast (October–December 2017) and non-monsoon (January–May 2018) seasons, moderate LST rates are observed. The net annual LST along the Karaikal coast is directed northwards. The monthly averaged net transport for the Karaikal coast ranges between 0.24×10^5 and $0.64 \times 10^5 \, \mathrm{m}^3/\mathrm{month}$.

The predominant direction of sediment movement is dictated by the wave approach angle and its deformed characteristics in the nearshore region. Several parameters, such as breaker wave height, wave breaker angle, surf zone width, sediment concentration, sediment grain size distribution, beach slope and wave–current interaction influence the magnitude as well as the direction of the sediment transport. For the months experiencing drift in southward direction, the beach slope is rather steep, and the observed significant wave heights are found to be higher in comparison with the other months. The principal direction of offshore wave approach angle drives the sediment along a similar trajectory. Generally, LST rate can vary depending upon wave climate, the sediment characteristics and the availability of sediments. The predominant direction of transport depends upon the wave approach direction and configuration of the coast. The net annual sediment transport rate along the Karaikal coast from the five different methods is shown in **Table 11.2**. All the methods except the method of Van Rijn (2001) yield similar results. The abrupt changes in the seabed slope during northeast and non-monsoon seasons yield extremely high LST rates using the formula of Van Rijn (2001), and thus is not considered for further discussion.

The estimated net LST rates along the Karaikal coast are compared with the earlier studies. In the present study, LST estimates are 0.23–$0.34 \times 10^6 \, \mathrm{m}^3/\mathrm{month}$ and 0.11–$0.3 \times 10^6 \, \mathrm{m}^3/\mathrm{month}$ towards north and south, respectively. These are much greater than the estimate of

Table 11.2.   Annual net transport rate estimated from five methods.

Method	Towards north $(\mathrm{m}^3/\mathrm{y})$	Towards south $(\mathrm{m}^3/\mathrm{y})$	Net $(\mathrm{m}^3/\mathrm{y})$
Komar (1978)	163,411	−102,628	60,783
CERC (1984)	170,202	−106,893	63,309
Komar distribution (1977)	164,272	−103,161	61,111
Kamphuis (1991)	210,911	−146,741	64,169
Van Rijn (2001)	316,956	−293,624	23,331

Chandramohan *et al.* (1990), in which the estimates were 0.05–$0.15 \times 10^6 \, m^3$/year towards north from March to October, and 0.05–$0.25 \times 10^6 \, m^3$/year towards south from November to February. Although a similar trend in a predominant direction of sediment movement is observed, the magnitude does not essentially match as the source of wave characteristics adopted by Chandramohan *et al.* (1990) was ship-observed. Natesan *et al.* (2015) studied LST along the Chennai coast with the variation of crest of berm, and the estimated LST rate varied from 0.03–$0.28 \times 10^6 \, m^3$/month towards north, and 0.01–$0.09 \times 10^6 \, m^3$/month towards south with a net transport of $0.7 \times 10^6 \, m^3$/year. Prabakar Rao (2002) reported the gross sediment transport rate of the southern Tamil Nadu coast between 0.6 and $0.7 \times 10^6 \, m^3$/year. From the studies reported along the Tamil Nadu coast, it is evident that the LST rates gradually increase from the southern to northern districts. The net LST rate along the India coast vary considerably depending upon the wave climate, sediment nature, beach slope, wave approach direction and configuration of the coast.

11.9. Summary

The WAM model has yielded reliable wave climate results for the regional study along the selected stretches along the east and west coasts of India. The simulated model results were used for the study period of 2017–2018 covering northeast, southwest and non-monsoon seasons. Nearshore wave parameters from the model results were compared well with measure wave data. To estimate the LST rate, five different empirical methods namely Komar, CERC, Komar distribution, Kamphuis and Van Rijn were employed to estimate the annual net sediment transport rate, the results of which are reported. The results imply that the predominant direction of sediment movement is towards north. Seasonal variability of dynamic nature of LST shows towards north during southwest monsoon season whereas transport towards south during northeast monsoon season. Monitoring and analysis of trends on LST along the Indian coast can play a major role in the assessment of erosion/accretion and designing of coastal structures.

References

Anand, K.V., Sundar, V., Sannasiraj, S.A., Murali, K., Rangarao, V. and Subramanian, B.R. (2011). Littoral transport estimate from the field measurement along north Chennai coast of Tamil Nadu, India. *Proceedings of the 6th International Conference on Asian and Pacific Coasts (APAC 2011)*, Hong Kong, China, December 14–16.

Bagnold, R.A. (1963). *Mechanics of Marine Sedimentation.* In: Hill, M.N. (Ed.), *The Sea*, Volume 3. New York: Wiley Interscience, pp. 507–528.

Brunn, P. (1954). Coast erosion and the development of beach profiles. Technical Memorandum No. 44, *Beach Erosion Board, Coastal Engineering Research Center, US Army Engineer Waterways Experiment Station, Vicksburg, MS*

Chandramohan, P., Nayak, B. U., Raju, V. S. 1990. Longshore-transport model for south Indian and Sri Lankan coasts. *ASCE Journal of Waterway, Port, Coastal and Ocean Engineering*, 116-4. DOI: 10.1061/(ASCE)0733-950X (1990)116:4(408).

Chandramohan, P., and Nayak, B.U. (1991). Longshore sediment transport along the Indian coast. *Indian Journal of Maritime Science*, 20(2), 110–114.

CERC (1984). *Shore Protection Manual.* Vols I and II. Coastal Engineering Research Center. U.S. Army Corps of Engineers, Washington, DC. U.S. Government Printing Office, Vicksburg.

Coastal Engineering Manual (2006). U.S. Army Corps of Engineers. 1110-2-1100, U.S. Army Corps of Engineers, Washington, D.C. (in six volumes).

Damgaard, J.S. and Soulsby, R.L. (1996). Longshore bed-load transport. *Proceedings of the 25th International Conference on Coastal Engineering*, ASCE, Orlando, pp. 3614–3627.

Exon, N.F. (1975). An extensive offshore sand bar field in the Western Baltic Sea. *Marine Geology*, 18, 197–212.

Holthuijsen, L.H. (2007). *Waves in Oceanic and Coastal Waters.* Cambridge University Press, Cambridge.

Inman, D.L. and Bagnold, R.A. (1963). Littoral processes. In: Hill, M.N. (Ed.), *The Sea*, Interscience, New York, Vol. 3, pp. 529 –533.

Kamphuis, J.W. (1991). Alongshore sediment transport rate. *Journal of Waterways, Port, Coastal and Ocean Engineering*, 117(6), 624–641.

Kamphuis, J.W., Davies, M.H., Naim, R.B. and Savo, O.J. (1986). Calculation of littoral sand transport, *Coastal Engineering Journal*, 10, 1–21.

Komar, P.D. (1976). *Beach Processes and Sedimentation.* Englewood Cliffs, NJ: Prentice-Hall, 429 pp.

Komar, P.D. (1977). Beach sand transport: Distribution and total drift. *Journal of Waterway, Port, Coastal and Ocean Engineering*, 103(WW2), 225–239.

Komen, G.J., Cavaleri, L., Donelan, M., Hasselmann, K., Hasselmann, S. and Janssen, P.A.E.M. (1994). *Dynamics and Modelling of Ocean Waves.* Cambridge, UK: Cambridge University Press.

Loveson, V.J., Gujar, A.R., Barnwal, R.P., Khare, R. and Rajamanikam, G.V. (2014). GPR studies over the tsunami affected Karaikal beach, Tamil Nadu, South India. *Journal of Earth System Science*, 123:1375–1385.

Ministry of Environment Forest and Climate Change (MoEFCC) (2019). Preliminary project report for integrated coastal zone management project. Government of India.

Mohapatra, M. (2015). Cyclone hazard proneness of districts of India. *Journal of Earth System Science*, 124(3), 515–526.

Natesan, U., Rajalakshmi, P.R., Ramana Murthy, M.V. and Ferre, V.A. (2015). Nearshore wave climate and sediment dynamics along southeast coast of India. *Journal of Indian Society of Remote Sensing*, 43(2), 415–427.

Nordstorm, K.F., Nancy Jackson, L., James, A. and Sherman, D.J. (2003). Longshore sediment transport on a microtidal estuarine beach. *Journal of Waterway, Port, Coastal and Ocean Engineering*, 129:1–4.

Prabakar Rao, K. (2002). Sediment Transport and Exchange around Rameswaram Island between Gulf of Mannar and Palk Bay. Berhampur, Ph.D. thesis, Berhampur University, India, 200 pp.

Ramana Murthy, M.V., Mani, J.S., and Subramanian, B.R. (2008). Evolution and performance of beachfill at Ennore seaport, southeast coast of India. *Journal of Coastal Research*, 24(1), 232–243. DOI:10.2112/05-0537.1

Sanil Kumar, V., Anand, N.M., and Chandramohan, P. (2004). Longshore sediment transport rate measurement and estimation on central west coast of India. *Coastal Engineering Journal*, 48, 95–109.

Suresh, P.K., Sundar, V., and Selvaraja, A. (2011). Numerical modelling and measurement of sediment transport and beach profile changes along south west coast of India. *Journal of Coastal Research*, 27(1), 26–34.

Sundar, V. (2002). Littoral drift. In: Narasimhan, S. and Kathiroli, S. (Eds.), *Harbour and Coastal Engineering*, Chapter 9, Vol. II, pp. 1–57.

Sundar, V., Noethel, H., and Holz, K.P. (1992). Wave kinematics in a groin field – frequency domain analysis, *Coastal Engineering Journal*, 18, 137–152. EID: 2-s2.0-0027790566.

Sundar, V., Noethel, H. and Holz, K.P. (1994). Wave directions in a Groin field. *Journal of Coastal Research*, 10(4), 839–849. https://doi.org/10.1016/0029-8018(93)90027-

Sundar, V. and Suresh, P. (2011). Comparison between measured and simulated shoreline changes near the tip of Indian peninsula. *Journal of Hydro-Environment Research*, 5(3), 157–167. DOI: 10.1016/j.jher.2011.02.002.

Van Rijn, L.C. (2001). *Longshore Sediment Transport*. Delft Hydraulics.

Vimala, J., Latha, G. and Venkatesan, R. (2014). Estimation of significant wave heights using numerical and neural techniques and comparison with buoy and satellite observations. *The International Journal of Ocean and Climate Systems*, 5, 223–236. DOI:10.1260/1759-3131.5.4.223.

Walton, T.L. and Bruno, R.O. (1989). Longshore transport at a detached breakwater, phase-II. *Journal of Coastal Research*, 5(4), 679–691.

Wang, P. and Kraus, N.C. (1999). Longshore sediment transport rate measured by short term impoundment, *Journal of Journal of Waterway, Port, Coastal and Ocean Engineering*, 125, 118–126.

Wellen, V.E., Chadwick, A.J., Lee, M., Baily, B. and Morfett, J. (1998). Evaluation of longshore transport models on coarse grained beaches using field data: A preliminary investigation, *Proceedings of International Conference on Coastal Engineering*, ASCE, Copenhagen, pp. 2640–2653.

Chapter 12

Numerical Modelling of Tidal Inlet Dynamics

Abstract

Tidal inlets are ecologically sensitive coastal features, where its opening along the shoreline permits free exchange of water through flooding
and ebbing flow. The shoreline openings may constitute bays, lagoons,
marsh or estuaries. The formation of sandbars, spits, and shoals is more
common at the inlet mouth particularly along the coasts dominated by
longshore sediment transport. These formations act as a barrier, blocking
the entrance of tidal inlet and thus restricting flow/exchange of water.
The sandbar and spit formation are usually minimized by construction
of a pair of training walls, prior to which monitoring the local hydrodynamics at the inlet vicinity is essential. In the case where the tidal inlet
meets an estuary, the volume of water exchanged at the inlet mouth are
governed by the inlet dimensions and rate of littoral transport. After a
comprehensive discussion on the hydrodynamics and morphodynamics
of estuaries, its application is considered through a detailed investigation
of a micro-tidal inlet at the Arasalar estuary, a tributary of the Cauvery
river, in Karaikal (10° 54′ 52″ N; 79° 51′ 09″ E), situated along the
southeast coast of the India. The numerical model using the finite volume method is applied to estimate the siltation rate and its distribution
within the domain, driven by the tide-induced currents and riverine discharge and validated through field measured data. This chapter provides
the procedures to be followed in identifying the problem of siltation and
arriving at a suitable mitigation measures to reduce siltation within the
inlet to facilitate smooth navigation.

12.1. General

12.1.1. *Tidal inlets*

Tidal inlets are the portion of the water body between the sea and a
river/backwater that is subjected to flooding during high tide and ebbing
during low tide. Such inlets are widely associated with river mouths or
streams and provide a regular exchange between the estuarine waters and
the open ocean. The exchange of water in an inlet is either maintained by

the cyclic ingress and flushing of tide or closed due to heavy longshore drift along the coast. The tidal flow deposits sand on either the landward or seaward side of an inlet. The sediments carried out by the flood tide will develop a shoal on the riverside or backwater side called the flood tidal delta or flood shoal. The tidal current moving towards the sea is called ebb current or ebb tide, which leads to the movement and settlement of sediments known as ebb shoals in the ebb-tidal delta. The dynamics at a tidal inlet involve major processes resulting in the highest magnitude of erosional-depositional changes along barrier islands.

The formation of ebb/flood shoals may increase over the years and result even in the closure of the inlet mouth. The sediment movements parallel and normal to the coast are driven by waves and tidal currents, respectively. Usually, the ebb-tidal delta comprises large quantities of sediments, comparable to the volume of sediments in the adjoining barrier islands. Any minor modification in the existing tidal prism could potentially affect the sediment supply to beaches in its vicinity. The presence of shoals in the ebb-tidal delta serves as a natural offshore submerged reef, reducing the magnitude/intensity of wave energy incident on the landward beaches; thus, the breaching of such shoals could accelerate erosion processes. The inlet sediment bypassing process across mixed energy shorelines may occur by three major mechanisms viz., inlet migration and spit breaching, stable inlet processes and ebb-tidal delta breaching, as illustrated in **Fig. 12.1**.

Inlets are maintained stable and open by the constant presence of strong tidal currents or gradually closed off due to sizeable longshore littoral currents and sediment transport. The wave forces are more predominant than the tidal currents along the micro-tidal environment, thereby resulting in temporary or permanent closure of the inlet mouth, whereas in a meso-tidal barrier island system, the existence of strong tidal flow enables to maintain an open inlet mouth, irrespective of the longshore sediment transport rate and wave-induced forces. However, meso-tidal inlets are also susceptible to migration of the inlets due to the movement of sediments by ebb and flood tidal currents.

The velocity of the flow at the inlet throat should be sufficiently large enough to keep the channel entrance clear of any sandbar, shoal or spit formation. The man-made interference to widen or deepen the channel will distort the equilibrium at the inlet, resulting in deposition or erosion of sediment, which is dictated by local hydrodynamics and sediment characteristics. The tidal prism is the measure of volume of water entering the

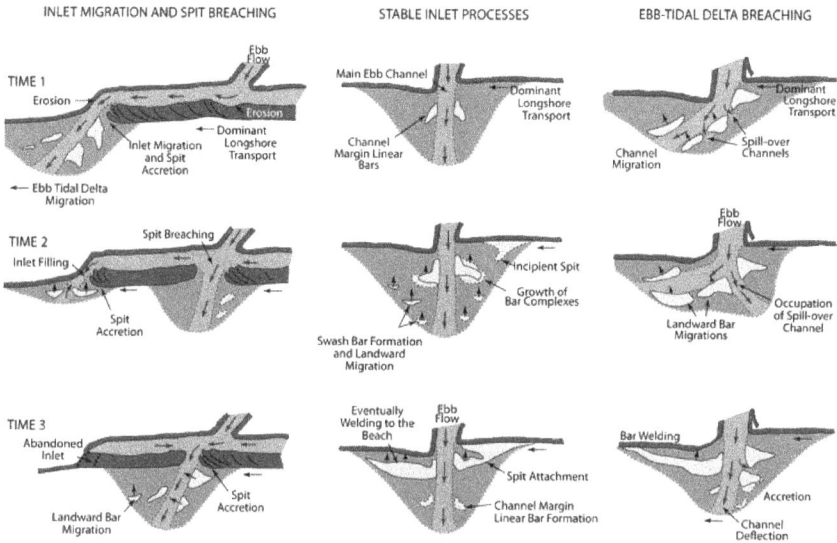

Fig. 12.1. Tidal inlet processes.
Source: Bruun and Gerritsen (1958).

channel during flooding and exiting the channel during ebbing, i.e., tidal prism volume, $P = hA$, where h is the average tidal range and A is the average surface area of the basin.

12.1.2. *Shoal and spit formation*

The sediments, sands or other unconsolidated materials deposited over the bed are referred to as shoals, which can also be described as a natural submerged ridge, bank or a bar. When these deposited materials build up, they pose a serious threat to navigation of vessels. In simple words, the existence of shoals indicates a relatively shallower region in a water body. Shoals can also refer to a rock mass on the water bed or any vegetation layer or any matter that leads to a reduction of water depth locally. A fully developed shoal could separate a smaller unit of the water body from sea, resulting in the formation of marine lagoons, brackish water estuaries, freshwater seasonal stream and river mouths and deltas. In certain coastal stretches, where the longshore drift dominates, the sand bar initially formed can build and propagate continuously over years to form a permanent structure. One such example is the formation of Hope Island (16.97° N; 82.35° E), in Kakinada along the east coast of India. The spit, gradually over 200 years,

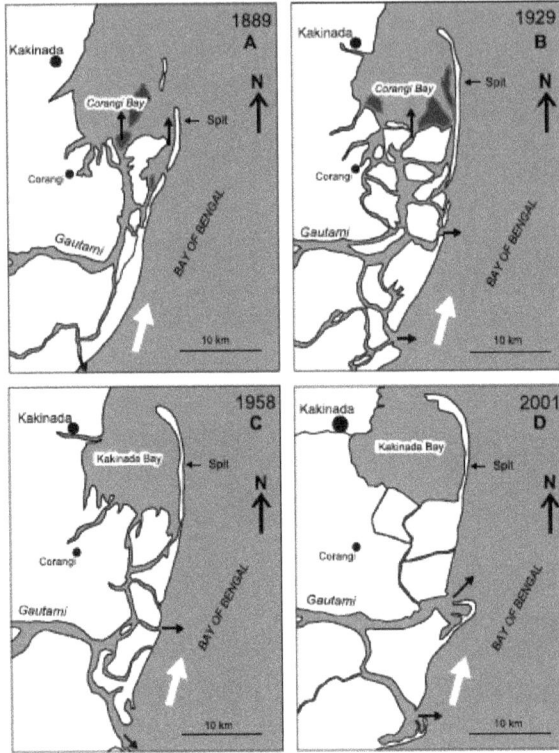

Fig. 12.2. Development of the Godavari spit (Hope Island).
Source: Nageswara Rao *et al.* (2005).

has propagated to a length of about 18 km towards the north (**Fig. 12.2**). The Kakinada Bay, measuring about 100 sq.km, is the area between the stretch of coast of Kakinada and Hope Island. Hope Island acts as a natural providing tranquillity to the ships anchored in Kakinada Bay.

On the formation of sandbars near the confluence of estuaries and the open ocean, Bruun and Gerritsen (1958) analyzed a large number of inlets and classified as per the ratio of Annual Littoral drift along a coast, M and the tidal discharge, Q (**Fig. 12.3**) as mentioned below.

(1) Bar bypassing [$M/Q > 200$], i.e., if M is very high and Q is low, the natural bypassing would be directly across the inlet over a bar.
(2) Tidal flow by passing [$M/Q < 7$], i.e., if M is low and Q is high, usually during monsoons, the drift would be bypassed by tidal flow

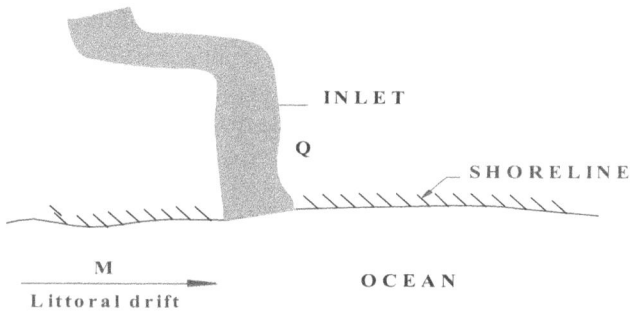

Fig. 12.3. Littoral drift and tidal discharge.

The investigations on the behaviour of tidal inlets are important in connection with the following.

- Development of tourism, fisheries and other related commercial activities.
- Constituting important public infrastructure such as jetties, ports and harbours serving the shipping industry.
- Provides an aesthetic environment and recreational space for the local community.
- Promotes the space for activities such as boating, fishing, swimming, diving, windsurfing and bird watching.
- Exchange of fresh and salt water promote the growth of mangroves, the roots of which stabilize the banks of the estuaries, which can act as buffers in reducing the inundation levels during extreme coastal hazards
- If the sandbar formation is not removed, no tidal flushing takes place, leaving the river to behave as stagnant pool of water, which will serve as excellent ground for breeding mosquitoes (health hazard).

12.1.3. *Stabilization of tidal inlets*

The broadly classified methods for stabilizing the tidal inlets are: (i) sand pump on a trestle up-drift of the inlet, (ii) sand pump at the tip of the up-drift groin (on a short trestle) and (iii) sand pump on a trestle down-drift of the inlet (the approach channel acts as a sand trap) bypassing through a pipeline across channel or bypassing through a reclamation berth.

   Along a littoral-drift–dominated coast, the widely adopted practice is to carefully plan and implement a pair of training walls, one on either side of the mouth of the estuary. If the direction of the drift is of equal magnitude on both sides of the mouth, the length of training walls could be of same length, whereas, if the coast is experiencing the net drift dominant towards a particular direction, the length of the training wall on the updrift side will have to be longer. The length of the training wall is equal to or twice the surf width in which distribution of the drift is concentrated.

12.2. Strategy To Be Adopted for Stabilization of Inlets

Tidal inlets are fairly sensitive to the near-field environmental changes, which can drastically influence the stability of an inlet. To decipher whether the inlet under consideration is stable or to assess the need for any protective measures, the flowchart in **Fig. 12.4** prescribes the sequence of considerations to be undertaken before planning the stabilization of a tidal inlet. The following classification is made in the initial framework.

- Morphologically important regions
  - Inlet
  - Regions that need developmental activities

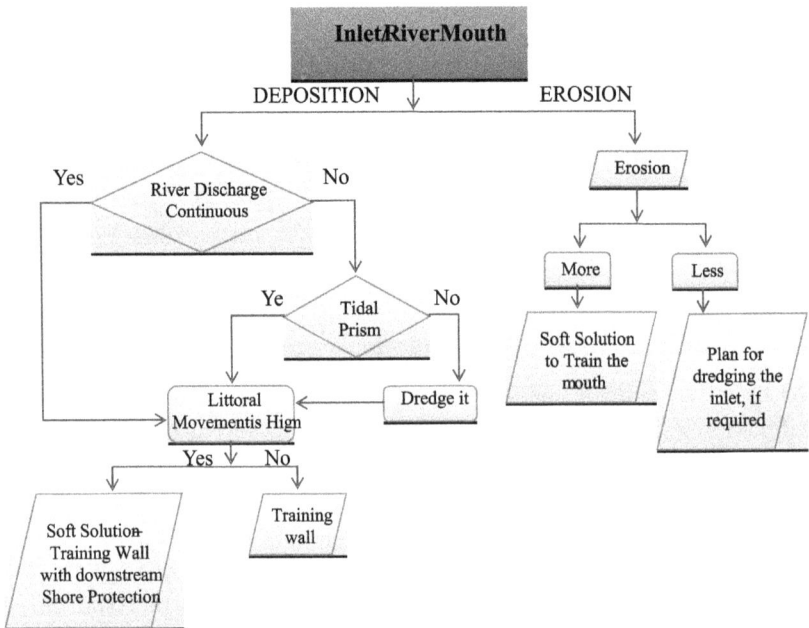

Fig. 12.4. Stabilization of tidal inlets.

- Highly populated
- Tourism and recreation
- Industrial sector
- Socio-economic aspects
- Ecologically sensitive regions

12.3. Literature Review

The inlet morphodynamic and hydrodynamic processes are complex and highly unpredictable. A stable inlet should possess an equilibrium, where the cross-sectional area of the inlet can fluctuate about a mean value. It is stated that the maximum current speed or shear stresses generated over a spring tidal cycle influences the cross-sectional area of the inlet (Escoffier, 1940). The generalized $P-A$ relation derived by O'Brien (1931, 1966) does not apply to all tidal inlets, where P is the spring tidal prism and A is the closure curve area. Styles $et\ al.$ (2016) state that the same tidal forcing in a much larger cross-sectional area will produce smaller velocities to preserve the continuity, whereas the weaker flows will reduce sediment transport capacity and cannot maintain a large inlet cross-section.

The strong seasonal variations in the river flow and wave climate in the wave-dominated coastal environment makes the inlet to migrate and eventually close (Tung $et\ al.$, 2009). The tidal inlet in an estuary has various parameters that contributes to modify the local morphodynamics, typically characterized by waves, tides, sediment supply and river discharge (Hayes, 1980; Boothroyd, 1996; Fitzgerald, 1996; Chang, 1997; Fitzgerald $et\ al.$, 2002; de Swart and Zimmerman, 2009). The three natural mechanisms of sand bypassing in an inlet are wave-induced transport, tide-induced transport, erosion and accretion in the tidal channel (Bruun and Gerritsen, 1959). The riverine tidal inlets serves as an inland waterway, where the variations in position and dimensions of the inlet can lead to potential navigational problems.

The estuary mouth interaction may have various morphological change over the season and the cyclic changes in position of the inlet channel and growth and migration of shoals (Oretel, 1977; Fitzgerald, 1984; Hume and Hendendorf, 1992; Gaudiano and Kana, 2001; Fitzgerald $et\ al.$, 2002; Siegle $et\ al.$, 2004; Elias and van der Spek, 2006; Cheng $et\ al.$, 2007; Cooper $et\ al.$, 2007). The morphodynamics of the tidal inlet is typical of barrier coasts formed during a period of continuous sea-level rise. The local littoral sediment transport rate and the evolution of the shoreline, geometry and hydrodynamics of tidal inlets (Tomlinson, 1991; Bruun, 2001; French, 2001; Hanson and Kraus, 2001; Castelle *et al.*, 2007; Fontolan *et al.*, 2007; Buonaiuto and Bokuniewicz, 2008; Kraus, 2009; Pacheco *et al.*, 2010; Wang and Beck, 2012). The field-measured data from the Karaikal inlet indicates that it is not in an equilibrium state. The exchange in the volume of the water during a tidal cycle is attained for a trained inlet, and the sediment transport is continuous from the up to downdrift side along the outer ebb-tidal zone, as opposed to seasonal splitting and merging of ebb shoals (Dean, 1988; Fitzgerald *et al.*, 2000).

The sediment size decreases upstream of the river, and it is also finer, with an increasing elevation over the intertidal flats. The flow is unsteady and non-uniform due to the tidal current variation in the estuary, and the sediment transport direction is observed to be predominant in the lateral direction along the width of the channel. The flow zone in the channel is separated by the upper zone (affected by turbulence), the central zone (motionless water in the lee side of the bedform) and the lower zone (no flow). The flow separates the bottom zone, and the finer grains are carried downstream as suspended load, the erosion point of flow eradicates from the channel (Davis, 1978) and the coarser grains stay back. The higher sandy portions (ebb shields) are usually coarser than the lower flood ramp areas on the flood tidal delta (Daboll, 1969).

When the current velocities are higher, the bottom bed shear increases or the shear stress is increased due to wave action, which leads to erosion. The shoals form due to decrease in the current velocity below the threshold value. The stable inlet would retain its cross-section either by settling sediments further away from flood and ebb delta or within the sediment traps over the bed (Bruun, 1986, 1989). The various dynamic and static parameters for morphological changes at inlet are net longshore sediment transport, tidal prism, riverine sediment supply, dredging of the channel, flood shoal evolution, wave refraction and diffraction over shoals, bathymetry, sediment size, etc. (Carr and Kraus, 2001). The current vortices are generated when tides move over the irregular topography within the inlet basin, and these currents are advected by the tidal flow resulting in a complex flow (Zimmerman, 1980; Ridderinkhof, 1989; Stanev *et al.*, 2003, 2007).

12.4.   Case Study

12.4.1.   *General*

To have better understanding of the foregoing discussion, a case study by considering the inlet of the Arasalar River located at Karaikal (10° 54′ 52′N;

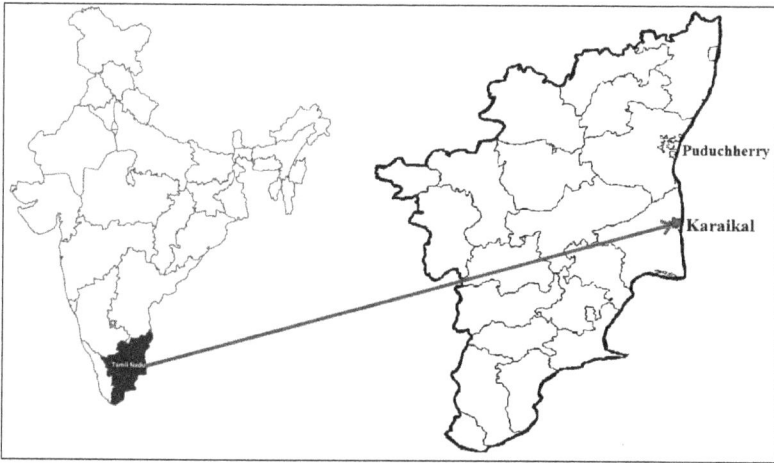

Fig. 12.5. Location map (Karaikal).

79° 51′ 09′E) along the southeast coast of the Indian peninsula is presented and discussed herein. The inlet is about 110 m wide exposed to a tidal range of less than 1 m. According to Davies (1964) and Hayes (1979), the Arasalar inlet can be classified as a micro-tidal inlet. The Arasalar River is one of the minor tributaries from the major river Cauvery, which drains into the Bay of Bengal. The location map is projected in **Fig. 12.5**. A pair of rubble mound training walls was constructed at the entrance of the channel in 2005 to prevent the sandbar formation; later in 2007, piers were built to create berthing facilities, about 800 m upstream of the inlet along the southern bank of the river. The beach profile slope is flat and almost in line with the mean sea level, due to which the coast is highly prone to coastal flooding during storm surges. The net littoral drift experienced in the coast is about 0.13 million m^3 per annum of littoral sediments heading northward and a gross transport rate of 0.24 million m^3 per annum (Kumar *et al.*, 2006). The study area along with the locations of field measurement points are projected in **Fig. 12.6**.

The local study reports that the recurrent problem in the entrance basin is excessive silt/shoal formation upstream of the river. Excessive siltation hinders the navigation of vessels, mandating frequent dredging activities to be commissioned. The main source of siltation in the inlet channel is tidal current and riverine discharge flows, and the tidal prism, inlet dimensions/area, bottom boundary friction, sediment concentration, etc., are the parameters dictating the rate and pattern of siltation. The training

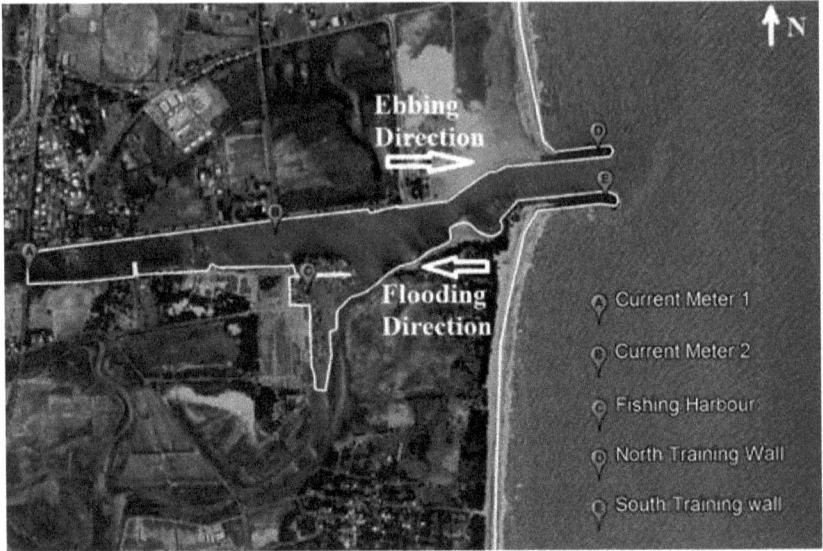

Fig. 12.6.   Study area (Arasalar River, Karaikal).

walls effectively serve as barriers against spit and sandbar formations, thus enabling the river mouth to remain open across all seasons. The training walls extend into the surfing zone, preventing coastal currents from depositing or removing sediments near the mouth. The seasonal closure of an inlet leads to two main problems, viz., the ocean access for boats gets limited and the water quality in the landward-side waterbody will deteriorate during the months of inlet closure (Senthilkumar *et al.*, 2017).

12.4.2.   *Field data*

A detailed field investigation was carried out for the different seasons of a typical year to serve as an input for the numerical model and to validate the output results. The field study includes the bathymetry survey conducted within the channel and the open coast up to -20 m water depth using an Echo Sounder and a GPS device. The measured bathymetry charts are projected in **Fig. 12.7**. A pair of current meters are fixed in the Arasalar River, one near the fishing harbour (10° $54'$ $41.74''$N 79° $50'$ $19.77''$E) and another near the railway bridge (10° $54'$ $44.87''$N 79° $50'$ $44.57''$E) about 2 km upstream from the river mouth. The highest magnitude recorded is 0.5 m/s from the unidirectional measurement as seen in **Fig. 12.8**. The local

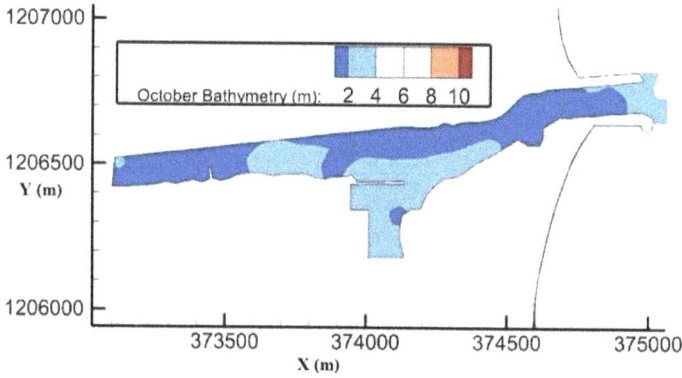

Fig. 12.7(a). Bathymetry survey chart for October 2019.

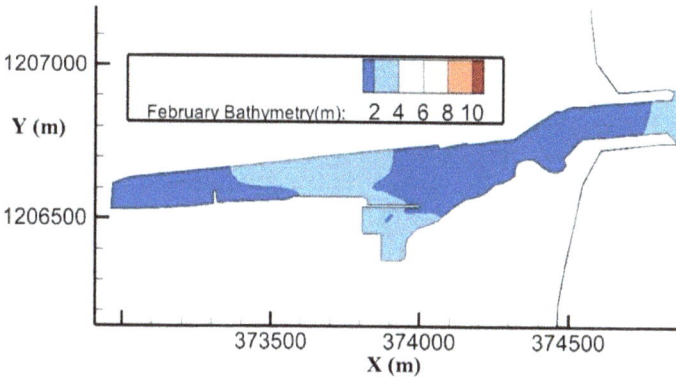

Fig. 12.7(b). Bathymetry survey chart for February 2020.

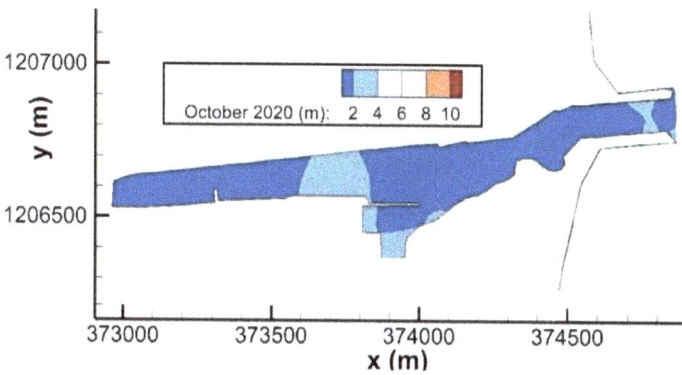

Fig. 12.7(c). Bathymetry survey chart for October 2020.

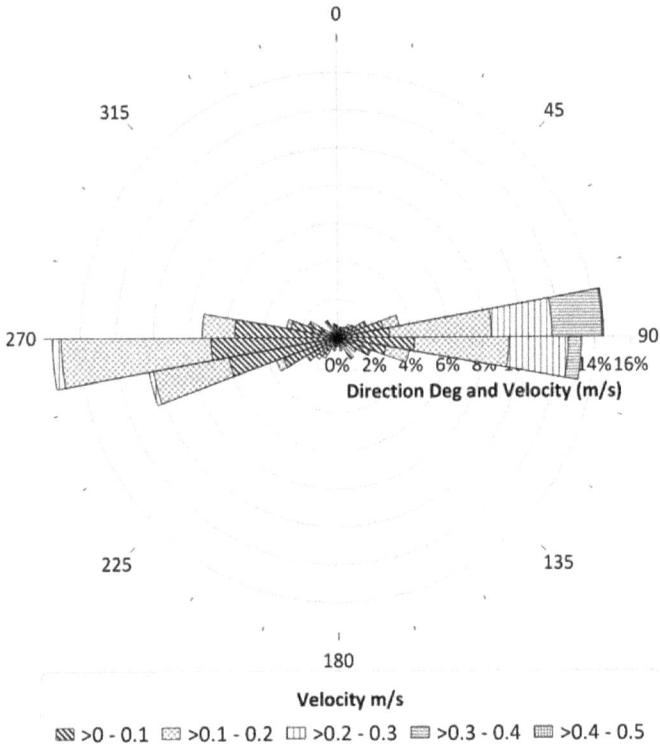

Fig. 12.8(a). Current velocity near railway bridge.

bathymetry influences bottom shear stresses and current flow parameters, which in turn affects the morphodynamic evolution in the nearshore region and river or tidal inlets (Pascolo *et al.*, 2018). The tide levels variations near the site were extracted from WX-Tide32, the average tidal range recorded is about 0.8–1.0 m. The soil sediment sample as well as the suspended sediment concentration at the site were measured to be in the range of 0.1–0.5 mm (finer sand).

12.4.3. *Problem identification*

The collection of data aforementioned was the basic requirement for understanding the dynamics of the estuary selected for the study. It was understood that the construction of training walls prevented the formation of

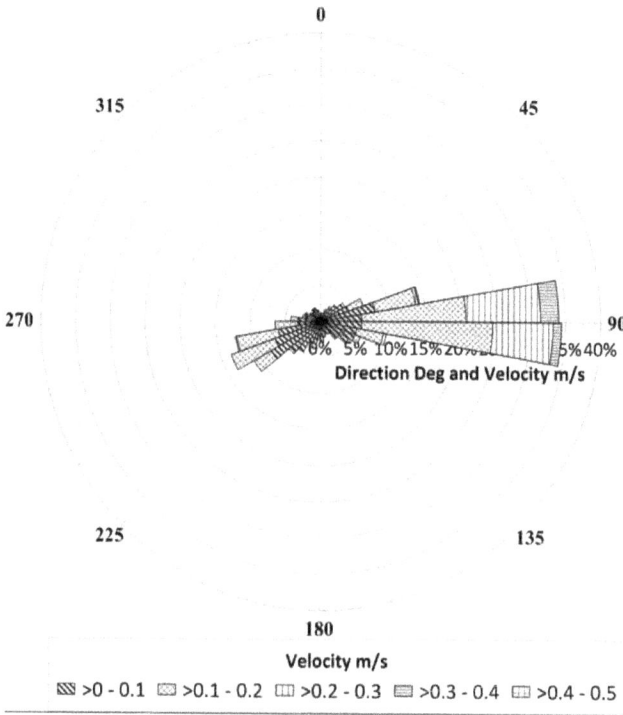

Fig. 12.8(b). Current velocity near fishing harbour.

sandbars and spits at the mouth of Arasalar inlet. The perennial rivers that travel from the source emerging to numerous tributaries carries heavy sediments along with the flow. The progressive build-up of sediments modifies the bed bathymetry, which in turn influences the local hydrodynamics and morphodynamics. The siltation in the channel leads to the reduction of water depth in the entrance channel, where the fishing boats draft reduces and hinders smooth navigation. The channel boundaries upstream of the trained inlet experience sediment erosion and deposition due to the varying current flow, which in turn is a long-term effect of the training wall construction. The stability is understood to have been disturbed due to the varying discharge from the river and dynamic tide-induced currents from the ocean. The instabilities along the north boundary is observed between 2011 and 2015 (**Fig. 12.9(a)**) and later from 2016 to 2020, extreme changes were

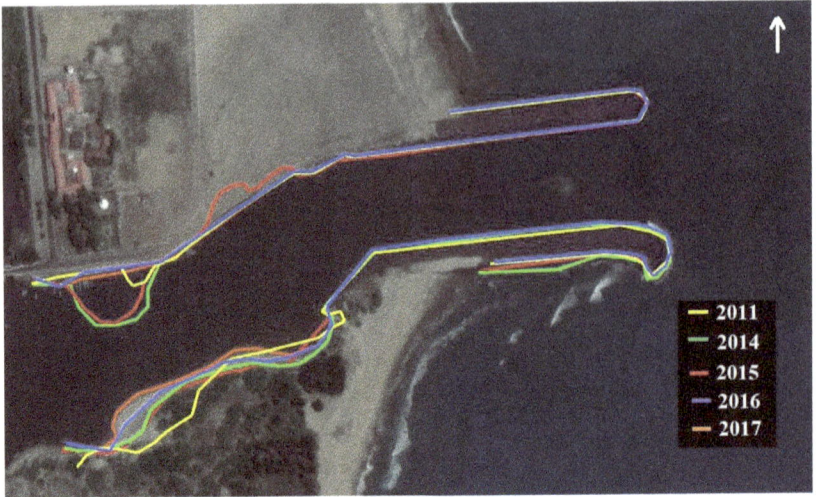

Fig. 12.9(a).   North boundary stability (2011–2017).

Fig. 12.9(b).   South boundary changes in the river channel (2018–2020).

Table 12.1.   Dredging quantity in the channel.

	2018–2020	Average for a year
Deposition	5732.34 m^3	1,910 m^3
Erosion	3088 m^3	1,029 m^3

Fig. 12.10. Revetment on north boundary.

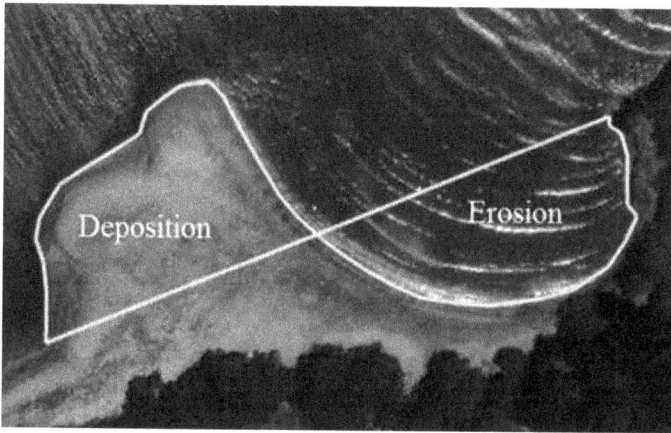

Fig. 12.11. Erosion and deposition in south boundary.

observed at the south boundary (**Fig. 12.9(b)**), causing not only a recurrent reduction in the local water depth within the inlet channel but also in its width at certain critical reaches along the entrance channel. Despite the frequent maintenance dredging, the reduction of channel width is compromised. The continuous erosion along the north boundary were resolved by constructing a stone revetment as shown in **Fig. 12.10**. The quantity of sediments eroded and deposited along the south boundary was calculated from the field-measured data as shown in **Fig. 12.11** between 2018 and 2020 and provided in **Table 12.1**. To arrive at a feasible and permanent solution for the siltation on the south boundary, a proposal was made to construct a revetment for a length of 450 m upstream of the river channel. The morphodynamic changes at the inlet location due to the construction of the revetment structure are being discussed in Section 12.6.2.

12.5.   Methodology Adopted

12.5.1.   *General*

The numerical model adapted for the study was developed over the years by Chitra *et al.* (1996), Murali *et al.* (2002) and Murali *et al.* (2014), where the local hydrodynamics are modelled and calibrated with the help of field-measured data. A vertically integrated form of shallow water equations (SWE) discretized by the finite volume method (FVM) is employed to mimic the real-field conditions in the model. The tide-induced current velocities can be obtained by solving the SWE (Eq. (12.1)), in the Cartesian coordinate system, where the z-axis represents the water depth.

$$q_{x,x} + q_{y,y} + \eta_{,t} = 0 \tag{12.1}$$

$$q_{x,t} + \frac{q_x}{H}q_{x,x} + \frac{q_y}{H}q_{x,y} - fq_y = -\rho^{-1}Hp_{a,x} - gH\eta_{,x} + \rho^{-1}(\tau_{ax} - \tau_{bx})$$

$$q_{y,t} + \frac{q_y}{H}q_{y,z} + \frac{q_y}{H}q_{y,y} + fq_x = -\rho^{-1}Hp_{a,y} - gH\eta_{,y} + \rho^{-1}(\tau_{ay} - \tau_{by})$$

$$\tag{12.2}$$

where η is the free surface elevation, u and v are the mean velocity vector components, h is the total depth ($h = d + \eta$), with d being the still water level. ε is the eddy viscosity, τ_{wi} surface stresses, τ_{bi} bed friction stresses ($i = x, y$), and f is the Coriolis parameter. S_{ij} are the components of the radiation stress tensor that represent the excess momentum fluxes associated with the oscillatory wave motion.

The mass and momentum conservation within the study domain area is ascertained by employing an FVM solver over an unstructured triangular mesh domain (Ashford, 1996; Roache, 1998). The domain is discretized into a set of non-overlapping control volumes, each with an integral form, which ensures arriving at a continuous and strong solution. The bottom boundary friction has arrived as a function of the Manning roughness coefficient. A Courant number varying between 0.35 to 0.5 is adopted to ensure the stability of the numerical scheme. Chezy's co-efficient for open channel flow is adopted for the riverine discharge with a uniform friction factor of 0.023 throughout the domain area. An eddy viscosity coefficient of 1 is adopted by calibrating the numerical model through the trial-and-error method. This parameter is essential to account for turbulence in the governing shallow water equations and aids in simulating the diffusion characteristics of turbulence in the flow field/domain. The numerical solver works the model with input from tide elevation, riverine discharge and open boundaries, and the domain area and concentrated mesh are shown in **Fig. 12.12**. Each of the numerical simulations was set to account for monthly variations, which could include two spring tides and one neap tide or two neap tides and one spring tide.

The arrived hydrodynamic model serves as a crucial input for the morphodynamic model. The numerical solver computes the bed level changes

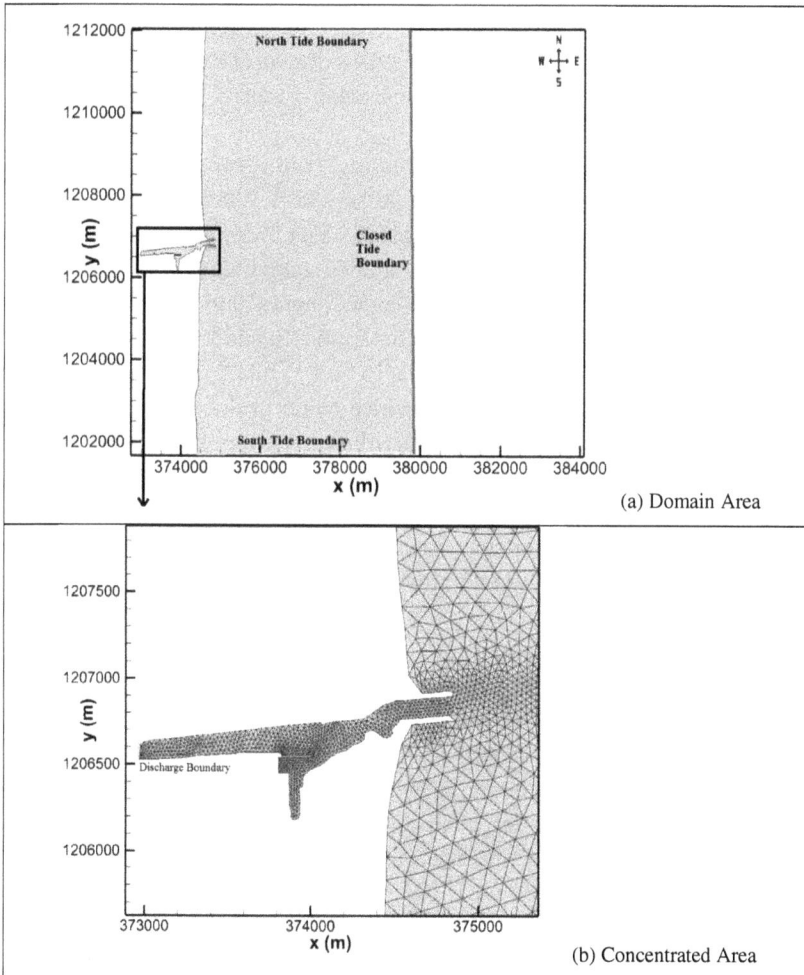

(a) Domain Area

(b) Concentrated Area

Fig. 12.12.   Tide and discharge boundaries.

within the domain through the mass balance of sediments at given control volumes. Equation (12.3) gives the expression to compute the rate of siltation.

$$\frac{\partial h}{\partial t} + \frac{1}{1-n} \nabla_h \bullet (q_x, q_y) = 0 \qquad (12.3)$$

where n is the porosity, h is the bed level, and q_x and q_y are the total sediment transport rate in x and y directions, respectively. As per above equation, it is assumed that the sediment entrainment and deposition rates are at equilibrium, under a balanced condition (Tambroni et al., 2010). Several researchers have proposed several sedimentation formulae over the past six decades to compute the sedimentation quantity and its distribution within a controlled study area.

To understand the inlet morphodynamics, three sedimentation formulae were chosen, viz., the formulation of Ackers and White (1973), Van Rijn (1984), and Engelund and Hansen (1967). The Ackers and White (1973) formula (Eq.(12.4)) compute the suspended and bedload sediments as a total load for assessing the fineness of sediments under a unidirectional current flow. It is claimed to be best applicable in shallow waters and for a wide range of sediment sizes. The Ackers and White formula overestimate the sedimentation for finer particles, which could probably lead to negative sediment concentration and result in a large scatter in the prediction of bed load sediments (Bisheng et al., 2008). The Van Rijn (1984) formula (Eq. (12.5)) has principally arrived from fundamental physics and empirical relation for computing sediment transport rate in rivers, where the bed and suspended load sediments are computed separately. The limitations of the application are that the flow depth should be greater than 1 m and grain size to be greater than 0.5 mm.

The suspended load computed using the Van Rijn formula was found to have a huge scatter compared to the field measured load (Nakato, 1990). The Engelund and Hansen (1967) formula based on the shear stress approach (Eq. (12.6)) was originally developed to compute the bedload transport of sediments under a unidirectional current flow regime. Yang (1973) claims the formula to be reliable for sandy sediments and is suitable to be applied for river channels. Bayram et al. (2007) stated that these formulae underestimate the quantity of bedload sediments. It employs the Shields parameter, the results of which do not depend on the skin friction calculation method. The uncertainties associated with employing various sedimentation formulae are discussed with examples by Pinto et al. (2006).

If the grain size distribution in the study area is unknown, the application of Ackers and White or Van Rijn formula is not suitable; if the bed sediment details are constrained, the application of Engelund and Hansen formulation is preferred (Silva *et al.*, 2009).

$$Q = VDB; \quad D_{gr} = D_{50}[g(S_s - 1)/\nu^2]^{1/3}$$

$$u_* = (g\,D\,S)^{1/2}; \quad n = 1 - 0.56\log(D_{gr})$$

$$A = [0.23/d_{gr}^{1/2}] + 0.14; \quad \alpha = 10$$

$$F_{gr} = \{u_*^n/[gd_{50}(S_s - 1)]^{1/2}\}\{V/[32^{1/2}log(\alpha D/d_{50})]\}^{1-n} \tag{12.4}$$

$$m = (6.83/d_{gr}) + 1.67; \quad C = 10^{2.79\log(D_{gr})-0.98[\log(D_{gr})]2-3.46}$$

$$G_{gr} = C[(F_{gr}/A) - 1]^m; \quad X = G_{gr}S_sD_{50}/[D(u_*/V)^n]$$

$$G = \gamma_w QX$$

where F_{gr} is the mobility number, D_{gr} is the dimensionless grain diameter and G_{gr} is the dimensionless sediment transport function.

$$\frac{q_b}{\bar{u}d} = 0.005 \left(\frac{\bar{u} - \bar{u}_{cr}}{[(s-1)\,gD_{50}]^{0.5}}\right)^{2.4} \left(\frac{D_{50}}{d}\right)^{1.2}$$

$$\frac{q_s}{\bar{u}d} = 0.012 \left(\frac{\bar{u} - \bar{u}_{cr}}{[(s-1)\,gD_{50}]^{0.5}}\right)^{2.4} \left(\frac{D_{50}}{d}\right) (D_*)^{-0.6} \tag{12.5}$$

where q_b and q_s are the bed and suspended load volume per width, respectively, $\bar{u}_{cr}$ is the critical mean flow velocity based on Shield's criterion, $\bar{u}$ is the mean flow velocity, D^* is the particle parameter, and s is the specific density.

$$q = 0.05\gamma_s V^2 \left(\frac{D_{50}}{g(\frac{\gamma_s}{\gamma} - 1)}\right)^{0.5} [\tau_o/(\gamma_s - \gamma)D_{50}]^{3/2} \tag{12.6}$$

where g_s is the sediment transport weight and V is the mean flow velocity.

12.5.2. *Boundary conditions*

The variations in the tidal ranges between Nagapattinam and Paradip were found to be from 0.62–1.87 m. The tide propagates from deep water to the coast, with their fronts nearly parallel to the coast. The current magnitudes were observed to increase slowly from south to north, i.e., 0.075 m/s at Nagapattinam to 0.249 m/s at Paradip. The tide data is extracted from

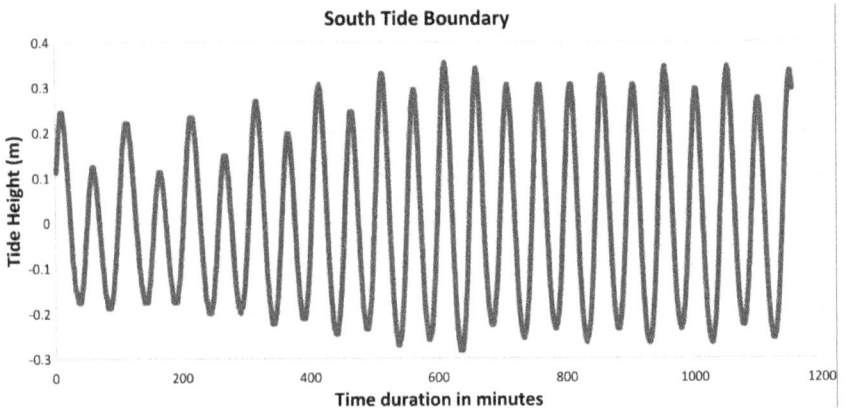

Fig. 12.13(a). Tide for south boundary.

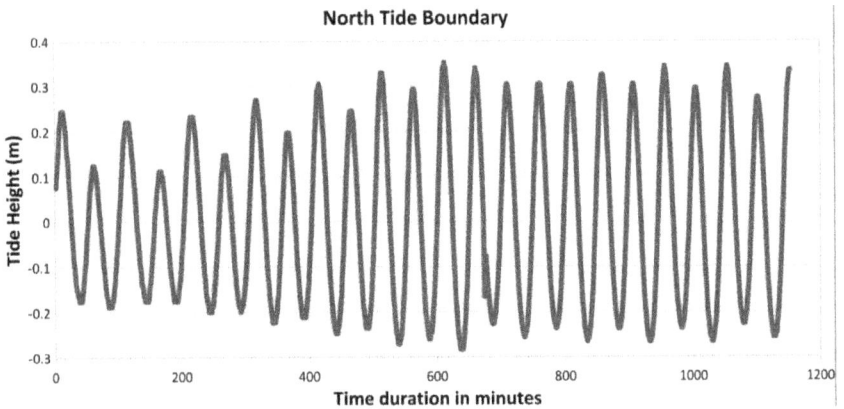

Fig. 12.13(b). Tide for north boundary.

WX-Tide 32 for the Karaikal site location as shown in **Figs. 12.13(a)** and **12.13(b)**. The tide boundaries for Karaikal have been set for 5 km north and south shore, and the boundaries have been extended up to 5 km in the ocean from the inlet of the channel for numerical modelling. The tide boundary has been being calculated from the south direction and the tide ranges have been set at 45 min delay from the south to north tide propagation, and the boundary parallel to the shore is considered as the closed boundary. For the modelling, the tides are set from south to north, and

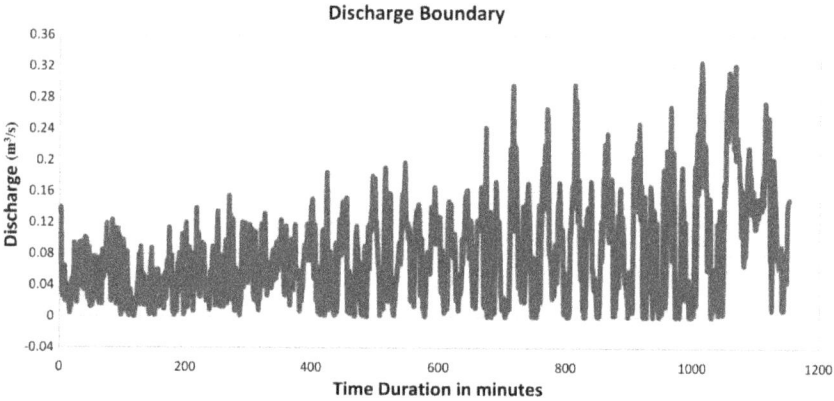

Fig. 12.13(c).   Discharge boundary.

the discharge has been provided inside the channel for the hydrodynamic study.

The current velocity measurements under the railway bridge is used to compute the quantity of discharge from the river as shown in **Fig. 12.13(c)**.

Similar studies were also conducted for other inlets of India. These are Kosasthalaiyar inlet, (South of Kattupalli port, 13° 18′57.26″N 80° 20′49.00″E), an inlet under unstable condition, and Cochin backwater inlet, which comes under the stable category. Salient results for the Kosasthalaiyar inlet are also brought out toward the end of this chapter for a better understanding.

12.6.   Results and Discussion

12.6.1.   *Hydrodynamic results and validation*

The current metre velocity and its directions are computed for a period of four months (October 2019–February 2020) to visualize the effects of northeast monsoon season. The average tidal range at Karaikal coast is observed to be 0.7 m for all seasons, and the coast experiences semi-diurnal tidal variations (i.e., two high tides and two low tides observed in a calendar day). The flooding flow heading towards the inlet of the channel from the ocean is forced from the tides and deformed wave forces. From the hydrodynamic modelling results, it is observed that the ebbing current velocities possess higher magnitude, and it is further forced by the riverine discharge during monsoon. The average discharge is measured as 0.2–0.3 m/s within the inlet

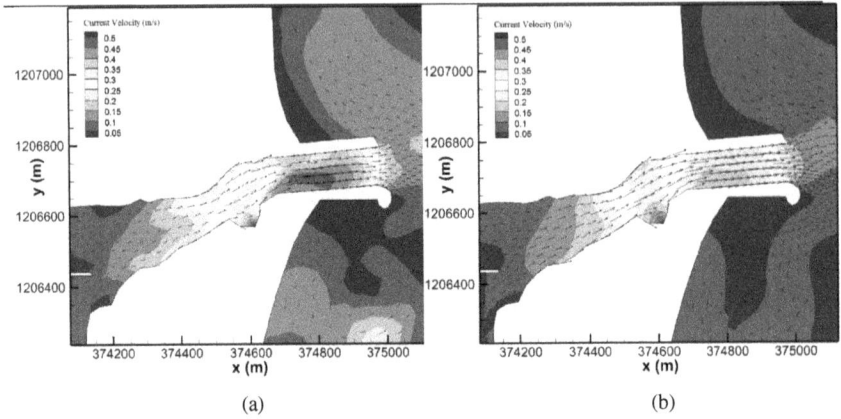

Fig. 12.14. Simulated local current velocities during spring tide: (a) October 2019 and (b) November 2019.

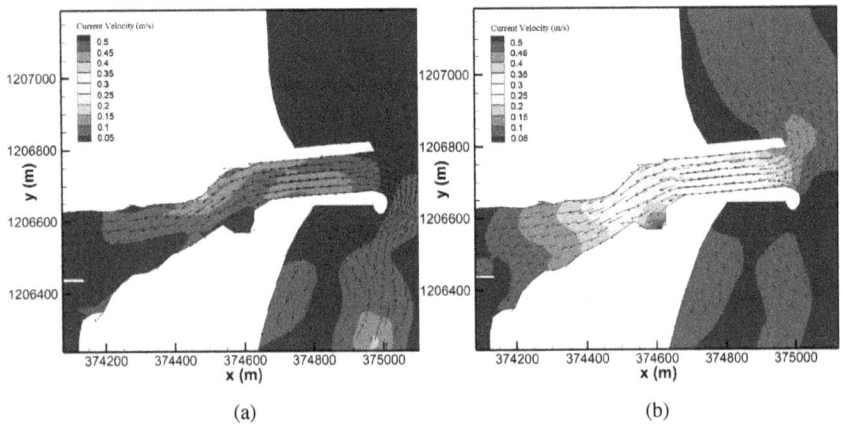

Fig. 12.15. Simulated local current velocities during neap tide: (a) October 2019 and (b) November 2019.

basin during monsoon. The current velocities observed for typical months are projected in **Figs. 12.14** and **12.15** during spring and neap tides. The current velocity at the mid-section of the inlet is steady and streamlined, compared to north and south wall boundaries. The hydrodynamic model results are calibrated at a macroscopic level, considering that the domain area extents over 37 km^2. The current meter and siltation numerical results is compared with the field-measured data. The deviations are observed at certain time intervals, the general trend between the two and the agreement is found to be satisfactory. The comparison between field-measured and simulated current velocities during high and low tides is projected in **Fig. 12.16**.

The maximum current velocity during the neap tide near the fishing harbour (October 2019) is measured as $0.35\,\text{m/s}$. The maximum current velocity during the spring tide near the fishing harbour (October 2019) is measured as $0.48\,\text{m/s}$.

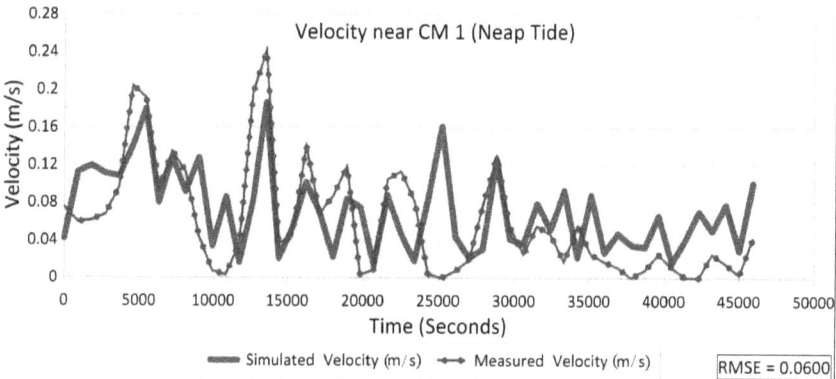

Fig. 12.16(a). Comparison between field-measured and simulated velocities at CM1 during neap tide.

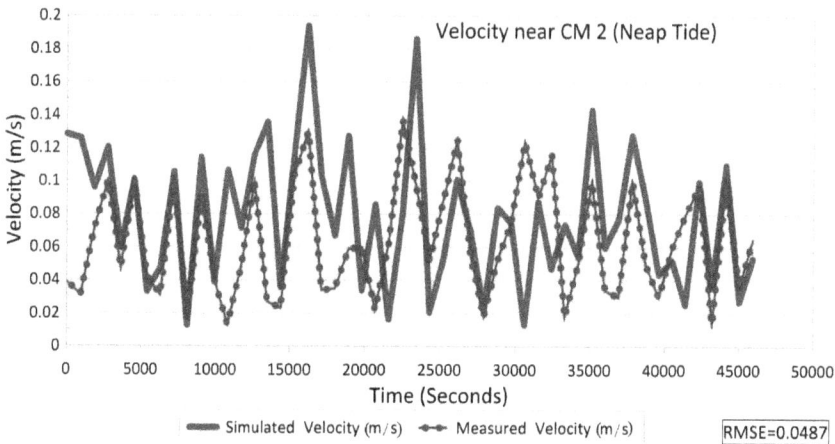

Fig. 12.16(b). Comparison between field measured and simulated velocities at CM2 during neap tide.

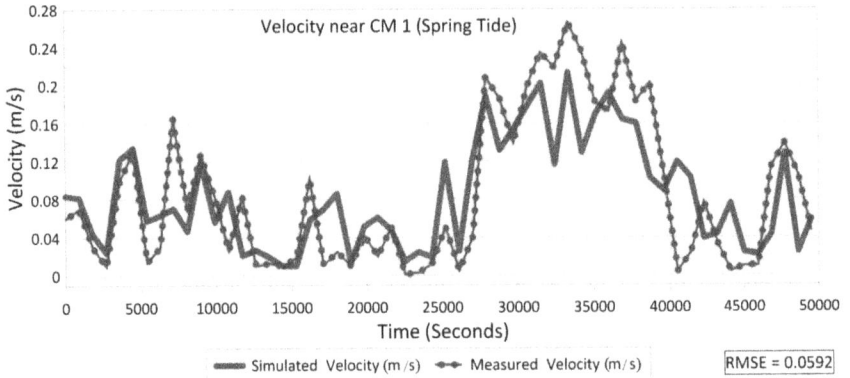

Fig. 12.16(c). Comparison between field measured and simulated velocities at CM1 during spring tide.

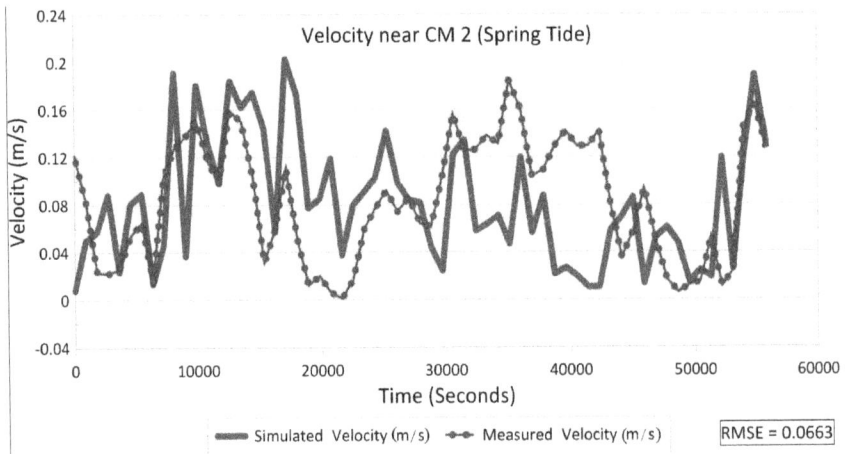

Fig. 12.16(d). Comparison between field-measured and simulated velocities at CM2 during spring tide.

The hydrodynamic model is calibrated for its flow parameters before proceeding with the morphodynamic studies. The sediment distribution within the domain area is dictated by flow limits and current gradients, and it is mandatory to calibrate and validate the hydrodynamic model. The computed root-mean-square error (RMSE) values (as shown in Eq. (12.7)) between the two current velocities are found to be below 0.2 for the monthly current velocities across two months (October 2019 and February 2020).

According to Williams and Esteves (2017), the obtained RMSE value for a hydrodynamic model to assess its accuracy demonstrates a statistically significant fit.

$$RMSE = \sqrt{\frac{i}{N_i} \sum_{i=1}^{N_i} (S_i - O_i)^2} \qquad (12.7)$$

where S_i is the simulated data and O_i is the observed data. Thus the performance of the hydrodynamic model is successfully validated.

12.6.2. Morphodynamic results and validation near the fishing harbour

The study explains the simulated siltation and bed level changes results in the Arasalar River, Karaikal, as illustrated in **Fig. 12.17**.

The morphodynamic study is carried out in the riverine inlet at Karaikal, and the field measured-data has been used as the input to determine the monthly bed changes. Through simulation, the hydrodynamics and morphodynamic changes in the domain area are observed. For the analysis of bed level changes in the channel, the three most popular formulas are adopted; Van Rijn, Ackers and White, and Engelund and Hansen. For model validation, field-measured and simulated siltation results are compared, near the fishing harbour area, and the siltation pattern is noted in the horizontal direction, west to east as shown in **Fig. 12.18**. The compared data points are projected in **Fig. 12.19**. The bed level changes have been observed near the fishing harbour for October 2019 to February 2020. The comparison of simulation resulted with the three formulae and the field-measured data enables to identify the best-suited formulae, i.e., Engelund and Hansen, to compute the morphodynamic changes in the inlet of the channel.

12.6.3. Morphodynamics results at inlet of the channel (problematic location)

A comparative study has been conducted between the said three sedimentation formulae and the field-measured data. The simulated siltation results using the Engelund and Hansen formula were found to closely follow the field-measured data for the northeast monsoon season between October 2019 and February 2020 (validated through field-measured bathymetry charts taken between corresponding time intervals). A similar model is executed for a period between February and October 2020, and results are

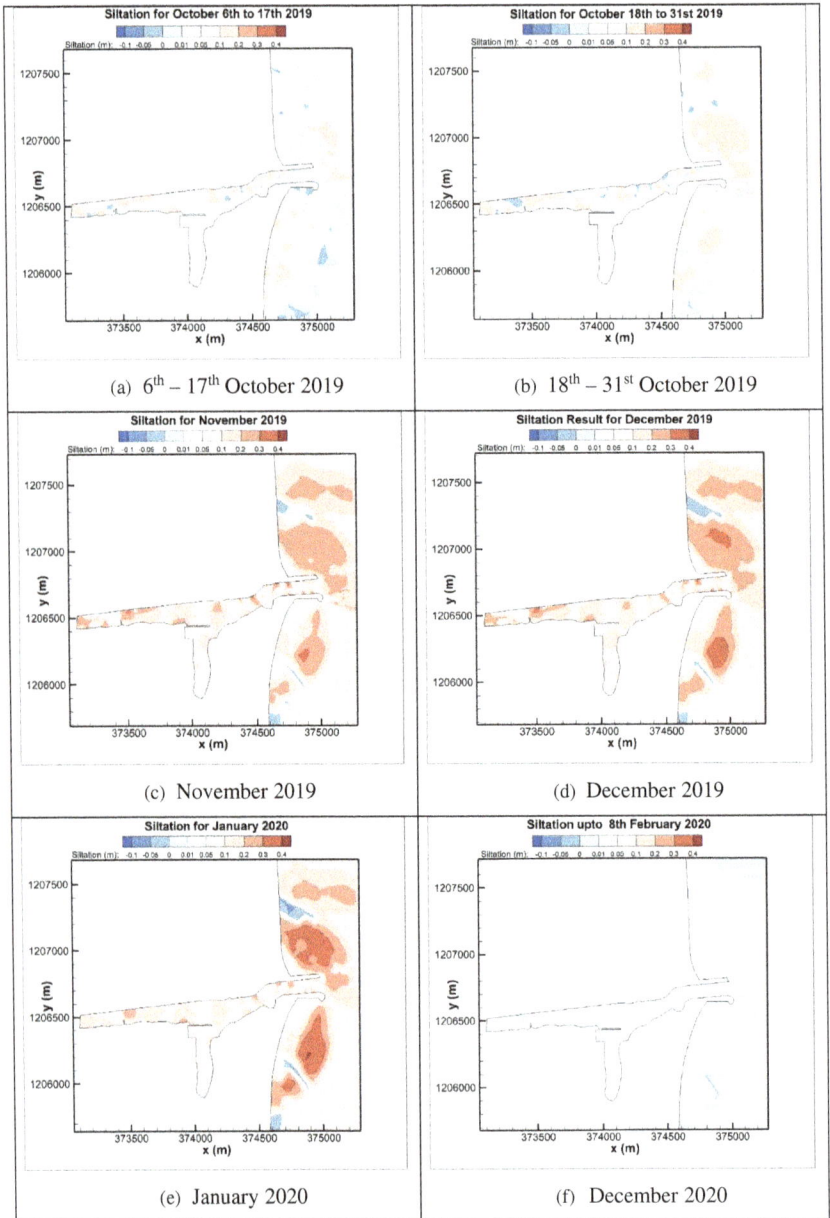

(a) 6th – 17th October 2019

(b) 18th – 31st October 2019

(c) November 2019

(d) December 2019

(e) January 2020

(f) December 2020

Fig. 12.17. Morphodynamic changes from October 2019 to February 2020.

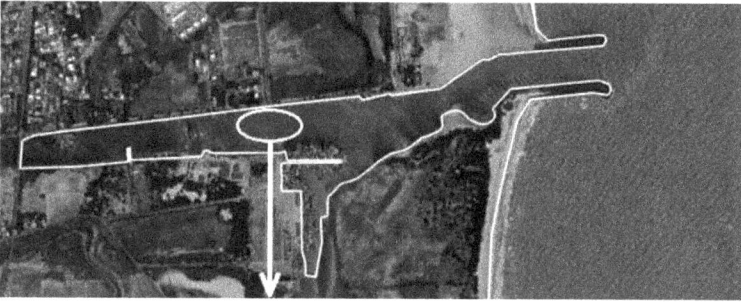

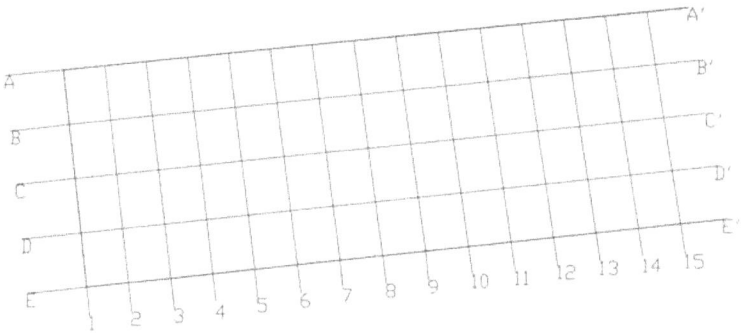

Fig. 12.18. Cross-section area near the fishing harbour.

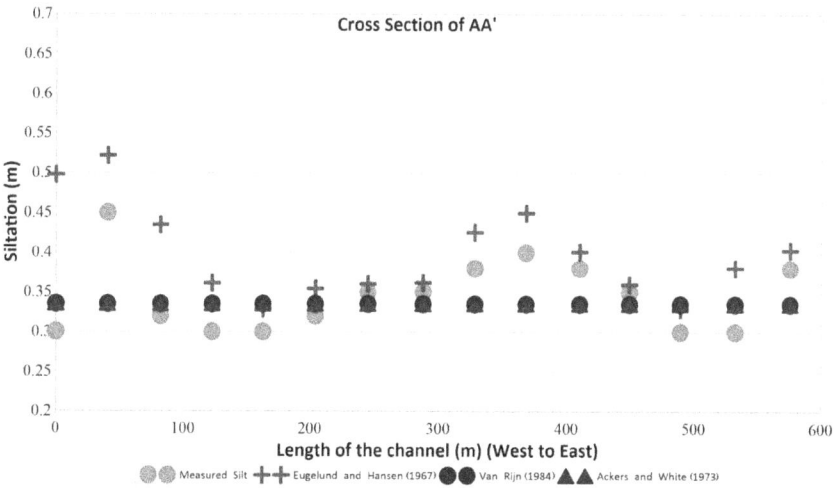

Fig. 12.19(a). Comparison for measured and simulated siltation at Section A-A' near the fishing harbour (October 2019–February 2020).

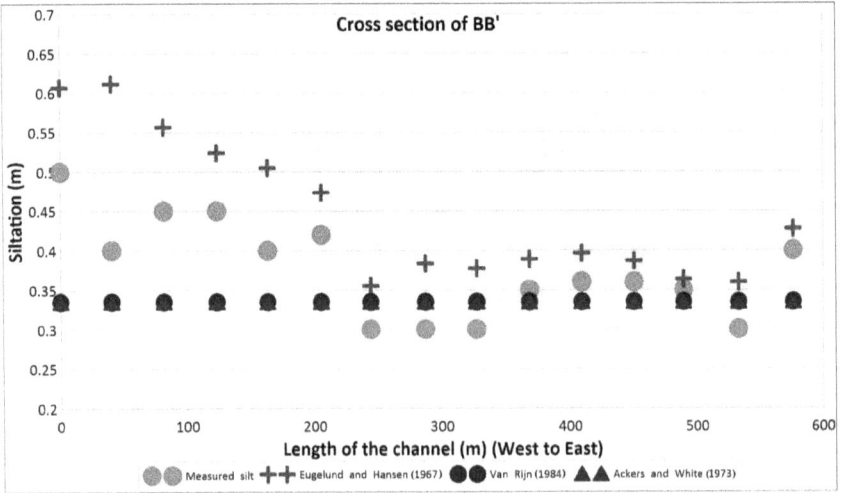

Fig. 12.19(b). Comparison for measured and simulated siltation at Section B-B' near the fishing harbour (Oct 2019–Feb 2020).

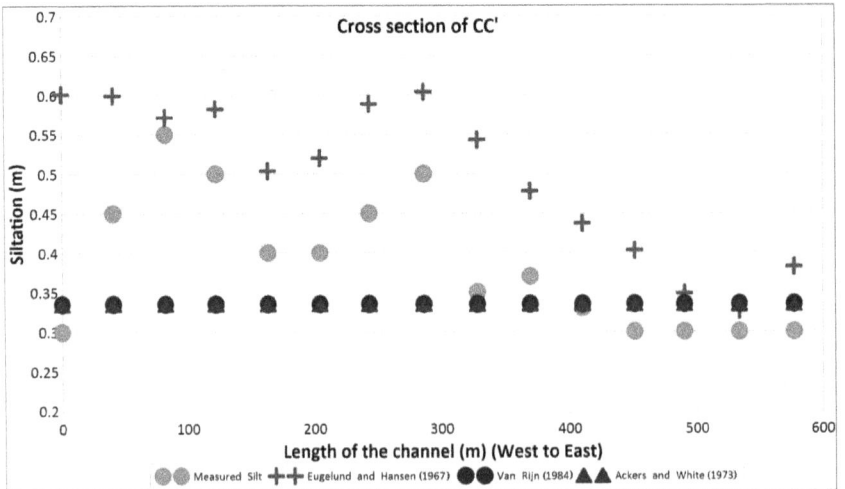

Fig. 12.19(c). Comparison for measured and simulated siltation at Section C-C' near the fishing harbour (October 2019–February 2020).

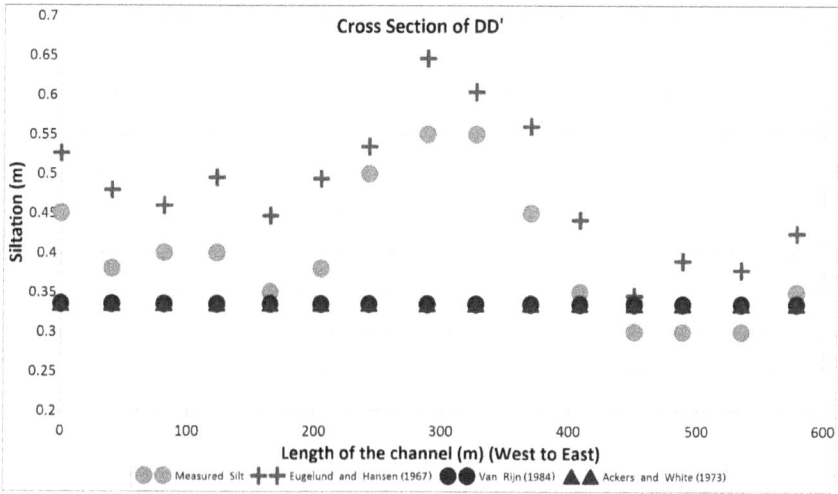

Fig. 12.19(d). Comparison for measured and simulated siltation at Section D-D' near the fishing harbour (October 2019–February 2020).

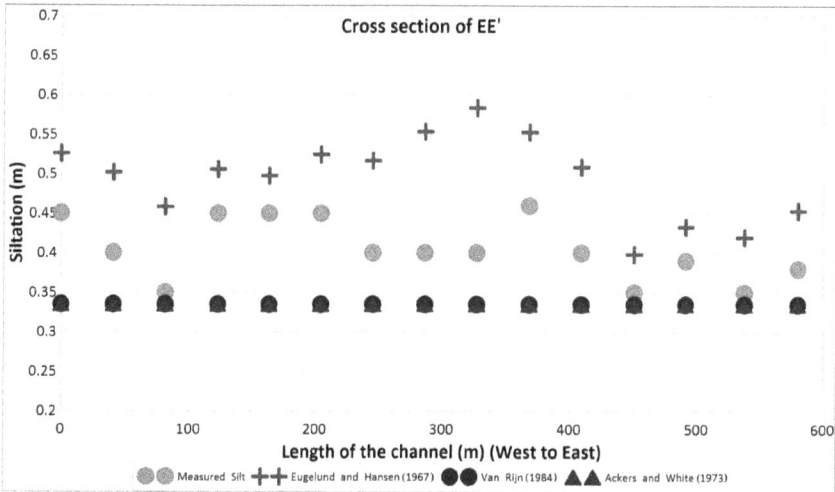

Fig. 12.19(e). Comparison for measured and simulated siltation at Section E-E' near the fishing harbour (October 2019–February 2020).

arrived at for real-field conditions as well as for the proposed solution of revetment construction is computed. The results of the siltation studies for the real-field conditions are projected in **Fig. 12.20**. To achieve better clarity, only a portion of stretch that is problematic for navigation is presented while projecting the results.

12.6.4.   *Simulated morphodynamics results at inlet of the channel (proposed revetment on south boundary)*

The siltation experienced at the river channel is more at the inlet of the channel at Arasalar River, Karaikal. The loosely packed sand on either side of the boundary tends to close the channel by erosion. The eroded sediment

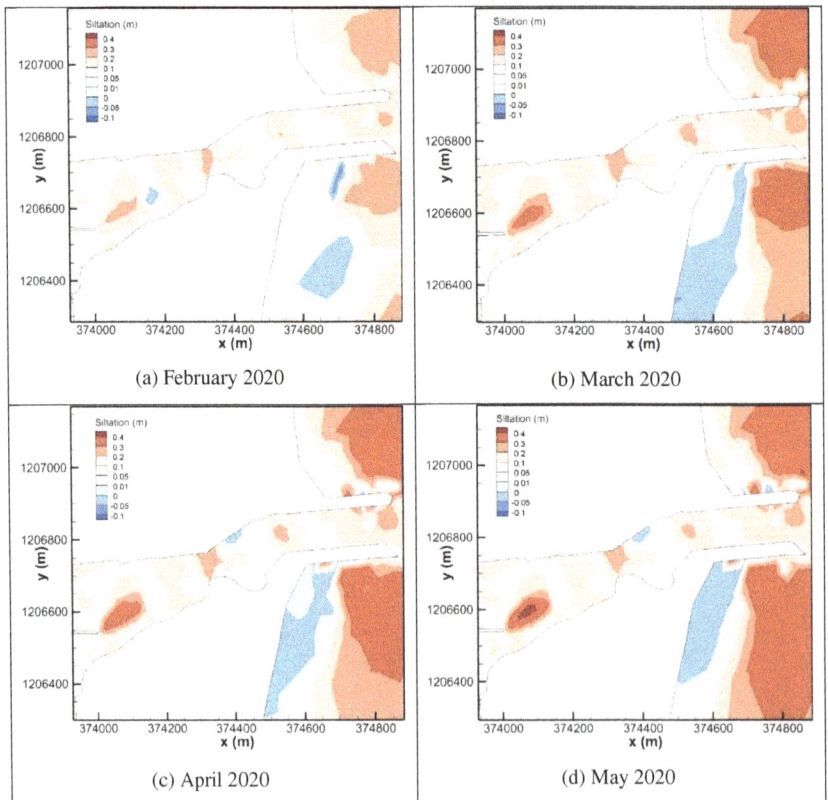

(a) February 2020   (b) March 2020

(c) April 2020   (d) May 2020

Fig. 12.20.   Morphodynamic changes from February to October 2020 (Without Revetment).

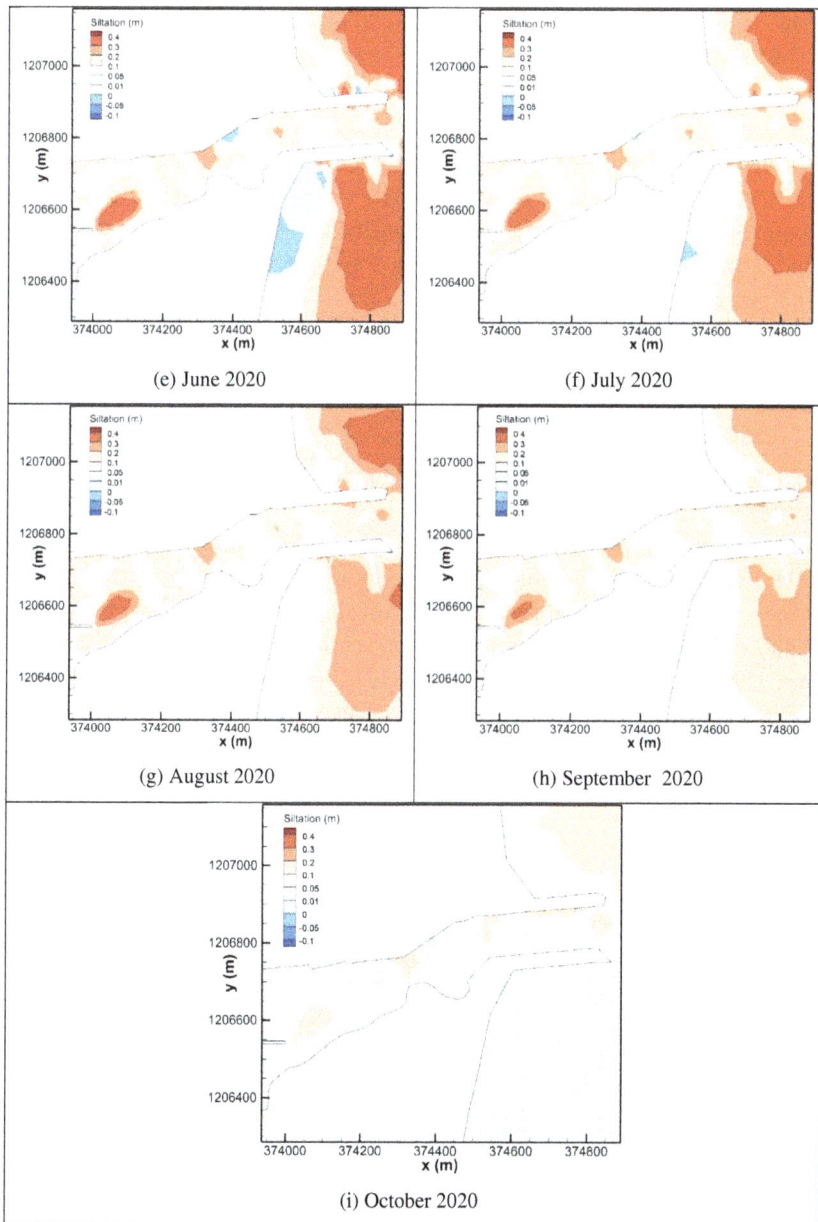

(e) June 2020

(f) July 2020

(g) August 2020

(h) September 2020

(i) October 2020

Fig. 12.20.   (*Continued*)

in the middle of the channel is removed by dredging and later the erosion at the north was controlled by constructing stone revetment covered by mesh. The revetment in the north region stays strong and protects the river from siltation. To arrive at a viable and permanent solution multiple options and layouts were considered and a proposal to construct a revetment over a stretch of 600 m westward of the existing southern training wall was proposed to make hassle-free navigation as shown in **Fig. 12.21**. The numerical study has been projected in **Fig. 12.22** for best-fit Engelund and Hansen formula.

From the results in the above figures, it is inferred that sizeable changes in the siltation patterns are observed for the same input data, in the domain area for the model in the presence and absence of a revetment structure. The comparison for the aforesaid results are projected in **Fig. 12.23**. The variation in bathymetry computed within the study area for the said cases is projected in **Fig. 12.24(a)**.

The quantity of siltation is more pronounced along the western end of the channel since the eastern end is closer to the seaward opening. In general, the measured and real field condition (i.e., simulated without revetment) are of comparable values across all sections. Invariably, across all sections, the morphodynamic model with a revetment structure has yielded results that project a lesser quantity of siltation, thus averting the need for periodic maintenance dredging. The future scope is to study the long-term effects of the channel boundary subjected to storm surges and various cyclonic events.

Fig. 12.21. Proposed revetment layout.

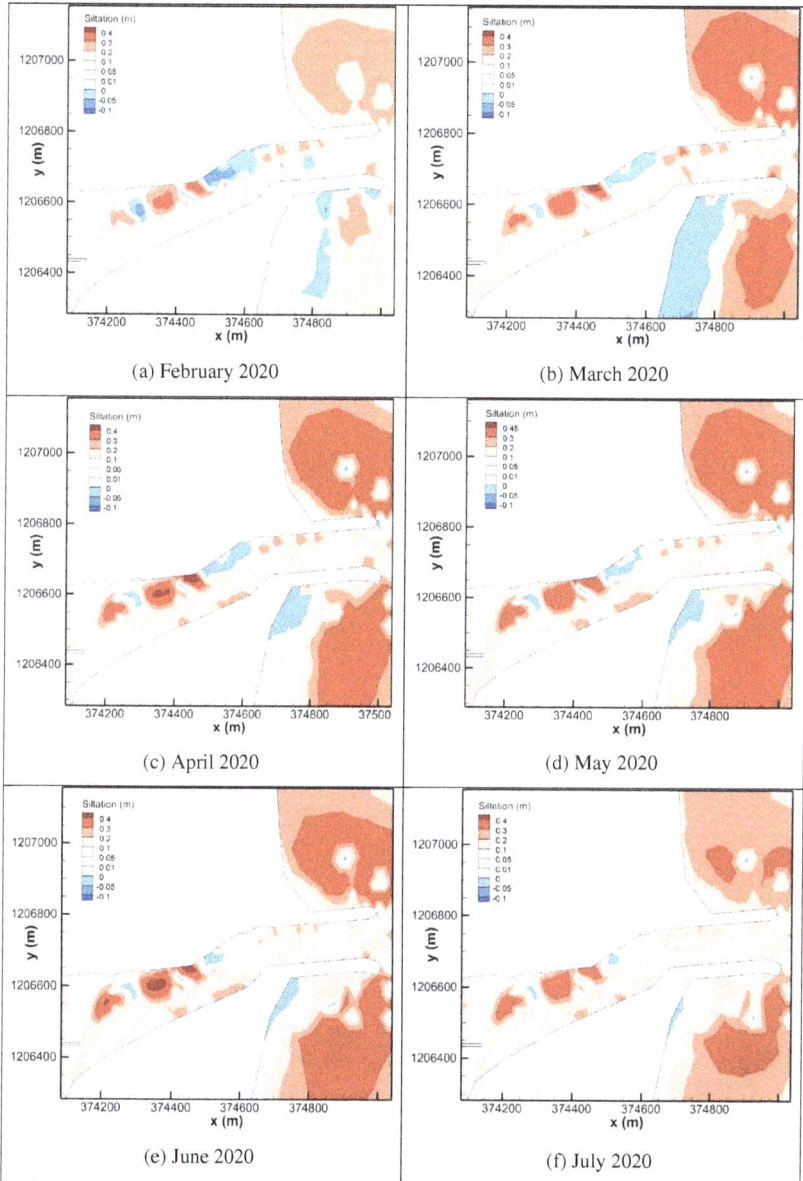

(a) February 2020

(b) March 2020

(c) April 2020

(d) May 2020

(e) June 2020

(f) July 2020

Fig. 12.22.   Morphodynamic changes from February 2020 to October 2020 (with revetment).

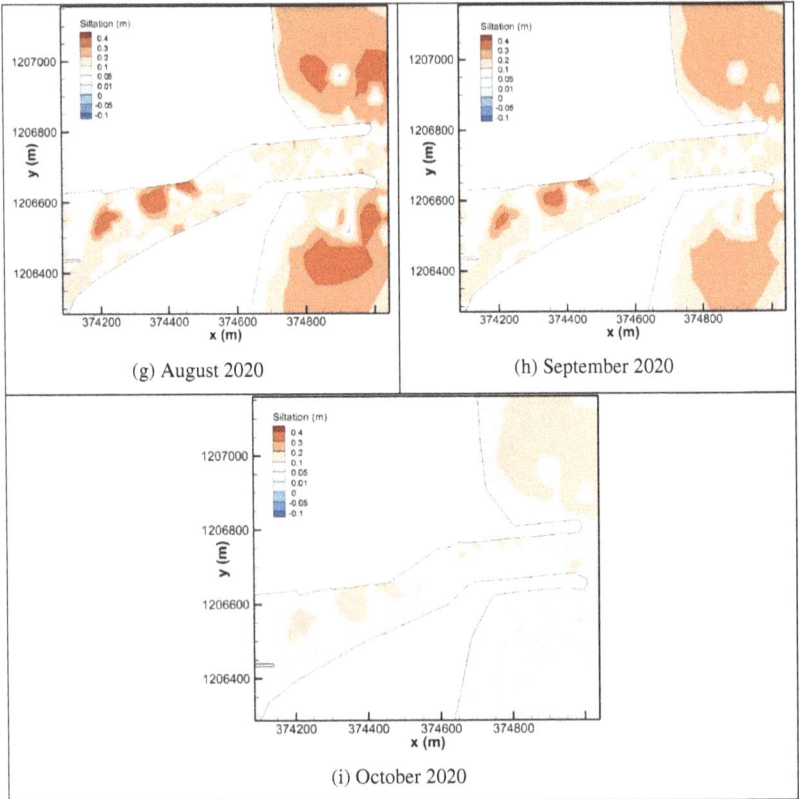

(g) August 2020　　　　　　　(h) September 2020

(i) October 2020

Fig. 12.22. (*Continued*)

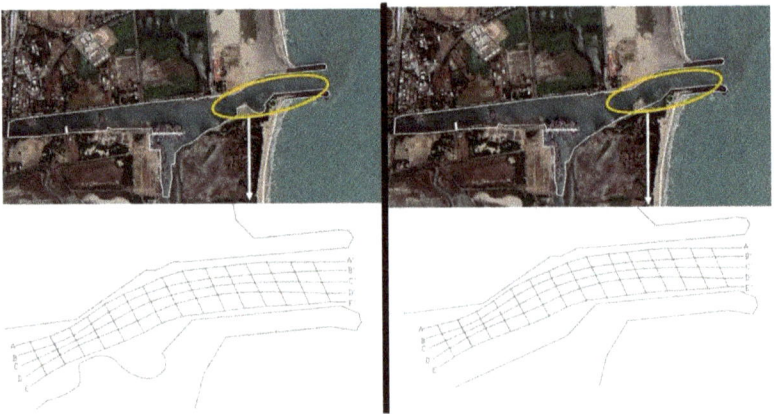

Fig. 12.23. Siltation study area (approximate area of 90,000 m^2).

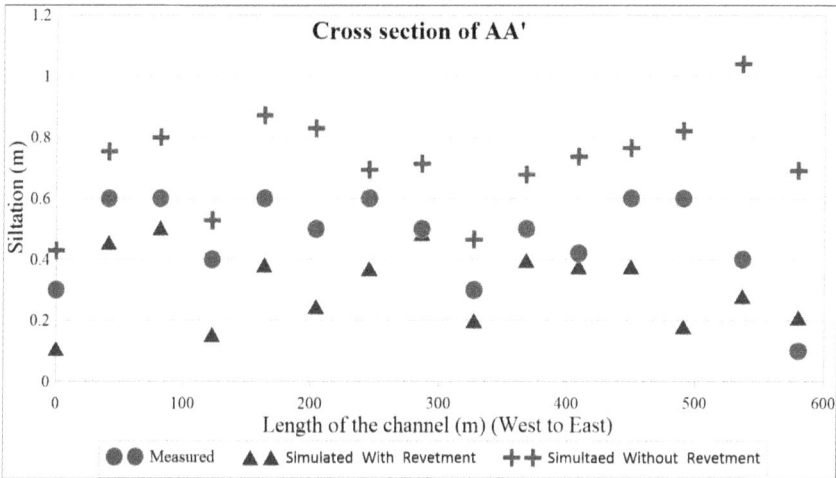

Fig. 12.24(a). Comparison of measured and simulated siltation at Section A-A' at inlet basin.

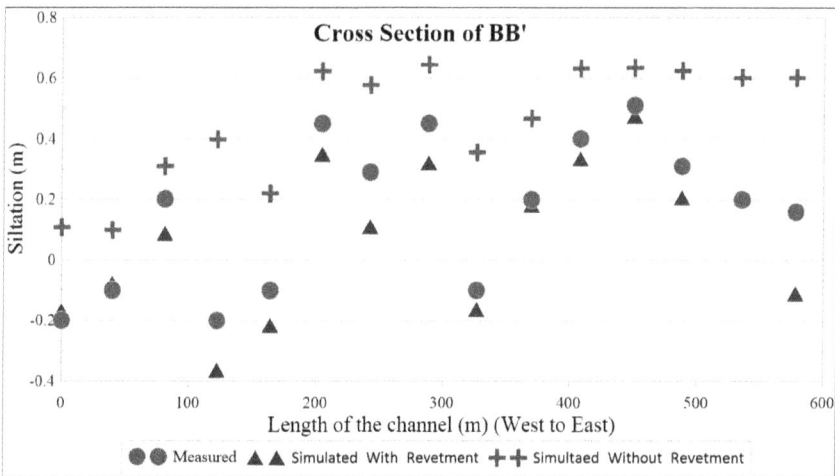

Fig. 12.24(b). Comparison of measured and simulated siltation at Section B-B' at inlet basin.

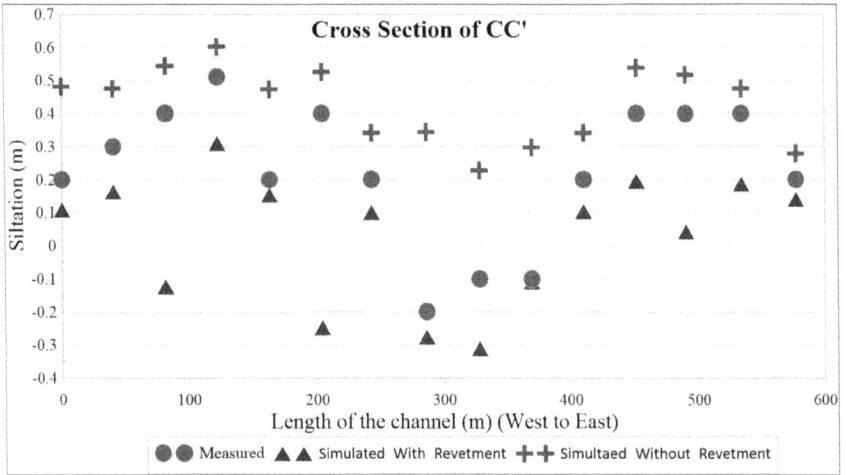

Fig. 12.24(c). Comparison of measured and simulated siltation at Section C-C' at inlet basin.

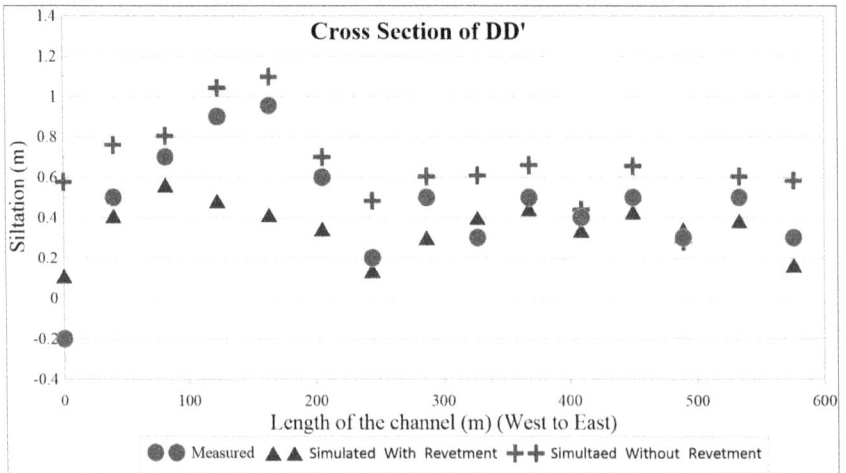

Fig. 12.24(d). Comparison of measured and simulated siltation at Section D-D' at inlet basin.

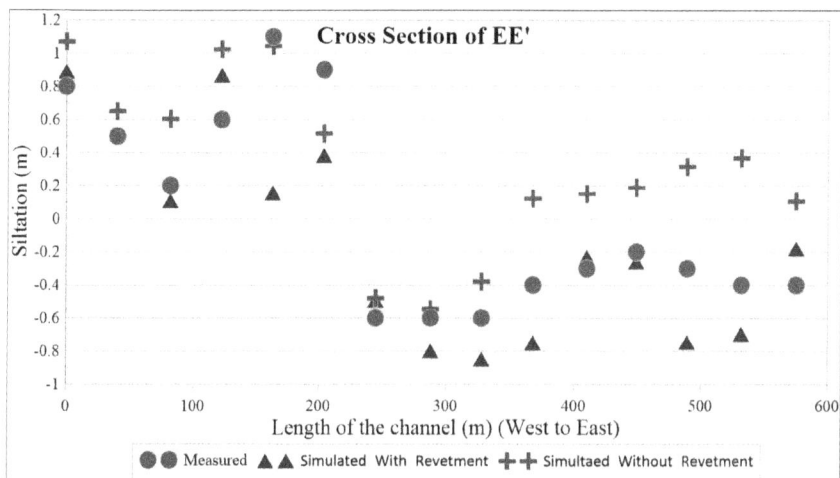

Fig. 12.24(e). Comparison of measured and simulated siltation at Section E-E' at inlet basin.

12.7. Summary

The primary focus of this chapter is to discuss the problems related to the hydrodynamics and morphodynamics of tidal inlets as their stability plays an important role in the livelihood of coastal communities. The occurrence of siltation hinders the free movement of vessels through inlets. To understand the application of numerical models and field measurements in identifying such problems and to suggest remedial measures, the siltation in the Arasalar riverine inlet along Karaikal situated on the southeast coast of Indian peninsula is considered.

The field measurements such as bathymetry charts, current velocities and sediment characteristics have been carried out for various seasons and applied in the numerical model along with the calibration constants to replicate the real-field event as close as possible with the means of an FVM solver. The hydrodynamic modelling for precise current velocity estimates was achieved with a minimal RMSE error as low as below 0.02. The simulated hydrodynamic and morphodynamic results in terms of current velocities and bed level changes are found to be comparable with the field-measured data, after calibration of flow parameters. Through their analysis, Engelund and Hansen (1967) predict the sedimentation inside the channel with better accuracy, since the flow carries sandy sediments within the riverine channel. The study suggests that the mid-portion of the channel

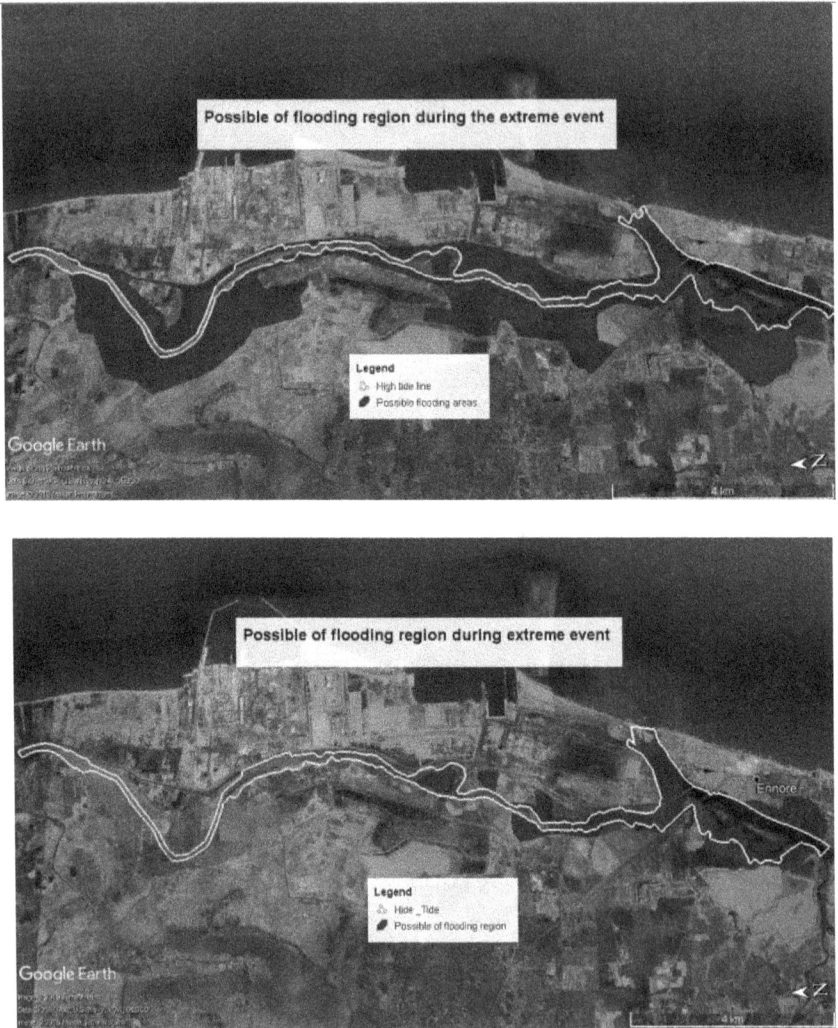

Fig. 12.25. Comparison flooding regions under extreme events indicating usefulness of coastal information monitoring. (Top: present condition; Bottom: with inlet training).

is relatively siltation-free in the nearby the fishing harbour, whereas the region near wall boundaries of the riverbanks is heavily silted since the current velocities, on interaction with the wall boundaries, lose energy and suspended sediments carried along in the flow drops down here after attaining its fall velocity. This results in free passage of fishing boats or smaller vessels docking along the jetties, and the maintenance dredging activities are being carried out at regular intervals. Other measures such as the construction of spurs, revetments, etc., can be proposed after a complete analysis of the design. The solution for the periodic siltation problem was considered to be the construction of a revetment structure. Morphodynamic modelling using the Engelund and Hansen (1967) formula generated the most accurate siltation prediction within the inlet channel of the domain. Thus, the same formula was adopted to study the siltation pattern with the proposed revetment structure along the southern wall boundary. Finally, through graphical comparison of the siltation levels along the navigation length between the mouth and the fishing harbour jetties, it is deduced that construction of the proposed revetment would minimize siltation levels and improve the navigational efficiency without frequent maintenance dredging activities.

A comprehensive planning and execution of coastal information and systematic modelling can provide an in-depth knowledge on coastal vulnerability, inlet morphodynamics and training of inlets. An example is shown in **Fig. 12.25**, wherein the vulnerability of the coastal community along the banks of Kosasthalaiyar river inlet has been quantified for the purpose of better management and evacuation practices near the inlet.

References

Ackers, P. and White, W.R. (1973). Sediment transport: New approach and analysis. *Journal of the Hydraulics Division, ASCE*, 99(11), 2041–2060.

Ashford, G.A. (1996). An unstructured grid generation and adaptive solution techniques for high Reynolds number compressible flows. PhD Thesis, the University of Michigan, Ann Arbor, MI.

Bayram, A., Larson, M. and Hanson, H. (2007). A new formula for the total longshore sediment transport rate. *Coastal Engineering*, 54(9), 700–710.

Bisheng, W.U., van Maren, D.E. and Linggun (2008). Predictability of sediment transport in the Yellow River using selected transport formulas. *International Journal of Sediment Research*, 23, 283–298.

Bruun, P. (1986). Morphological and navigational aspects of tidal inlets on littoral drift shores. *Journal of Coastal Research*, (2), 123–146.

Bruun, P. (1989). *Port Engineering*, Vol III. IV. Houston, TX: Gulf Publishing Company.

Bruun, P. (2001). The development of down drift erosion. *Journal of Coastal Research*, 17(1), 82–89.

Bruun, P. and Gerritsen, F. (1958). Stability of coastal inlets. *Journal of Waterways and Harbors Division*, ASCE, 84(3), 1644.

Bruun, P. and Gerritsen, F. (1959). Natural by-passing of sand at coastal inlets. *Journal of Waterways and Harbors Division*, WW4, 75–107.

Buonaiuto, F.S. and Bokuniewicz, H.J. (2008). Hydrodynamic partitioning of a mixed energy tidal inlet. *Journal of Coastal Research*, 24(5), 1339–1348.

Boothroyd, J.C. (1985). Tidal inlets and tidal deltas. In: Davis Jr., R.A. (Ed.), *Coastal Sedimentary Environments*. Springer-Verlag, New York, pp. 445–532.

Carr, E.E. and Kraus, N.C. (2001). Morphologic asymmetries at entrances to tidal inlets. *Coastal and Hydraulic Engineering Lab,* Engineering Research and Development Center, Vicksburg, MS.

Castelle, B., Bourget, J., Molnar, N., Strauss, D., Deschamps, S. and Tomlinson, R. (2007). Dynamics of a wave-dominated tidal inlet and influence on adjacent beaches, Currubin Creek, Gold Coast, Australia. *Coastal Engineering*, 54(1), 77–90.

Chang, H. (1997). Modeling fluvial processes in tidal inlet. *Journal of Hydraulic Engineering*, 123(12), 1161–1165.

Cheung, K.F., Gerritsen, F. and Cleveringa, J. (2007). Morphodynamics and sand bypassing at Ameland inlet, the Netherlands. *Journal of Coastal Research*, 106–118.

Chitra, K., Murali, K. and Mahadevan, R. (1996). Simulation of storm surges along east coast of India using an explicit FEM. In: *International Conference on Ocean Engineering'96*. Ocean Engineering Centre, IIT, Madras, India, pp. 17–22.

Cooper, J.A.G., McKenna, J., Jackson, D.W.T. and O'Connor, M. (2007). Mesoscale coastal behavior related to morphological self-adjustments. *Geology*, 35(2), 187–190.

Daboll, J.M. (1969). Holocene sediments of the Parker River estuary. Massachusetts: Cont. No. 3, Coastal Research Group, Dept. of Geology, University of Massachusetts, pp. 137.

Davies, R.A. (1964). What is a wave dominated coast? *Marine Geology*, 60(1–4), 313–329.

Davis, R.A. (1978). Coastal Sedimentary Environments. Springer, New York, NY. https://doi.org/10.1007/978-1-4684-0056-4

Dean, R.G. (1988). Sediment interaction at modified coastal inlets: Processes and policies. In: *Hydrodynamics and Sediment Dynamics of Tidal Inlets*. Springer Verlag, New York, pp. 412–439.

de Swart, H.E. and Zimmerman, J.T.F. (2009). Morphodynamics of inlet systems. *Annual Review of Fluid Mechanics*, 41(1), 203–229.

Engelund, F. and Hansen, E. (1967). A monograph on sediment transport in alluvial streams. *Hydraulic Engineering Reports.*

Elias, E.P.L. and van der Spek A.J.F. (2006). Long- term morphodynamic evolution of Texel Inlet and its ebb-tidal delta (The Netherlands). *Marine Geology*, 225 (1–4), 5–21.

Escoffier, F. (1940). The stability of tidal inlets, *Shore and Beach*, 8(4), 114–115.

Fitzgerald, D.M. (1984). Interactions between the ebb-tidal delta and landward shoreline: Price Inlet, South Carolina. *Journal of Sedimentary Petrology*, 54(4), 1303–1312.

Fitzgerald, D.M. (1996). Geomorphic variability and morphologic and sedimentologic controls on tidal inlets. *Journal of Coastal Research* [Special Issue] 23, 47–71.

Fitzgerald, D.M., Buynevich, I.V., Davis, R.A Jr. and Fenster, M.S. (2002). New England tidal in-lets with special reference to riverine-associated inlet systems. *Geomorphology*, 48(1–3), 179–208.

Fitzgerald, D.M., Kraus, N.C. and Hands, E.B. (2000). Natural mechanisms of sediment bypassing at tidal inlets. US Army Corps of Engineers.

Fontolan, G., Pillon, S., Delli Quadri, F. and Bezzi, A. (2007). Sediment storage at tidal inlets in northern Adriatic lagoons: Ebb-tidal delta morphodynamics, conservation and sand use strategies. *Estuarine, Coastal and Shelf Science*, 75(1–2), 261–277.

French, P.W. (2001). Coastal defences: processes, problems and solutions. In: *Earth Surface Processes and Landforms*, Vol. 28. John Wiley & Sons, Ltd., London.

Gaudiano, D.J. and Kana, T.W. (2001). Shoal bypassing in South Carolina tidal inlets: Geomorphic variables and empirical predictions for nine mesoscale inlets. *Journal of Coastal Research*, 17(2), 280–291.

Hanson, H. and Kraus, N.C. (2001). Chronic beach erosion adjacent to inlets and remediation by composite (T-Head) groins. US Army Corps of Engineering, Vicksburg, MS.

Hayes, M.O. (1979). *Barrier Island Morphology as a Function of Wave and Tide Regime*. New York: Academic Press, pp. 1–29.

Hayes, M.O. (1980). General morphology and sediment patterns in tidal inlets. *Sediment Geology*, 26 (1–3), 139–156.

Hume, T.M. and Hendendorf, C.E. (1992). Factors controlling tidal inlet characteristics on low drift coasts. *Journal of Coastal Research*, 8(2), 355–375.

Kraus, N.C. (2009). Engineering of tidal inlets and morphologic consequences. In: Kim, Y.C. (Ed.), *Handbook of Coastal and Ocean Engineering*, World Scientific, Singapore, pp. 867–901.

Kumar, V.S., Pathak, K.C., Pedenkar, P., Raju, N.S.N. and Gowthaman, R. (2006). Coastal processes along the Indian coastline. *Current Science*, 91(4), 530–536.

Murali, K., Lou, J. and Kumar, K. (2002). An unstructured model simulation for Singapore strait. *Maritime Port Journal Singapore*.

Murali K., Sundar, V. and Sannasiraj, S.A. (2014). Mathematical modelling of sedimentation within an intake basin at Krishnapattinam. *Proceedings of 19th Congress of the Asia and Pacific Division of the international Association for Hydro-Environment Engineering and Research (IAHRAPD)* September 21–24, Thuyloi University, Hanoi, Vietnam.

Nageswara Rao, K., Sadakata. N., Hema Malini. B. and Takayasu, K. (2005). Sedimentation processes and asymmetric development of the Godavari delta, India. *Society for Sedimentary Geology*, pp. 433–449.

Nakato, T. (1987). Discussion of "Modeling of River Channel Changes" by H. H. Chang (February, 1984, Vol. 110, No. 2). *Journal of Hydraulic Engineering*, 113(2), 262–265.

O' Brien, M.P. (1931). Estuary tidal prism related to entrance areas. *Civil Engineering*, 1(8), 738–739.

O'Brien, M.P. (1966). Equilibrium flow Area of tidal inlet on sandy coasts. *Proceedings of the 10th Coastal Engineering Conference*, Vol. I, pp. 676–686.

Oretel, G.F. (1977). Geomorphic cycles in ebb deltas and related patterns of shore erosion and accretion. *Journal of Sedimentary Research*, 47(3), 1121–1131.

Pacheco, A., Ferreira, O., Williams, J.J., Garel, E., Vila-Concejo, A. and Dias, J.A. (2010). Hydrodynamics and equilibrium of a multiple-inlet system. *Marine Geology*, 274(1–4), 32–42.

Pascolo, S., Marco, P. and Silvia, B. (2018). Wave current interaction: A 2DH model for turbulent jet and bottom-Friction dissipation. *Water*, 10(4), 392. http://doi.org/10.3390/w10040392.

Pinto, L., Anfre B.F. and Paula, F. (2006). Sensitivity analysis of non-cohesive sediment transport formulae. *Continental Shelf Research*, 26, 1826–1839.

Roache, P.J. (1998). Verification and validation in computational science and engineering. *Hermosa Publication*, 895, Albuquerque, NM.

Ridderinkhof, H. (1989). Tidal and residual flows in the Western Dutch Wadden Sea III: Vorticity balances. *Netherlands Journal of Sea Research*, 22(1), 9–26.

Siegle, E., Huntley, D.A. and Davidson, M.A. (2004). Physical controls on the dynamics of inlet sandbar systems. *Ocean Dynamics*, 54(3–4), 360–373.

Silva, P.A., Bertin, X., Fortunato, A.B. and Oliveira, A. (2009). Intercomparison of sediment transport formulas in current and combined wave current conditions. *Journal of Coastal Research*, SI 56, 559–563.

Senthilkumar, R., Murali, K. and Sundar, V. (2017). Stability of a micro-tidal inlet using semi-numerical approach. *The International Journal of Ocean and Climate Systems*. 8(3), 113–125. DOI: 10.1177/1759313117736747

Stanev, E.V., Floser, G. and Wolff, J.O. (2003). Dynamical control on water exchanges between tidal basins and the open ocean. A case study for the East Frisian Wadden Sea. *Ocean Dynamics*, 53, 146–165.

Stanev, E.V., Burghard, W. Flemming, A., Joanna, B., Staneva, V. and Wolff, J-O. (2007). Vertical circulation in shallow tidal inlet and back barrier basins. *Continental Shelf Research*, 27, 798–831.

Styles, R., Brown, M., Brutsché, K., Li, H., Beck, T. and Sánchez, A. (2016). Long-term morphological modeling of Barrier Island tidal inlets. *Journal of Marine Science and Engineering*, 4(4), 65. https://doi.org/10.3390/jmse4040065

Tambroni, N., Ferrarin, C. and Canestrelli, A. (2010). Benchmark on the numerical simulations of the hydrodynamic and morphodynamic evolution of tidal channels and tidal inlets. *Continental Shelf Research*, 30(8), 963–983.

Tomlinson, R.B. (1991). Processes of sediment transport and ebb tidal delta development at a jettied inlet. *Proceedings of Coastal Sediments'91*, pp. 1404–1418.

Tung, T.T., Walstra, D.J.R ., Graaff, J. van de. and Stive, M.J.F. (2009). Morphological modelling of tidal inlet migration and closure. *Journal of Coastal Research*, SI 56, *Proceedings of the 10th International Coastal Symposium*, pp. 1080–1084.

Van Rijn, L.C. (1984a). Sediment transport, Part I: Bed load transport. *Journal of Hydraulic Engineering ASCE*, 110(10).

Van Rijn, L.C. (1984b). Sediment transport, Part II: Suspended load transport. *Journal of Hydraulic Engineering ASCE*, 110(11).

Wang, P. and Beck, T.M. (2012). Morphodynamics of an anthropogenically altered dual-inlet system: John's Pass and Blind Pass, west-central Florida, USA. *Marine Geology*, pp. 291–294 (0), 162–175.

Williams, J. and Esteves, L. (2017). Guidance on setup, calibration, and validation of hydrodynamic, wave, and sediment models for shelf seas and estuaries. *Advances in Civil Engineering*. https://doi.org/10.1155/2017/5251902

Yang, C.T. (1973). Incipient motion and sediment transport. *Journal of Hydraulic Division*, ASCE, 99(10), 1679–1704.

Zimmerman, J.T.F. (1980). Vorticity transfer by tidal currents over an irregular topography. *Journal of Maritime Research*, 38(4), 601–631.

Index